LIVING BEACHES
of the Gulf Coast

[signatures]

October 2023

LIVING BEACHES
of the Gulf Coast

A Beachcomber's Guide
Including Texas, Louisiana, Mississippi, Alabama,
and Florida's Panhandle

Blair and Dawn Witherington

Pineapple Press
Palm Beach, Florida

Photos Not Captioned

Front Cover

Main: Hatchling loggerhead sea turtle

Insets:
Sharkeye moonsnail
Lined sea star
Spotted ground squirrel
American oystercatcher
Texas pricklypear cactus

Back Cover, Common sundials

Front Flap, Willet

Fly Page, Boca Chica, Texas

Title Page, Long-billed curlew

p. 1, Wave ripples
p. 49, American oystercatcher
p. 269, Seaoats and beach croton
p. 347, Mineral sands
p. 363, Galveston storm memorial
p. 404, Sanderling

Pineapple Press

An imprint of Globe Pequot, the trade division of
The Rowman & Littlefield Publishing Group, Inc.
4501 Forbes Blvd., Ste. 200
Lanham, MD 20706
www.rowman.com

Distributed by NATIONAL BOOK NETWORK

British Library Cataloguing in Publication Information Available

Library of Congress Cataloging-in-Publication Data

Names: Witherington, Blair E., 1962- author. | Witherington, Dawn, author.
Title: Living beaches of the Gulf Coast : a beachcomber's guide including Texas, Louisiana, Mississippi, Alabama and Florida's Panhandle / Blair and Dawn Witherington.
Description: Sarasota, Florida : Pineapple Press, 2022. | Includes bibliographical references and index. | Summary: "Includes beach anatomy, phenomena, beach animals, plants, minerals. Also includes man-made structures (i.e. lighthouses and other historical structures), and beach art. Suggested beach scavenger hunts and hints for success. There is a section on conservation. Finally, there is a section on resources and suggested reading"— Provided by publisher.
Identifiers: LCCN 2021053944 (print) | LCCN 2021053945 (ebook) | ISBN 9781683340560 (paperback) | ISBN 9781683340577 (epub)
Subjects: LCSH: Beaches—Gulf Coast (U.S.)—Guidebooks. | Coastal ecology—Gulf Coast (U.S.)—Guidebooks. | Gulf Coast (U.S.)—Guidebooks.
Classification: LCC GB459.4 .W57 2022 (print) | LCC GB459.4 (ebook) | DDC 551.45/70976—dc23/eng/20211208
LC record available at https://lccn.loc.gov/2021053944
LC ebook record available at https://lccn.loc.gov/2021053945

Printed in New Delhi, India

Contents

Beach Animals and Microbes

Beach Plants and Fungi

Beach Geology

Hand of Man

Acknowledgments

For contributions, reviews, and advice we are indebted to Kraig Anderson, Roger Birkhead, Dennis Bonal, Mike and Sam Burnett, Ryan Chabot, Donna Lee Crawford, Robert Deans, Carly DeMay, Kevin Edwards, the Gulf Specimen Marine Lab, Cullen Hanks, Shigetomo Hirama, Ashleigh Holden, the Houston Museum of Natural Science, Steve Johnson, Gary Kidder, Kim Matheny, Adrienne McCracken, Kate McElvaney, D.J. McNeil, Kim Mohlenoff, the Padre Island National Seashore, Ed Perry, Tina Petway, Steven Pinker, Tom Pitchford, Terry Ross, the Schauman Family, Donna Shaver, Derke Snodgrass, the Texas Marine Mammal Stranding Network, Jace Tunnell, and Ryan Welsh.

Top Forty Living Beaches of Texas and the Northern Gulf

Every Gulf beach has life, but some beaches stand out as splendid examples of natural processes free to run their course. These are not beaches devoid of humans (many are among the most visited shores). But they are beaches where our influence has been more casual than insistent. In geographic order a list includes:

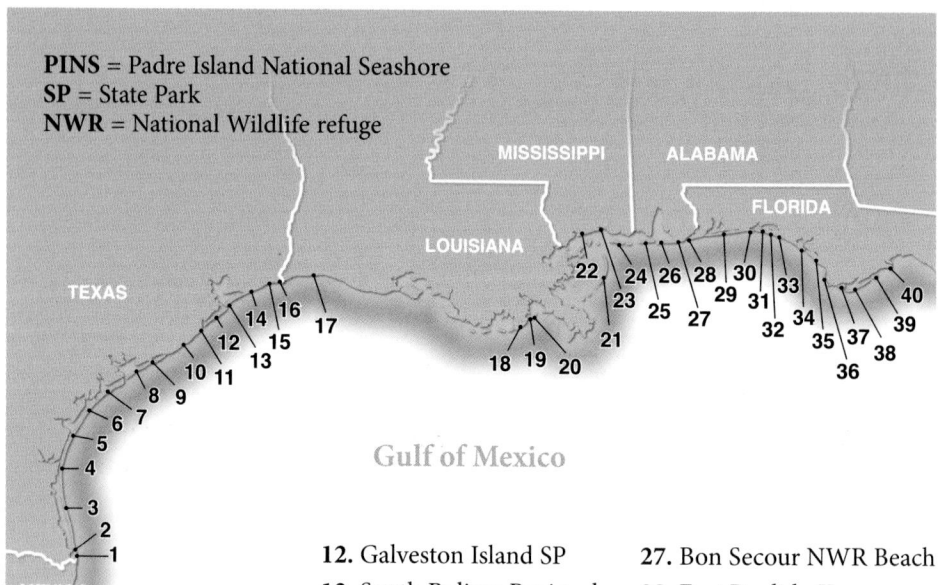

1. Boca Chica SP
2. Brazos Island SP
3. Padre Island, South
4. Padre Island, North and Malaquite Beach (PINS)
5. Mustang Island SP
6. San José Island
7. Matagorda Island
8. Greek Island
9. Matagorda Bay Nature Park and Matagorda Beach
10. San Bernard NWR
11. Follett's Island Beach
12. Galveston Island SP
13. South Bolivar Peninsula
14. McFaddin NWR
15. Sea Rim SP
16. Texas Point NWR
17. East Holly Beach
18. Elmer's Island Wildlife Refuge
19. Grand Isle SP
20. Grand Terre Island
21. Chandeleur Islands
22. West Pass Christian Beach
23. Deer Island
24. Cat, Ship, Horn, and Petit Bois Islands
25. West Dauphin Island
26. Fort Morgan Beach
27. Bon Secour NWR Beach
28. East Perdido Key
29. Gulf Islands National Seashore
30. Eglin Beach Park
31. Henderson Beach SP
32. Topsail Hill Preserve SP
33. Grayton Beach SP
34. St. Andrews SP and Shell Island
35. Crooked Island
36. St. Joe Peninsula SP
37. St. Vincent Island
38. Cape St. George and St. George Island SP
39. Dog Island
40. Alligator Point and Bald Point SP

The Beaches are Alive!

Yes, our Gulf beaches are alive. Some of this life is obvious. Stroll onto the beach, sink your toes in the sand, and look around you. On the dune-front, gulls glide above seaoats flagging in the sea breeze. On the open beach, crabs toss sand from their burrows. And at the tide line, shorebirds busily poke and turn the clumps of seaweed. Look closer and you'll see even greater evidence of life. The seashore is vibrant with dozens of dune-plant species; a diverse array of stranded seashells; birds that dive, run, wade and soar; and the wrack, that ever-changing line of formerly floating drift-stuff from faraway.

Clearly, beaches attract, foster, and collect life … and the testament of life. But in an important way, beaches are also alive themselves. Beaches and dunes grow, diminish, evolve over years, and shift with the seasons. To pulsate with change is the very nature of a sandy sea coastline. This change is the essence of what makes beaches so fascinating.

In the long term, beaches are dynamic, tumultuous, even dangerous places. Yet, a short-term visit allows a pleasant acquaintance with the beauty generated by all that turmoil. These wonders both originate on beaches, and arrive each day, produced by the most primeval and spectacular wilderness on Earth—the sea. A beach visit allows us to peer into that wilderness, and even examine it closely, for much of the sea's mysterious nature ends up on its beaches.

We hope that this book will provide some helpful interpretation for the curious seashore visitor. In part, it is a guide to critters, plants, formations, and stuff that might be puzzling enough to go nameless without a little assistance. But an additional objective we have is to share the intrigue inherent to the common things … those easily identified but little known. From the elegant to the plain, from the provocative to the mundane, everything on a beach has a story to tell.

A lined sea star lighted by the sunset seeks the sea at low tide

How the Story Unfolds

This book is organized into sections dividing major groups of beach stuff—*Beach Features, Beach Animals and Microbes, Beach Plants and Fungi, Beach Geology,* and *Hand of Man.* Within each section, groups of related items are presented together and share an identifying icon at the top corner of the page.

   Most items have a map showing where and when one might find it on a Gulf beach. These ranges pertain specifically to an item's beach distribution, which may be different from the places it occupies when not on a beach. For example, many of the plants that produce drifting seeds known as seabeans live far away within inland tropical rainforests. Few of these plants live locally, but their attractive floating seeds show up on our beaches at particular places and times (due to ocean currents and weather). Coastal lines on the maps are **red** if they describe a warm-season distribution and **blue** if they describe a range mostly during cooler months. **Purple** lines represent all seasons. Lines are **solid** to show a relatively common occurrence (relative to other locations, not to other items), and **open** to show the item as relatively uncommon. Because the range maps are not absolute, a gap may indicate either rarity or uncertainty. The maps show gaps along marshy, submerged shorelines with little or no beach.

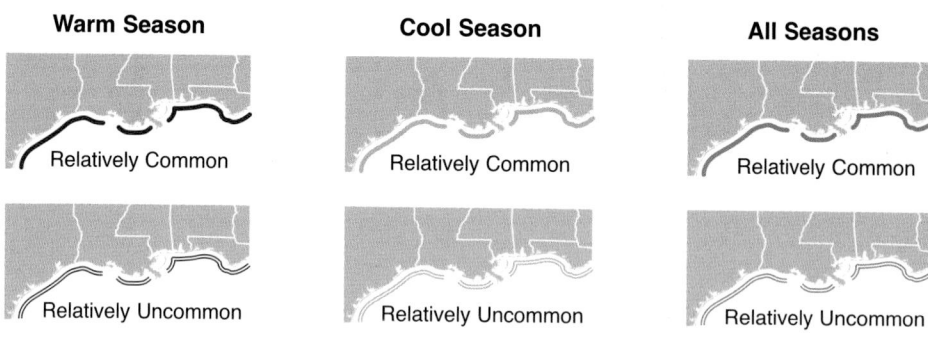

Because this is a guide to beach-found things, depictions are of things as one might find them on a beach. That is, many items tend to show a beach-worn look. Although we've tried to represent the living elegance of creatures, some are merely deceased lumps and blobs by the time they reach a beach. Images with a catalog number are of specimens graciously loaned by the Houston Museum of Natural Science.

   Note that where an item's size is given, the measure refers to maximum length or height, unless otherwise indicated. Also note that a few featured items hold the potential for an unpleasant encounter. These will have a watch-your-step (X) or don't-touch symbol (🚫), which we hope you will see before you engage the thing with your bare feet or put it in your pocket.

   Rather than simply set the scene, introduce a cast of characters, and leave you hanging, we've ended the book with motivation. We propose a short list of "quests." These target a selection of rare, beautiful, or otherwise compelling prizes that can provide a blanket excuse for beach adventures.

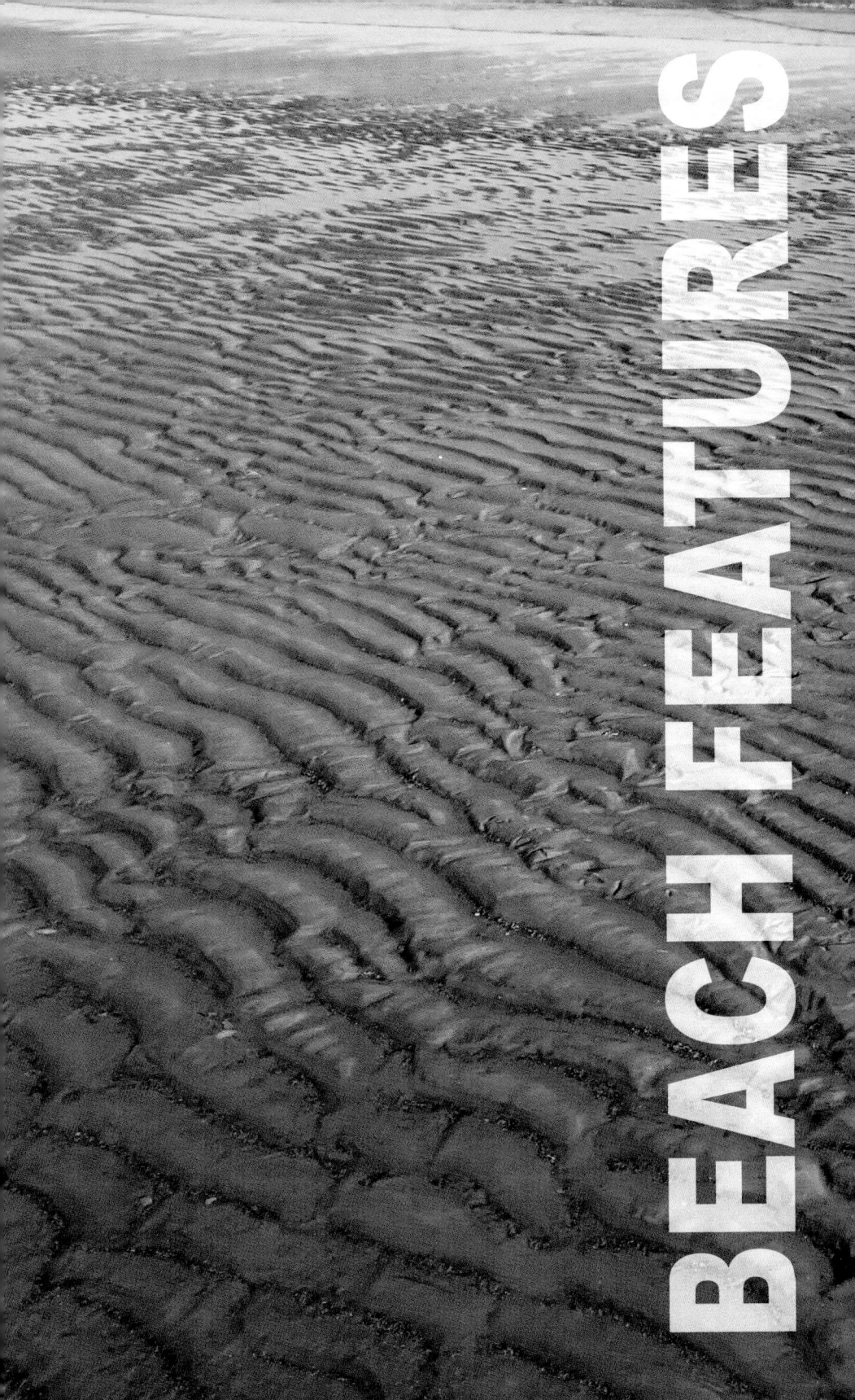

BEACH FEATURES

What are Beach Features?

A beach's features show its life signs. These traces reveal a beach's relationships, history, growth, retreat, and other aspects of restlessness. Although much of this book describes beach things, the following section on *Beach Features* deals mostly with processes and the evidence they leave.

Beach features show us the intimate relationship between land and sea. To some, this relationship seems like a battle. But a more fitting metaphor describes an economic relationship with mutual exchange. The most dynamically traded element in this land-sea commerce is sand. The land stakes claim to former seabottom blown into dune mounds and covers its claim with pioneering greenery. But the sand is just a loan. In the economy of sand, this currency shifts between land and sea, between undersea features, between beach elevations, and between adjacent beaches. The dynamics of this bustling economy can be read in a beach's features. To an educated eye, beach features are both telling evidence of past events and indicators for prediction.

The give-and-take between land and sea also involves energy. As you'll see, beach features include exchanges of heat, wind, waves, and biological actors that further define the beach economy.

To a true beach aficionado, an otherwise pedestrian pile of sand offers abundant confirmation of a beach's beating pulse. Sensing that pulse is basic to understanding what beaches are all about and can add intrigue and wonder to a coastal visit.

Sandbars, surf currents, and tides determine where fishers can make their catch

Beach Anatomy

Each visit to a Gulf beach is unique. No two beaches are exactly alike, and at any given beach, every day is different. But despite their dynamic and fluid nature, our beaches tend to share a common anatomy that is predictable based on location and season.

You may have noticed that Gulf beaches have a lot of sand. Most of this sand is quartz, eroded from rocks millions of years ago and carried by rivers into the Gulf. These rivers also carried feldspar grains and heavier sediments like metal oxides and other minerals. On a few Gulf beaches, shells and shell bits are an important fraction of piled sediments. The origins of Gulf sands span the majority of the North American continent east of the Rockies, and include animal bits from waters near and far. Because river drainages and directional nearshore currents carry different sediments, sand characteristics vary between beaches. How wind and currents move sediment also drives variation in sand character between seasons and locations within the beach-dune system.

Beach anatomy is largely determined by elevation above the sea. Beach zones range from high, dry, and occasionally wave-washed (the **dune**), to frequently wave-washed (the **backshore**), to constantly wave-washed (the **foreshore**). Seaward of the foreshore are two zones critical to the beach, and formerly beach themselves: the **nearshore** and **offshore** zones.

Within each zone are the lumps, dips, ripples, and wave-washed stuff that further describe a beach's structure. The **dune scarp**, if present, marks the elevation where recent storms have swept away dune sands. Between the dune base and the daily high-tide mark lie one or more **wrack lines.** These are piles of marine organisms (mostly seaweed) that, in their death, bring life to the beach. The highest average tide generally reaches the

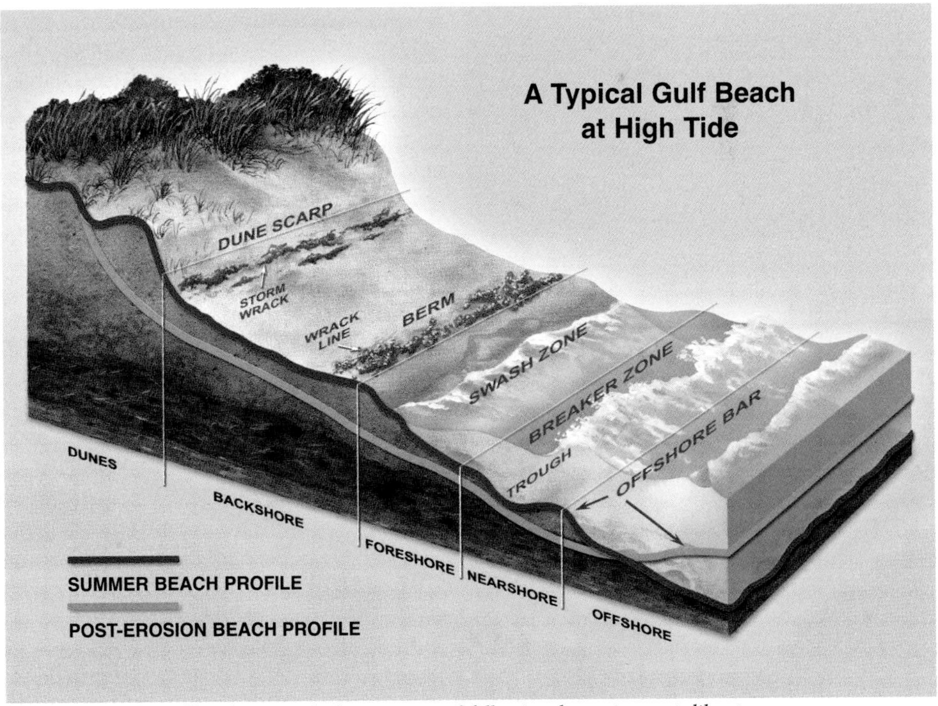

A Typical Gulf Beach at High Tide

DUNE SCARP

STORM WRACK

WRACK LINE

BERM

SWASH ZONE

BREAKER ZONE

TROUGH

OFFSHORE BAR

DUNES

BACKSHORE

FORESHORE

NEARSHORE

OFFSHORE

SUMMER BEACH PROFILE

POST-EROSION BEACH PROFILE

Dune and beach profiles change with the seasons and following dynamic events like storms

berm, a sandy platform between the flat backshore and the sloped foreshore. Beach meets the sea at the **swash zone**, where waves rush the sandy incline and wash back into the following breaker. Often, this final pounding of wave energy creates a step, down into a **trough** landward of the **breaker zone**. The breakers begin where the **offshore bar** presents a rise shallow enough to trip incoming waves. As many as three, deeper, offshore bars lie seaward of the first bar. The deeper sand-bars are built by sand movement during storms and are where large waves break during these tumultuous events.

Beach profile and anatomy change between periods of storm and calm. Compared to a typical midsummer beach fronting calm seas, beaches enduring high winds and waves tend to have a narrow berm and more distant offshore bars. High winds and waves often occur after cold fronts, and following tropical cyclones in late summer and early fall.

Artificially nourished (manmade) beaches seldom fit the natural profiles described above … at least in the beginning. These beaches start out with an engineered anatomy, generally wider and flatter than nature would produce, and equilibrate over a period of years as the sea sculpts the foreshore, then backshore, then dune.

Beaches around the Gulf vary in anatomical features according to how much and what kind of sand supplies them, and by the extent of wind and waves that act on those sediments. These factors produce some Gulf beaches that are a thin veneer of fine sand over clay or peat fronting a low marsh, and other beaches that have enormous dunes, with beach sands that extend more than 50 ft (15 m) deep.

A beachcomber who knows a little beach anatomy and who pays attention to local tides and sea conditions can find some amazing beach stuff. The ocean scatters its varied treasures in different

Matagorda Island, Texas, on a calm day

beach zones depending on its mood. For example, combing the swash zone (**A**) at low tide is the best way to find small and delicate seashells. When a stiff onshore wind is blowing, this is also the place to find ocean-drifting "blue animals." The recent high-tide line at mid-beach (**B**) is normally the best place to find large or fluttery shells, buoyant items like seabeans (pp. 318–341), and invertebrates like sponges and soft corals. Keep in mind that the high-tide wrack from previous days may have been higher up the beach (**C**), where drift treasures can be found if they have not been covered with windblown sand. The largest waves during the highest tides sweep up the beach to the storm wrack (**D**), which is often at the base of the dune. Although wrack-hunting is fruitful for almost anything immediately after a storm, even months-old storm wrack yields persistent, storm-stranded items like big shells, seabeans, driftwood, and lost cargo. Storm wrack on remote, infrequently combed beaches is filled with rare finds.

Beach anatomy becomes less esoteric and more evident to visitors who experience features forming before their eyes. **Aeolian (wind-driven) transport** of sand is but one of these watchable developmental processes. To experience sand flowing over a beach on a breezy day is to witness the origins and pulsation of many beach features. As sand flows across the beach, some features are exposed, others are buried, sand ripples march with the wind, and the growth of a dune from seed begins at a seaweed clump that had drifted at sea for hundreds of miles. This drama is common on breezy Padre Island, Texas, which has wind velocities capable of significant sand transport about 85 percent of the time.

Lettered areas show where to find beach treasures

Wind waves wash over Bryan Beach, Texas

Sand in the process of aeolian transport

5

Tall dunes on wind-dominated Brazos Island, Texas

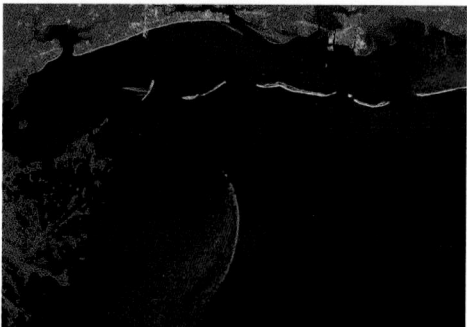

The "delta lobe" Chandeleur Islands, LA (bottom center)

A wave-dominated, central Texas barrier island

Wave-dominated, St. Joseph Peninsula, Florida

Beach Profiles

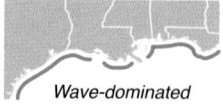

Wave-dominated

Wind-dominated

Delta Lobe

WHAT ARE THEY? A beach's profile is its shape in cross section, from the dune to the water. Gulf beach profiles follow three natural patterns, which are determined by wave and wind forces, and by sand supply. A fourth, artificial pattern, is determined by us (p. 393). Nearly all Gulf beaches are **wave-dominated, dissipative** beaches, which occur where tidal ranges are less than three times average wave height. These beaches are relatively wide and low, and have multiple offshore bars where the largest waves break, dissipating their energy before reaching the beach. Wave-dominated beaches are generally on long barrier islands with few inlets, but some are on the mainland, and some are on relatively short, **delta lobe islands**, which rim the low-gradient, marshy shorelines with abundant sediment surrounding the Mississippi River Delta. Southern Texas beaches are sub-categorized as **wind-(and wave-)dominated**, and show greater effects from wind waves and aeolian sand movement (pp. 5, 20).

HOW COME? Wave energy, tidal range, sand coarseness, and sand abundance are all connected to a beach's profile. Fine sands contribute to a flat beach because they do not drain well. Waterlogged sands allow waves rushing up the beach to rush back down instead of soaking in. This backwash carries sand with it, filling the lower beach and flattening the profile.

DID YOU KNOW? Exactly what determines beach profile is extremely complex, hard to model, and a bit of a mystery.

Beach Lifespans

Circles of life turn at varied rates. For some features of a living beach, mere seconds pass between birth and death. Others may persist for a century or more. But in comparison to life-spans elsewhere on our planet, few beach features ever get very old.

Even the least patient among us have attention spans that would allow appreciation of multiple generations of some beach features. Walk through the wet sand of the swash zone (pp. 23, 24) and look behind you at your **footprints**. Within a few seconds, a wave will end its life by crashing onto the beach and rushing up the slope in a froth-margined sheet of water. The final energy of the wave's life is spent suspending the sand that fills your tracks. As the leading edge of the wavewash reaches its peak, a swash mark (p. 23) is born where grains that were floating on the water's surface tension come to rest in a tiny ridge. The wavewash then slips back into the surf, leaving bubble streams from air forced out of the beach as water soaks into the sand. The escaping air leaves bubble holes (p. 27). These are either obliterated by the next wave or remain during a falling tide until a rising tide replaces them with a different set of bubble holes.

The incoming tide also marks life's end for **sandcastles,** some cusps and scarps (pp. 18, 19), and other topography of the beach berm such as wind ripples (p. 20). On the open beach, ghost crabs take diurnal refuge in burrows that may last only until the next high tide. But their burrows in the dune may allow the crabs to remain protected for months above the wash of winter storms. Wrack lines (pp. 16, 17) pushed high on the beach are testament to the highest tides of the month, and may remain for the weeks it takes for component seaweeds and marsh reeds to decompose. Just above this recent high-water mark, sea turtle nests become unrecognizable following a few days of wind or rain, although a successful nest will contain

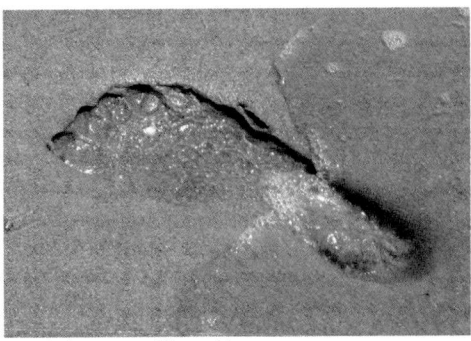

Beach footprints fill quickly

Sandcastles seldom survive high tide

Tidal cycles drive daily, consequential, beach changes

Beach changes shorten lives of structures meant to last

7

viable eggs for about two months until hatchlings escape from their eggshells and emerge from the nest (pp. 205–209).

On the dune, life is at the mercy of periodic storms. For many pioneering dune plants, life strategies anticipate periodic catastrophes. In late summer and early fall, when the risk of erosion heightens due to tropical cyclones, these plants rush to bear dispersing seeds that will germinate on some future dune. Many of these seeds benefit from dispersal as storm tides wash away the dunefront.

The same events that define lifespans of living things near the dune crest, also define the useful lives of our own beachfront structures. Only well behind the dunes does one find lives spanning more than a decade or so. Dune buildings, and the features we construct to defend them from the sea, last only as long as the next storm will allow. Retreating beaches (p. 31) foreshadow the future of coastal forests, and of buildings placed where forests once stood, as entire landforms move. Features anchored to the land are inevitably swept away in their own time. Nothing near beaches is permanent, and yet, cycles of birth and death persist through ages.

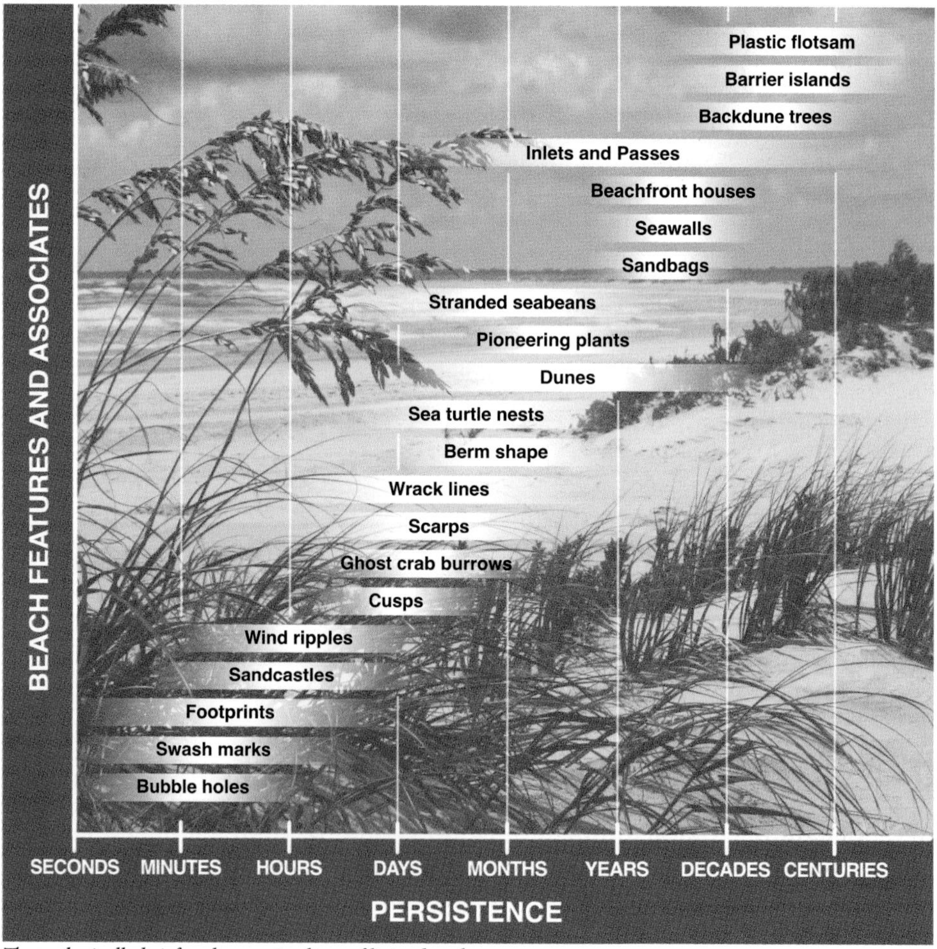

BEACH FEATURES AND ASSOCIATES

Plastic flotsam
Barrier islands
Backdune trees
Inlets and Passes
Beachfront houses
Seawalls
Sandbags
Stranded seabeans
Pioneering plants
Dunes
Sea turtle nests
Berm shape
Wrack lines
Scarps
Ghost crab burrows
Cusps
Wind ripples
Sandcastles
Footprints
Swash marks
Bubble holes

SECONDS MINUTES HOURS DAYS MONTHS YEARS DECADES CENTURIES

PERSISTENCE

The geologically brief and recurring lives of living beaches

Dunes and Coppice Mounds

WHAT ARE THEY? Dunes are piles of sand, typically stabilized by fast-growing, salt-tolerant, "pioneer" plants. Plants also create their own dunes by gaining a foothold on the upper beach where they intercept wind-blown sand. Over the backshore, plants or wrack form the nucleii of **coppice mounds** (embryonic dunes), which may or may not develop into larger sets of shore-parallel ridges making up the **primary dune**. Although the **foredune** (seaward face) of the primary dune may be eroded by storm tides every few years, the backdune is often stable enough to support semi-permanent woody vegetation between severe storms. In many dune systems, secondary dunes are also observed. These dunes were once primary dunes, or they formed behind the primary dune. This formation may have been dramatic, as when severe storms breached primary dunes and deposited sand inland. Or, they may have formed gradually, as winds blew fine-grained sand inland past the primary dune. Secondary dunes may lie within an extensive dune field that spreads hundreds of yards behind the beach. **Active dunes** move at a rate that outpaces plant growth. On windy Padre Island, Texas, active dunes can move 35 feet (10 m) in a year. These dunes are likely to have a bare crest and steep slipface opposite the prevailing wind direction. Dunes are part of a sand "banking" system within which the beach makes continuous deposits and withdrawals.

SIZE: Small mounds to ridges more than 50 ft (15 m) high.

HOW COME? Dunes form when sand that was moved by water or wind slows down enough to settle into a pile. Storm waves push sand onto the upper beach,

Grasses cover a foredune, backed by woody plants

Coppice mounds in front of a primary dune ridge

The Florida Panhandle has impressive dunes

9

A bare, active dune moving right to left

Dune overtakes grassland, Padre Island, Texas

A dune overtakes the road at Padre Island, Texas

and onshore winds blow beach sand into wind shadows behind old wrack lines and vegetation. Pioneer plants colonize these sand piles. As roots keep the sand from migrating windward, the dune grows into a shore-parallel dune ridge.

FOUND: All beaches. Dunes on delta lobe beaches (p. 6) tend to be small. Larger dunes form where there is strong directed wind and extensive sand supply, such as southern Texas and Florida's Panhandle. In Florida, swales behind the dunes provide woody plants shelter from salt spray, and are places where scrub communities and maritime forests can become established. The Texas climate is too dry for most woody dune plants. Beachfront development often levels the dune and prevents reformation. Coppice mounds and new dunes are unable to form where there is raking of the upper beach (p. 390).

SEASONS: All year. Rapid erosion and formation can occur during tropical storms and hurricanes.

DID YOU KNOW? A dune is far more than just a pile of sand. Sediments in a dune are complex and organized. Although winds move mostly lighter quartz sand grains, water can move heavier mineral grains into dunes. These darker sediments trace sand layers that record a dune's history. Dunes also contain a lot of biology, including the seeds of future dune plants, and symbiotic fungi that are critical to plant growth. The formation of dunes has made much of the Gulf coastline we see today. Dig a hole near the shore and you will likely find sand that was once blown from a beach into an ancient dune.

Salt Pruning

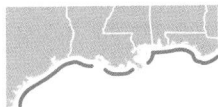

A dune wave made by salt aerosol from surf waves

WHAT IS IT? Salt pruning describes the trimming effects of wind and salt spray, which is the salty aerosol released from breaking waves. This process creates dune shrubs and trees with a sloping-hedge appearance.

SIZE: Salt spray can sculpt century-old live oaks into wavelike forms that are knee-high on their seaward side and well overhead on their protected side.

HOW COME? Salt spray from breaking waves settles on the outer leaves of dune plants. Evaporation of this spray leaves behind concentrated salt that can enter leaves through abrasions caused by wind-whipping. The salt gradually kills the most exposed, windward leaves. This trimming stimulates extensive branching, which produces dense, windward canopies. Salt pruning can result in bonsai-looking shrubs that lean away from the sea. Like bonsai, the beautiful forms of salt-pruned dune shrubs are acquired over decades.

FOUND: Beaches with mature dunes and woody vegetation. The Florida Panhandle has some of the most picturesque examples of salt pruning.

SEASONS: All seasons, although most of the actual pruning occurs during the driest months when there is little rain to wash salt from exposed leaves.

DID YOU KNOW? Salt-spray resistance dictates which plants exist on the dune. Survivors of this torture benefit by having few plant-competitors.

Decades of seaward leaf-pruning by salt spray

A salt-pruned sand live oak

11

Dune swales are wet troughs between dunes

Western Lake, Grayton Beach State Park, Florida

A dune-lake outfall, Florida Panhandle

Swales and Dune Lakes

Swales   Dune Lakes

WHAT ARE THEY? These are freshwater features close to the sea. **Swales** lie between dune ridges where rain and groundwater collect, and may be standing water, wet marsh, or sands just damp enough to allow water-loving plants to grow. **Dune lakes** are large coastal bodies of fresh water, tea-stained from plant tannins, that occasionally flow through an **outfall** into the Gulf across the beach.

SIZE: Swales range from narrow depressions to as wide as a football field. Dune lakes range from small ponds to over 250 acres (100 hectares), and average only neck deep.

HOW COME? Swales develop as dunes build seaward. Low areas between dunes collect fresh water and have less seawater intrusion with increasing distance from the sea. Fresh water and protection from salt spray allow a unique coastal ecosystem to thrive. Dune lakes formed roughly 10,000 years ago when erosion carved out their basins. Ground water and streams keep the lakes fresh or brackish.

FOUND: Swales are scattered among stable dune fields. Dune lakes occur only between the large dunes of the central Florida Panhandle.

SEASONS: Swales are prettiest with the blooming flowers and green vegetation of summer and early fall. Dune lakes often overflow through beach outfalls with the heavy rains of summer.

DID YOU KNOW? Coastal dune lakes are rare, occurring in only a few spots around the world. Fresh water near the beach brings diverse plant and animal life that would not be present without this mix of waters.

Ridges, Runnels, and Lagoons

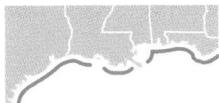

WHAT ARE THEY? Beach **ridges** are exposed sandbars or berms that have been breached by high tides. **Runnels** are mid-beach troughs that are filled with water at high tide, and that can persist through multiple tidal cycles as **"beach lagoons."**

SIZE: Ridges may be up to hundreds of feet wide at low water. Runnels may be equally wide and up to 6 ft (2 m) deep.

A beach ridge (A) and runnel (B) at low tide

HOW COME? These are features of beaches with moderate wave energy and tidal changes. Ridges and runnels indicate a beach in transition. Offshore sandbars become beach ridges when they have swelled with sand from the eroded beach. Runnels are troughs that were carved by longshore currents and are maintained by water flow after filling at high tide. If water can't exit through a rivulet (p. 22), it can persist when muddy sediment slows drainage through the sand. Some ridges grow into barrier islands, and some runnels grow into persistent lagoons.

FOUND: These features are most common on beaches that have endured a recent erosion event. Beach lagoons re-fill during the highest tides.

Beach lagoons flood during the highest tides

SEASONS: All seasons

DID YOU KNOW? Runnels that persist as lagoons become important areas for beach life. Receding waters concentrate trapped invertebrates and small fishes, which in turn attract shorebirds and wading birds.

A reddish egret haunts a beach lagoon at low tide

13

Saltmarsh and tidal channel behind a Gulf beach

An oyster reef behind a beach exposed at low tide

Beach erosion exposes layers of clay

An eroded peat outcropping on a Texas beach

Saltmarsh, Oyster Reef, and Outcroppings of Peat and Clay

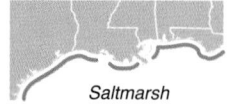

Saltmarsh

Oyster Reef

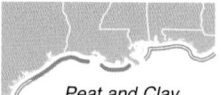

Peat and Clay

WHAT ARE THEY? A **saltmarsh** is a grassy estuary, periodically flooded by tides. An **oyster reef** is a mass of eastern oysters (p. 116) that have fused together to form rocklike shoals. **Peat outcroppings** are compacted, partially decayed plants from old marsh formerly behind the beach. **Clay layers** comprise extremely fine sediments under the peat.

SIZE: Saltmarsh rims coastal areas protected from waves, from a few feet to miles wide. Oyster reef can cover many acres of shallow estuary. Peat and clay outcroppings can extend for hundreds of yards down the beach, but are generally only about two feet (60 cm) in thickness.

HOW COME? Saltmarsh grasses and oysters need calm waters with a blend of fresh and salt. Barrier islands give protection, and coastal rivers provide fresh water. Peat forms where dead marsh grasses accumulate faster than they rot away in the oxygen-poor bottom. Many Gulf beaches lie over peat or clay, which is revealed where the sand supply is low. Peat and underlying clay remain because they erode more slowly than sand.

FOUND: At current and former estuaries.

SEASONS: All seasons

DID YOU KNOW? Underlying clay layers and peat outcroppings may be thousands of years old. Oysters clean estuarine waters by filtering up to a barrel of water per oyster per day.

14

Sand Layers and Shell Pavement

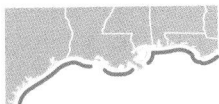

WHAT ARE THEY? Sand layers show up as contrasting horizontal lines containing sands of different densities, and are often seen in eroded scarps. **Shell pavement** remains where wind has scoured away sand covering shell lag deposits (layers of shells).

SIZE: Sand layers range from paper thin to the depth of an entire dune. Lag deposits can be one or many shells thick and may be exposed over flat areas as large as a football field.

HOW COME? Layers of sand mark events when wind or water moved sands of similar makeup onto the upper beach or dune. Light breezes move fine sands, stronger winds move coarse sands, and the heaviest, dark-mineral sands are moved into layers by storms and flowing water. Lag-deposited shells were pushed onto the upper beach by waves and later exposed as wind blew away surface sands. The exposed shell acts as a pavement that prevents further wind erosion.

FOUND: Sand layers occur in all dunes. Natural shell lag deposits lie on the upper beach where storm waves have reached, or where artificial sand pumping (p.393) has occurred.

SEASONS: All seasons

DID YOU KNOW? Layers tell the history of a dune, including when human artifacts were deposited. Nearly all the bivalve shells (e.g., clamshells) paving lag deposits rest concave-side down. In this most hydrodynamic orientation, they were least likely to be resuspended by the flowing water that brought them.

Layers of dark sand revealed in an eroded dune

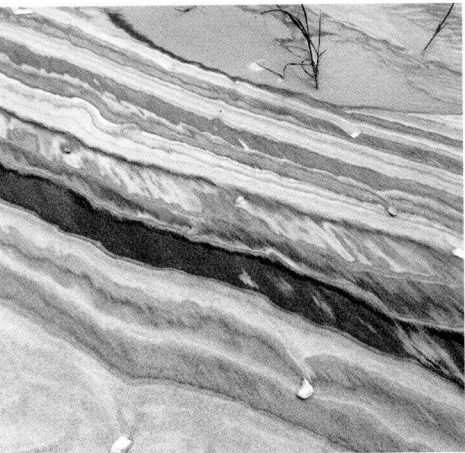

Layers in a beach scarp reveal a complex history

Shell pavement in a lag deposit

15

Driftwood in a Texas wrack line

Dense sargassum on a southern Texas beach

Spartina *reeds are common in wrack near inlets*

Wrack Lines

WHAT ARE THEY? Wrack lines, or strand lines, are lengthy piles of marine stuff that has washed in with the tide. Think of wrack as the ocean's bathtub ring. Wrack is often dominated by open-sea, drifting algae called **sargassum** (p. 310), uprooted seagrasses (pp. 316–317), and **reedy marsh plants**. But wrack can also include **driftwood**, seeds, **sprouting plants**, sea foam (p. 28), a variety of expired animals, and manmade litter. Wrack typically accumulates where waves from the highest tides of the month have swept up the beach. Old wrack can get pushed into the dunes during storms, and new wrack is sometimes spread over the lower beach as it gets reworked during each tidal cycle. The wrack line is a diverse place. Much of this book contains descriptions of the varied animals, plants, and debris that occur in the wrack.

SIZE: Although knee-high piles of wrack are common following rough weather, under calm conditions a beach may be starved of wrack for weeks. Elements of the wrack are as small as single-celled plankton and as large as a shipwreck.

HOW COME? Everything that goes around comes around. This is especially true for organisms floating at sea. After years of roundabout travel in surface currents, plants, animals, and debris originating far and wide often end up on a beach somewhere. But some wrack may be local. Seagrasses and some algae may come from the sea bottom near the beach, and reeds may have washed out a nearby inlet flushing the saltmarsh behind the barrier island. Examine elements of the wrack closely for signs of a lengthy sea voyage, such as animals that live only in the open

sea, and flotsam bearing encrusting hydroids, algae, barnacles, and bites from marine animals.

FOUND: All beaches, although the components of the wrack vary by location. Northern Gulf beaches are more likely to receive marsh plants. Central and southern Texas beaches get abundant sargassum. Benthic (bottom-growing) macroalgae (pp. 310–315) are most common on beaches close to jetties (p. 392) and lagoon inlets (p. 45). There may be many wrack lines on a beach indicating where the tides have reached. Old wrack is the highest, and can linger for weeks before disappearing under sand.

SEASONS: Wrack lines can be extensive in fall and spring following storms. Look for the greatest wrack deposits following days of strong onshore winds. All elements of the wrack arrive under these conditions, including collectible finds such as **seabeans** (pp. 318–341).

DID YOU KNOW? Wrack contains treasure. Although the most famous of these treasures are seabeans, wrack lines are also the best places to search for many other "drifters." This broad category of buoyant items includes interesting thorns, corky bark, surfworn wood, purple sea snails (p. 87), and the spiral shells of ram's horn squid (p. 147). But there is much more to wrack than collectible treasures. The plants and animals that wash onto beaches at the end of their life's journey provide the base of an important beach food web. This web includes easily overlooked life such as fungi and the tiny invertebrates that feed on decaying plants, as well as some of the most charismatic animals on the beach, like our shorebirds. Many of the beach's most appealing animals would be absent were it not for the lowly wrack.

Ruddy turnstones shop the wrack for groceries

A railroad vine sprouts from a drift seed in old wrack

A hamburger bean—just one of many wrack treasures

Cusps at regular intervals on Dauphin Island, Alabama

Mostly rhythmic, crescentic bars off Santa Rosa Is., FL

Rhythmic Shorelines

WHAT ARE THEY? Shoreline features often undulate. When these undulations become regularly spaced, they are described as "rhythmic." Rhythmic shoreline features include **beach cusps**, which are periodic sand hills on the lower beach, and **crescentic bars**, which are segments in the offshore sandbar. Both features are curved mounds convex to the sea.

SIZE: Each generally crests 1–6 ft (0.3–2 m) above the surrounding sand.

HOW COME? Rhythmic features form when a beach receives a steady interval of incoming waves. These waves reflect, and set up shore-parallel "edge waves" trapped within the nearshore trough. When edge waves of similar wavelength come together from opposite directions, a standing edge wave is formed. This wave, stationary relative to the beach and perpendicular to surf waves, has localized peaks and troughs. Peaks increase the height and local erosional power of incoming surf, creating the convex embayments of the beach cusp or bar crescent.

FOUND: Wave-dominated beaches (p. 6)

SEASONS: Cusps are common in summer as beaches build. Crescentic bars form anytime surf is strong but regular.

DID YOU KNOW? Cusps and crescentic bar segments move slowly in wavelike fashion down the beach. The spaces between bar segments are where rip currents (p. 36) flow out. These rips are weaker than those in longer, uninterrupted offshore bars.

Scarps and Slipfaces

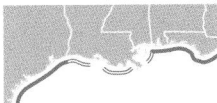

WHAT ARE THEY? Scarps (escarpments) are cliffs in the beach marking the recent line where erosion has taken sand. **Slipfaces** are the steep, downwind sides of active dunes.

SIZE: Scarps are generally 1–6 ft (0.3–2 m) high. Slipfaces are in this same height range, relative to dune size.

HOW COME? Scarps form due to rapid beach erosion, often within a single tidal cycle. They are common during storms or when moderately rough surf and a strong longshore current follow a period of sand buildup (accretion). Dune slipfaces mark "active" (moving) dunes, which have a curved slope toward the prevailing wind and are steep (34 degrees) to the point of slippage on the leeward side. Wind blows sand up to the crest, which periodically collapses down the slipface like a slowly breaking wave.

FOUND: All beaches form scarps during pronounced erosion events. Slipfaces are most common in southern Texas where strong, dry, directed winds blow.

SEASONS: Scarps are common in the stormy fall, winter, and spring. Dune faces slip most during dry days with persistent high winds.

DID YOU KNOW? Dunes can become active if their vegetation is removed and will march landward over anything in their path.

A lower-beach scarp shows recent erosion

Dune slipfaces point downwind

19

Plant arcs reveal recent wind changes

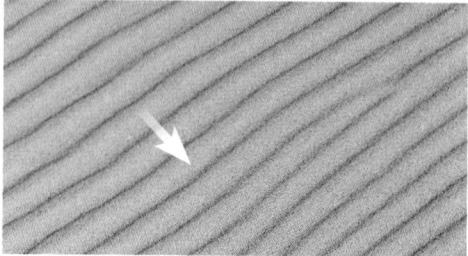

Wind ripples. Arrow shows air flow

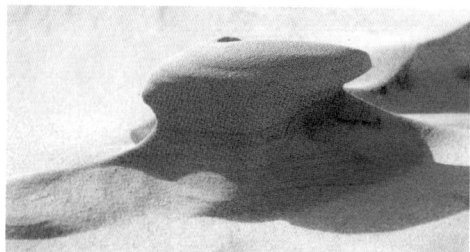

A mushroom pedestal on a windy beach

Harrow marks capped with shells

Sand in the wind shadow of a sea purslane sprout

Plant Arcs, Wind Ripples, Pedestals, and Harrow Marks

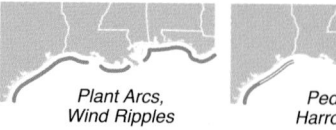

Plant Arcs, Wind Ripples

Pedestals, Harrow Marks

WHAT ARE THEY? Plant arcs trace where plants sweep sand as winds shift direction. **Wind ripples** are miniature dunes arranged perpendicular to the wind. **Pedestals** are mushroom-shape sculptures of salt-cemented sand with a rounded head atop a narrow base. **Harrow marks** are wind-eroded pedestals capped by a shell shard or seaweed scrap.

SIZE: A plant's size determines the arc it sweeps. Wind ripples are almost always about 3 in (7.5 cm) from crest to crest. Pedestals reach bowling-ball size. Harrow marks vary with the size of the shell-bit topping it; the smallest are capped by a single, coarse sand grain.

HOW COME? Ripples are spaced by how far a wind-blown sand grain can leap, in a process called saltation. Leaping grains are more likely to land where there is a slight rise, and these crests gain coarse sand. Finer grains fall into the wind shadows between crests. **Wind shadows** also occur behind objects on the beach. Pedestals and harrow marks occur where salts remain from evaporated wavewash. The brine cements sand grains together in a mix called "salcrete." Strong winds can erode the salcrete, except where the sand surface is protected by an object like a seashell.

FOUND: Most common on the upper beach along the windiest coastlines.

SEASONS: These features form when winds blow greater than 15 mph (24 kph).

DID YOU KNOW? Ripples move slowly downwind, with their crests collecting larger, denser, often darker grains.

Wave and Current Ripples

WHAT ARE THEY? Current and wave ripples are formed by moving water. **Wave ripples** result when sheets of water wash over the beach, as from incoming waves. These ripples have wave forms in cross section, with a steeper slope in the direction of water flow, and are in parallel sets. **Ladderback ripples** are from two perpendicular sets of superimposed wave ripples. **Current ripples** result from turbulent flowing water and are irregular and assymetrical.

SIZE: Most ripple crests are separated by about 3 in (8 cm), but strong currents can pile up much larger **sand waves**.

HOW COME? Assembly of ripples by flowing water is similar to the formation of wind ripples (previous page). Wave ripples are created by even water flow. Swift, deep, turbulent currents create undulating current ripples and larger sand waves that are similar to dunes. As current flow ends, parallel, wave ripples still under water erode first at their crests, making them flat-topped. Ladderback ripples begin as parallel ridges, and receive superimposed ridges when a new flow direction is perpendicular to the original current.

FOUND: Anywhere water moves over sand, especially in runnels (p. 13) and on tidal flats near inlets.

SEASONS: All year

DID YOU KNOW? Like wind ripples, water ripples move continually in the direction of the flow creating them until the tide drains water away. These formations are briefly suspended in time for us to admire until the next high tide arrives.

Wave ripples revealed at low tide

Ladderback ripples show two current-flow directions

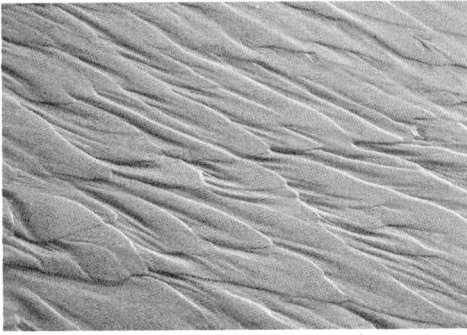

Current ripples in a drained beach runnel

Large, current sand waves from an inlet tidal flow

A set of tree-form rill channels at low tide

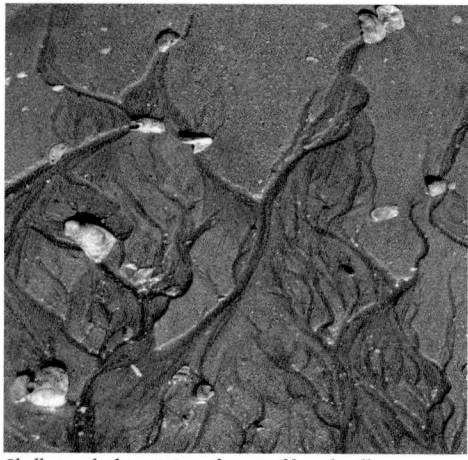

Shells mark the origins of a set of beach rills

A rivulet flowing from a beach runnel

Rills and Rivulets

WHAT ARE THEY? Rills are marks etched by water flowing out of the beach. **Rivulets** are ephemeral flows that drain tidal lagoons and runnels behind the beach.

SIZE: Rill channels have pencil-thin branches but may connect to tree-size systems originating up the beach. Most rivulets can be jumped across. The largest, semi-permanent rivulets might be considered inlets.

HOW COME? Rill marks form where groundwater seeps out from saturated beach sands and the flow is enough to erode fine sands into tiny channels. Rills appear where the water table intersects with the foreshore. Most often, this flowing water is salty, but following wet weather, it may be fresh. Rivulets flow back to the Gulf with water that was pushed into temporary lagoons and runnels during high tide. Rivulets often last just one tidal cycle and re-form along a slightly different path during the next falling tide.

FOUND: Rills form on fine-sand beaches as the edge of the water table is exposed at low tide. Rivulets form most prominently following the highest water of bi-monthly spring tides (p. 34).

SEASONS: All year.

DID YOU KNOW? Sands saturated with water often seep where shells or stones have primed a flow by wicking water through capillary action. The shell produces a tiny spring and the flow erodes fine sands into a miniature river delta.

Swash and Backwash

WHAT IS IT? Swash and **backwash** are the sweeps of water from incoming waves. Swash rushes up onto the beach and backwash flows down into the next wave. This wave-lapping moves sand, creates lines and bubble holes (p. 27), and defines an important home for critters in the sand.

SIZE: From a few feet (about 1 m) on steep beaches and during small waves, to over 100 ft (30 m) wide on flat beaches and during large surf.

HOW COME? The distant energy that produced a wave is finally spent in the swash zone. Sand in this zone varies dynamically from wet to dry in frequent cycles. **Standing (stationary) waves** in the swash form when backwash flows into the subsequent wave. The most landward travel of each swash wave leaves tiny ridges of swept-in sand grains and debris called **swash marks**. These marks often feature coarse grains and tiny shells that can "float" on the water's surface tension. The BB-size floats of sargassum algae (p. 310) are commmon there.

FOUND: All beaches. Larger waves create a wider swash zone.

SEASONS: All seasons.

DID YOU KNOW? The swash zone is not quite beach and not quite sea. It is probably the liveliest part of the beach, being home to crabs, clams, and worms that feed a host of birds, fishes, and other animals. See the *Beach Animals* section for tracks, burrows, and other evidence of these swash-zone critters. Swash-zone sands have unique properties. Seawater soaking into the porous beach transforms the sand into a liquidlike medium giving animals an opportunity to sand-swim.

The swash zone on a southern Texas beach

A sheet of backwash flows into a standing wave

Swash marks with sargassum floats

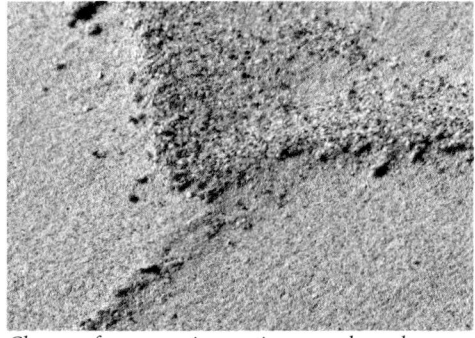

Closeup of coarse grains cresting a swash mark

23

Antidune ripples inscribed by a falling tide

Crescent marks at a sargassum clump

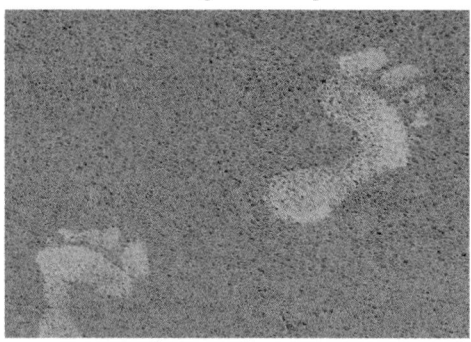

Footprints in hard, wet sand

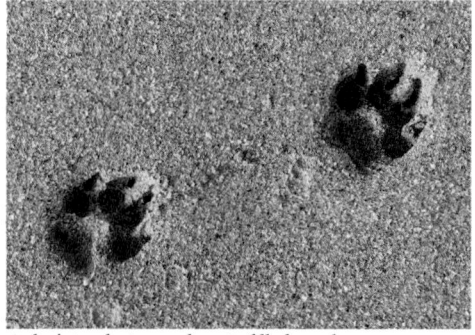

A dog's tracks in mushy, air-filled sand

Antidunes, Crescent Marks, and **Hard and Mushy Sand**

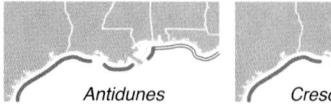

Antidunes   Crescent Marks

WHAT ARE THEY? Antidunes are low, parallel ridges in the swash zone. **Crescent marks** are isolated, swash-sand V marks with an apex at a shell or other object. **Hard sand** is compact and wet, and **mushy sand** is airy.

SIZE: A foot's length separates most antidune bands, and is the width of the average crescent mark. Prints in hard sand barely scuff the surface, but steps in mushy sand can sink down to the shin.

HOW COME? On flat forebeaches, backwash sheet flow against the incoming swash creates standing waves, which form bands of turbulence. This turbulence sorts sediments of varied density into antidunes. These sand features have their steepest face landward. Crescent marks form due to the turbulence presented by an object in the backwash flow. Sand hardness varies with how the sand was packed by swash action. In wet sand, water films surrounding each grain form surface-tension bridges linking grains together. In dry sand, only grain-to-grain friction keeps a foot from sinking. Mushy sand has air spaces within a matrix held by weak salt-bridges.

FOUND: Swash zone and nearby. The hardest sand is on the lower foreshore and mushy sand is near the high-tide line.

SEASONS: All year.

DID YOU KNOW? Footprints on wet sand are light against darker sands because the quartz grains detached by the step are free to refract incident light. Wet sand absorbs light within the beach.

Dark Sands

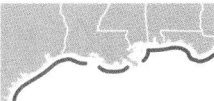

WHAT ARE THEY? Dark sand contrasts against more abundant quartz and feldspar sands. These layers contain the heaviest (densest) mineral grains on the beach, and may be slate gray, greenish, or bronze.

SIZE: Dark sand layers are typically less than 1/4 in (0.6 cm) thick, but can cover the entire lower beach.

HOW COME? Dark sands are revealed following erosion of lighter sands due to rough surf. The sand appears "dirty," but is merely composed of variously dark colored minerals (pp. 351, 352) that make up a tiny fraction of a beach's sand. The darkest mineral layers contain mostly iron and titanium oxides, whereas greenish and bronze layers also have rarer minerals like green hornblende, and gemstone grains like garnet, tourmaline, and zircon. Dense sands tend to settle together in sheets and are last to suspend in wavewash.

FOUND: All beaches. Mixtures of mineral sands differ between locations due to varied sediment contributions from regional rivers.

SEASONS: All seasons, but most common following rough surf.

DID YOU KNOW? Swash-sand features like ripples and crescent marks become more conspicuous where mineral sand layers are exposed. This is because of the correlations between sand color, density, and erosion resistance. To explore the world of tiny but abundant semi-precious gemstones (p. 351) use a jewelers loupe to examine samples of greenish and bronze sand layers.

A storm erodes lighter sand to uncover a dark layer

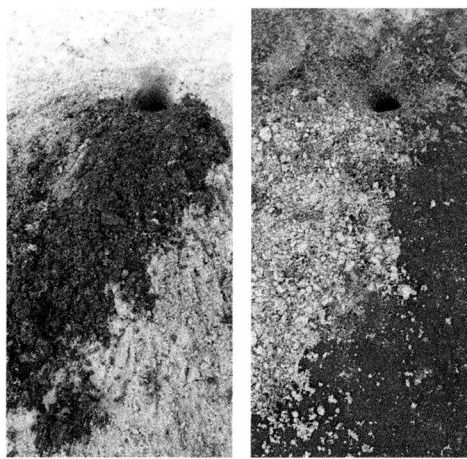

Crab burrows reveal contrasting light and dark layers

Dark magnetite with swirls of other mineral sands

25

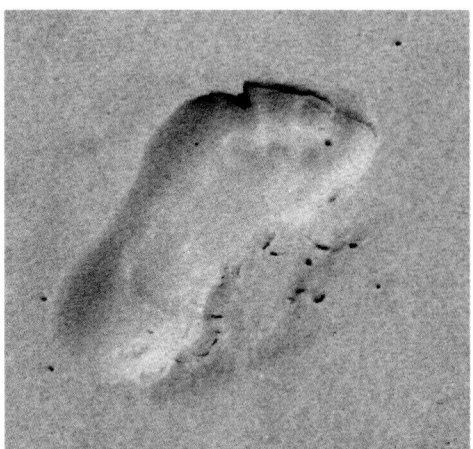

Footsteps in powdery backshore sands often squeak

Shell hash contains well-worn shell bits

Big Shell Beach, Padre Island, Texas

Beach Squeaks and Shell Hash

Beach Squeaks

Shell Hash

WHAT ARE THEY? Beach squeaks (also whistles and barks) are the unique sounds made by forceful footsteps in sugary sand. **Shell hash** describes surface patches of mollusk shell-shards.

SIZE: Footsteps in squeaking sands produce staccato notes with a pitch equivalent to a high female voice (about 1,000 Hz). This makes them heard (and appreciated) best by children. Shell hash may be a foot (30 cm) or more thick and cover thousands of square feet.

HOW COME? Each sand grain rubbing past another makes a sound. Compound this action by many thousands and you have a footstep squeak. More forceful steps make barking sounds. These sand sounds require dry, clean, uniform (well-sorted), quartz sand. Shell hash comprises mollusk shells broken into shards. Rounded and polished pieces of roughly similar size are assembled by waves and currents and remain in the swash zone after a period of rough surf.

FOUND: Squeaks and barks can be made in the upper beach and dunes of natural beaches (without artificially placed sand), especially northern Gulf beaches with predominantly quartz sand. Shell hash can assemble anywhere there are sufficient shells. Little Shell and Big Shell Beaches on Padre Island, Texas, have abundant shell hash due to a convergence of currents there (p. 37).

SEASONS: Sugar-sand beaches squeak any time the sand is dry. Shell hash is most common during calm periods after rough surf.

DID YOU KNOW? Shell hash is the best place to find seaglass (p. 383).

Bubble Holes, Blisters, Pits, and **Volcanoes**

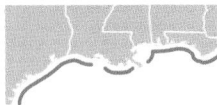

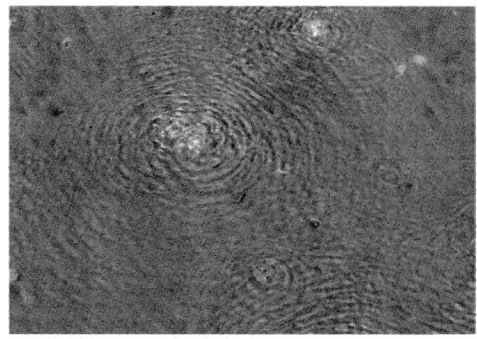

Air bubbling out of a fluffy beach

WHAT IS IT? All these features are perforations in the lower beach just above the swash zone. **Bubble holes** are punctures, **blisters** are bumps over air pockets, **pits** are holes within a depression, and **volcanoes** are bubble holes surrounded by a tiny rim.

SIZE: These air outlets are pin- to nail-size holes. Blisters are generally the size of a quarter.

HOW COME? Sand pushed up the beach by wavewash drains to leave air trapped within a mushy matrix. When a sheet of wave-water surges up the beach, water drains into the sand and forces air to bubble out through holes or to erupt forcefully like a tiny volcano of watery sand. Air trapped beneath a skin of salcrete (salt-cemented sand) may merely push up a sandy "blister." If the blister pops, a pit remains.

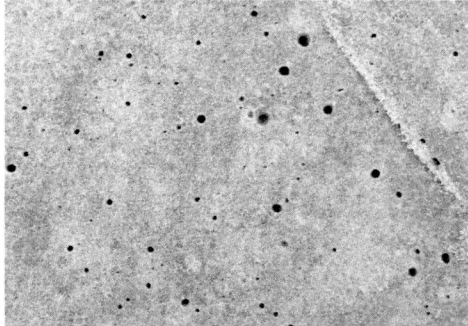

Bubble holes left after swash waters soaked in

FOUND: All beaches in areas of mushy sand, typically in the beach berm at the upper reach of the recent swash.

SEASONS: All seasons, although these holes form most during dry weather. Look for these features forming on a rising tide and remaining after a falling tide.

Blisters and pits in an air-puffed beach

DID YOU KNOW? Sand drained of water without bubbling can retain air as a third of its volume, which makes it very mushy. These swash holes are commonly mistaken for animal burrows. Most critters in this zone sand-swim rather than construct persistent homes. The fortified tubes of ghost shrimp (p. 164) are an exception.

A volcano bubble hole shows forceful air expulsion

A line of sea foam left by receding backwash

A ruddy turnstone searches for snacks in sticky foam

Sanderlings forage near brownish sea foam

Sea Foam

WHAT IS IT? Sea foam is the white, greenish, or brownish froth whipped up by breaking waves and brought to the beach by onshore winds.

SIZE: Most sea foam probably comes from diatoms, which are microscopic single-celled plants in glasslike capsules. The foam they generate in the wavewash is generally the consistency of a good stout beer, but airy sea foam can occasionally roll from the swash in thick suds rivaling any bubble bath.

HOW COME? Sea foam originates from a wide range of mostly planktonic (small and drifting) plants and animals. Wind and surface currents bring billions of these tiny critters to the surf where their cells are pulverized and their proteins and fats are whipped into suds.

FOUND: All beaches, but the choppy surf of breezy southern Texas seems to produce the Gulf's thickest sea foam.

SEASONS: Sea foam is generated when there are nearby "blooms" of plankton and when local sea conditions are rough enough to whip the plankton into froth. Late spring through early fall have the largest blooms and the foamiest surf.

DID YOU KNOW? The oceanic stuff that creates sea foam provides one of the largest active carbon reservoirs on Earth, more than all the rainforests put together. Most sea foam is harmless. However, bursting sea-foam bubbles during a red-tide event (p. 43) can release algal toxins that severely irritate the nose and throat.

Sand Erosion

WHAT IS IT? Erosion is beach sand loss. Of course, the sand isn't really lost. It just goes someplace else. The term "critical erosion" is used to describe sand loss that threatens buildings and other human interests on the dune.

SIZE: It is common to lose 1–2 ft (30–60 cm) of sand depth after a moderate storm. During hurricanes, all the dry sand on the upper beach and dune can move away in a matter of hours.

This standing pine serves as an erosion "dipstick"

HOW COME? Erosion comes from high water, waves, and currents. Waves suspend the sand and currents carry it away. High water brings this erosion to the upper beach and dune. Erosion is constant, even on beaches that seem to be growing. When accretion (sand build-up) is out-paced by erosion, beaches become steeper and more narrow. This net erosion can occur rapidly during storms that drive rough surf and strong longshore currents. Chronic erosion can occur where updrift inlets and jetties intercept the longshore flow of replacement sands.

FOUND: All beaches. Occasionally, net erosion can "sink" beaches at the ends of barrier islands. Where dunes are replaced by coastal armoring, narrow, eroded beaches become former beaches.

Upper beach and dune following an erosion event

SEASONS: Most moderate erosion occurs between late fall and spring. Profound erosion occurs during intense storms.

DID YOU KNOW? Sand that erodes from beaches generally goes no farther than the offshore bar. After severe storms, this bar widens to become exposed beach at low tide.

Moderate erosion takes place each high tide

29

This buried boat is testment to accretion

The simplest objects can record sand movement

A sign of accretion

Sand Accretion

WHAT IS IT? Accretion is beach sand build-up. Accretion and erosion (sand loss) are the yin and yang of Gulf beaches. Their dynamic balance maintains beaches as open and sandy places.

SIZE: Beach accretion is noticeable when stationary objects on the upper beach become buried by sand. Logs and other large items on the lower beach may be buried in sand within a single tidal cycle.

HOW COME? Accreting sand comes from eroding updrift beaches (up the longshore current stream) and from the eroding offshore sandbar. Accretion is typically more gradual than erosion. During calm periods, breakers suspend sand and carry it up the beach-face where the sand falls out of suspension. To understand the perpetual nature of a beach's dynamic sand balance, watch your footprint disappear after a single wavewash.

FOUND: All beaches. Some beaches, such as those updrift from inlets and passes, may experience years of net accretion (more accretion than erosion).

SEASONS: Most accretion occurs gradually during summer. Rapid accretion can occur during or after intense storms.

DID YOU KNOW? Although beach erosion makes headlines, beach accretion takes place in obscurity. Judging only by news reports, there should be nothing left of Gulf beaches.

Retreating Beaches

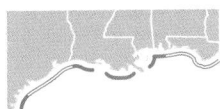

WHAT ARE THEY? Beaches retreat when the shoreline moves landward. **Boneyard beaches** occur where the sea has eroded through the dune into forest, killing the trees with salt water and leaving their woody "skeletons" on the beach. Other symptoms of retreat include exposed **underlying peat formations**, and remnants of **former marsh** where there was once sandy beach.

SIZE: Beach retreat happens across an enormous scale. For example, 80 percent of Texas beachfront is retreating, with an average rate of four feet (1.2 m) per year, and with some beaches moving landward at 55 feet (17 m) per year.

HOW COME? Beach retreat is complicated, although sea level rise due to climate change is the most obvious driver. Other contributors include (1) net sand loss due to sediment capture by dams on contributing rivers, (2) subsidence of sediments due to oil extraction, (3) localized tectonic subsidence (sinking of Earth's crust), and (4) increased storm activity that escalates erosion and landward sand flow at spillover fans (p. 32).

FOUND: Retreat is highest at mainland and delta lobe beaches between western Texas and the Mississippi Delta, but is also rapid on the Matagorda Peninsula, Texas, the barrier islands of Mississippi, and in the Florida Panhandle near tidal inlets.

SEASONS: All year

DID YOU KNOW? Beach retreat has resulted in barrier islands joining to the mainland and sinking into sandbars.

Boneyard beach pine forest, Cape San Blas, Florida

A beach retreats into underlying peat, Matagorda, TX

Stubble of former marsh reeds at a retreating beach

Blowouts leave only the highest plant tufts

Old overwash fans on a central Texas beach

A washout channel between dunes

Blowouts and Overwash

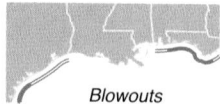

Blowouts

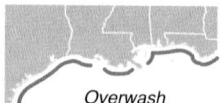

Overwash

WHAT IS IT? Blowouts are dune gaps created by wind. **Overwash** occurs when waves push water through dunes, creating **overwash fans** and **washout channels**. This water flow often crosses entire barrier islands.

SIZE: The smallest gaps are a few yards (meters) wide; the largest span hundreds of yards of former dune.

HOW COME? Blowouts begin as small vegetation gaps that focus winds, and widen from erosion as winds continue. During storms, offshore shoals focus wave energy toward particular beaches where low dunes allow wave-water to push landward. On thin islands, storm tides can flow overland to the backing estuary, creating fans of former beach/dune sand behind the island.

FOUND: These features are most common on wave-dominated barrier islands, but form any place storms strike.

SEASONS: Overwash flow and blowouts take place during storms, but traces of these processes persist for years.

DID YOU KNOW? Coastal geologists struggle to predict where major overwash events will take place based on models factoring offshore bathymetry, wave height, storm surge, and dune elevation. Often, overwash past is a prologue for both future overwash and inlet formation. Overwash flow is essentially a short-lived inlet. Tidal flow after passage of the storm surge can keep a passage open. Evidence for past overwash events includes layers of shells and mineral sediments.

Sandbars

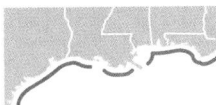

WHAT IS IT? A beach's offshore **sandbar** (also called the longshore bar) is a long, narrow shoal just off the beach and parallel to the coast. A trough separates the sandbar from the beach. Our wave-dominated beaches tend to have multiple offshore bars.

SIZE: From 50 ft (15 m) to more than 1,000 ft (300 m) off the beach, and in 3–20 ft (1–6 m) of water. The shallowest bars are often exposed at the lowest tides.

HOW COME? The sandbar is part of the beach's sand exchange system, which also includes the dune, updrift and downdrift beaches, and offshore shoals. The sandbar typically has the most sand when the beach has the least, and vice-versa. A bar that is removed will re-form at the expense of the beach.

FOUND: Nearly all beaches. Beaches with gentle slopes have the most pronounced sandbars.

SEASONS: The summer bar is typically smaller and closer to the beach than the winter bar.

DID YOU KNOW? After a severe hurricane, sandbars swell with the sand eroded from the beach and dune. Extensive beach-sand loss can be mirrored by a similar increase in sandbar elevation. Much of this sand returns quickly (in weeks) to nearby beaches. The sandbar is home to a varied array of small animals living within its shifting sands.

Shallow sandbar off a Panhandle beach on a calm day

Sandbars are favored loafing spots for seabirds

Surf breaking on a sandbar

33

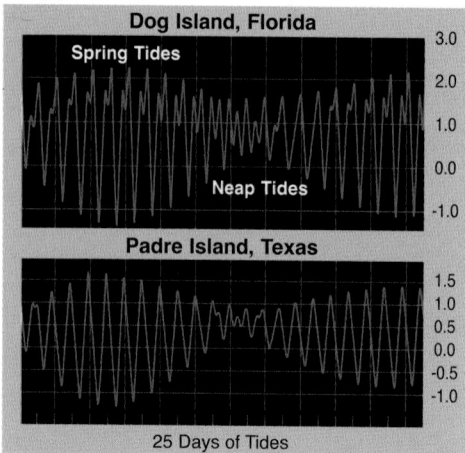

Tide levels (ft) at two Gulf beaches 800 miles apart

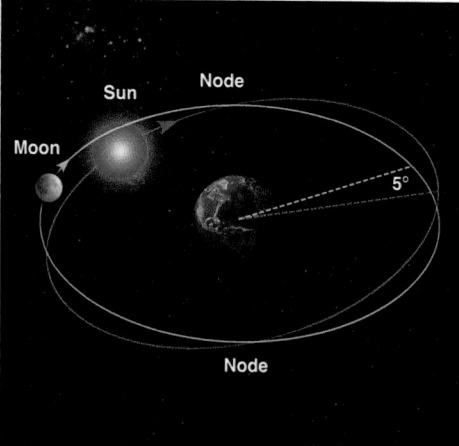

Gulf neap tides occur when the Moon is at a node

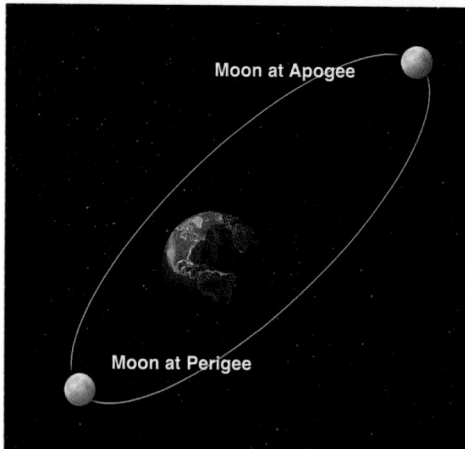

King tides occur when the Moon is at perigee

Tides

WHAT ARE THEY? Tides are periodic changes in local sea level. These levels flood until high tide then ebb to low tide. **Spring tides** have a large amplitude between highs and lows. **Neap tides** have small amplitudes.

SIZE: Average amplitudes between Gulf high and low tides are 1–2 ft (30–60 cm).

HOW COME? Gulf beaches have predominantly diurnal tides (one high and one low per day). The cadence of Gulf tides comes from rhythmic sloshing between the middle Gulf and its margins. This means that all Gulf beaches experience each high and low tide at about the same time. The dominant driver of this rhythm is gravitational tug from **the Moon**, which orbits Earth throughout a 27.3-day "tropical month." The plane of the Moon's orbit is inclined relative to **Earth's equatorial plane**. The lunar, tide-raising force is greatest when the Moon is at its maximum angle north or south of the equator, bringing spring tides to the Gulf. When the Moon's orbit is over the equator at a "**node**," we have neap tides. During neap tides, Gulf beaches get mixed semidiurnal tides, characterized by two sets of tides per day, with one low tide and one high tide being dominant. In the top figure, **Dog Island** in the eastern Gulf shows this trend toward neap, mixed tides.

FOUND: All beaches

SEASONS: "King tides" (perigean spring tides) occur when the Moon's elliptical orbit puts it at **perigee**, closest to Earth. King tides are greatest near summer or winter solstice, adding a couple of inches to high tide.

DID YOU KNOW? "Wind tides" are common on breezy southern Texas beaches where strong onshore winds blow, forcing water ashore in an apparent high tide that can last for days.

Waves and Surf

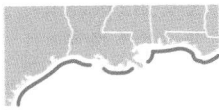

WHAT ARE THEY? Waves (swells) bring energy to the beach from far away. A wave "trips" in shallow water to become a breaker in the **surf**.

SIZE: From 0–12 ft (0–3.7 m) in height (measured base to crest). A large, 10-ft (3-m) wave from a distant hurricane with a 10-second swell-period will evolve into a 12-ft (3.7-m) breaker. Most Gulf surf is less than 4 ft and choppy, with a crest-to-crest wave period of 4–5 sec.

HOW COME? As a wave enters shallow water, it drags the bottom and slows down. This drag causes waves from many directions to gradually turn (refract) parallel to shore. At a depth of about 1.3 times wave height, a wave breaks. Breaking occurs when a wave slows down, gets higher and steeper, and its crest becomes turbulent. This crest slows last, pushes ahead of the wave face, and the wave breaks. Wave crests identify three kinds of breakers: **spilling, plunging,** and **surging.** Spilling breakers have frothy crests flowing down the wave-front and occur when waves advance on a gentle slope. Plunging breakers have crests that curl over the wave-base. They come from large waves approaching a moderate slope, and are the best waves for surfing. Surging breakers have crests that start to spill just as the leading wave-base slides up the beach. Surging breakers occur when the beach approach is steep relative to wave height.

FOUND: All beaches

SEASONS: August has the Gulf's lowest average surf (hip-high), and December has the highest (head-high).

A brown pelican surfs a spilling breaker

Plunging breakers show a moderately sloped approach

A surging breaker slides up the beach

Some currents pose a swimming risk

A small rip channel through an offshore bar

Escape a rip current by swimming parallel to shore

A strong tidal current exits an inlet channel

Nearshore Currents

WHAT ARE THEY? Nearshore currents include the **longshore current**, which flows parallel to the beach, **rip currents**, which flow perpendicular to the beach, and **tidal currents**, which flow through inlets and passes.

SIZE: Longshore currents run for miles along a beach. Rip currents end at a plume just outside the breakers. Tidal currents run the extent of inshore channels. Each can flow at a swift walking pace.

HOW COME? The longshore current runs inside the surf and is forced by waves driving toward the beach at oblique angles. Longshore currents tend to run counter-clockwise along the Gulf's margins, except in southern Texas, and for barrier islands that are not parallel with the Gulf's rim. The **convergence** of north and south longshore currents in Texas occurs at Shell Beaches on Padre Island, where Gulf waves focus their energy (diagram, facing page). Rip currents come from waves driving water into the trough between the beach and the offshore bar. This water rushes away from the beach along short-lived **rip channels** through the bar. Rip currents are strongest when large swells arrive near low tide. Tidal currents go slack a couple of hours after max flood or ebb tides.

FOUND: All beaches. Nearshore currents are strongest on beaches with high surf.

SEASONS: Longshore and rip currents are strongest fall through early spring. Spring tides (p. 34) drive swift tidal currents.

DANGERS: Panic induced by rip currents causes many drowning deaths. To **escape a rip**, don't swim into it. Swim parallel to the beach until you exit.

DID YOU KNOW? Although the **longshore current direction** can shift, its average flow is clearly revealed by net sand transport. Rivers that have supplied sand to western Gulf beaches have unique mineral-sediment signatures used to trace this flow. Analysis of sand signatures clearly shows net current flow from the Rio Grande north, from the Colorado and Brazos Rivers south, and from the Mississippi west. This net longshore flow also carries seashells. Where north and south longshore currents meet on **Padre, Island, Texas**, the shells carried by these currents form a vast accumulation of shell hash (p. 26) at Shell Beach (Big and Little Shell Beaches). The consistency of this convergence is due to Shell Beach's location along the curved Texas coast. North and south of Shell Beach, the predominant wave approach is at an oblique angle, which pushes longshore current flow toward the shelly location where wave front direction averages parallel to the shoreline."

Longshore current flow off a Florida Panhandle beach

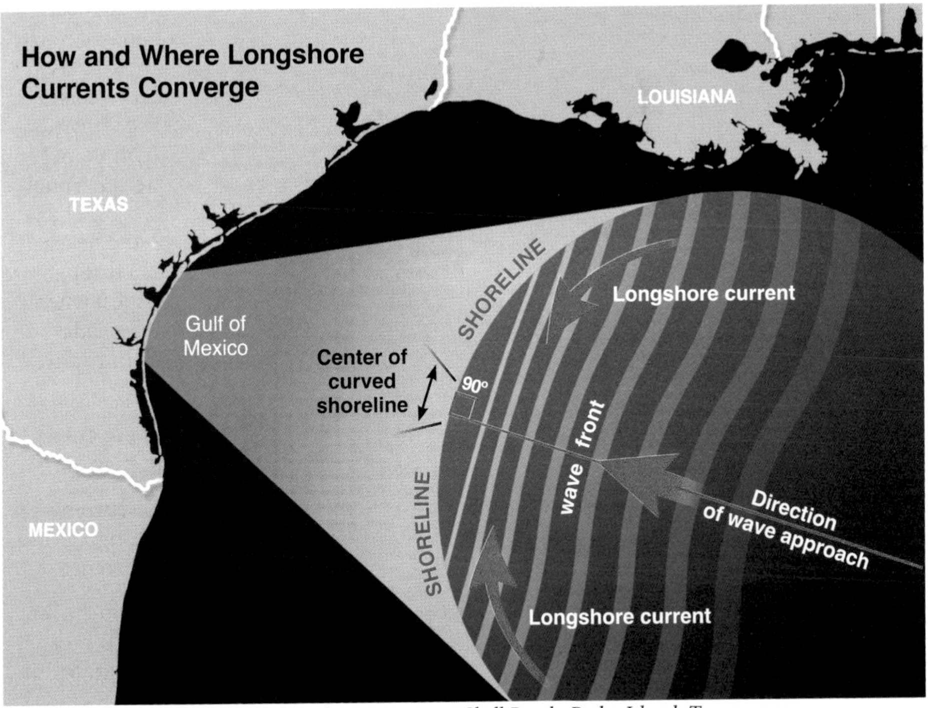

How and Where Longshore Currents Converge

LOUISIANA

TEXAS

Gulf of
Mexico

Center of
curved
shoreline

90°

SHORELINE

Longshore current

wave front

Direction
of wave approach

SHORELINE

MEXICO

Longshore current

Convergence of north and south longshore currents at Shell Beach, Padre Island, Texas

Offshore Currents

Open-sea currents link Gulf beaches to faraway events. The Gulf of Mexico receives water that has circulated around the North Atlantic in a system of surface currents known collectively as the North Atlantic Gyre—a clockwise swirl spanning the New and Old Worlds. These currents are pushed by easterly tropical tradewinds and include a continuous flood entering the Gulf between Yucatan and Cuba. This head of water drives the **Loop Current**, which makes a U-turn short of the Florida Panhandle. Periodically, the Loop Current's loop pinches off into an **eddy** that slowly drifts westward. Loop-Current water leaving the Gulf in the Florida Current leads into the Gulf Stream to circle the Atlantic Ocean and flow into the Caribbean Sea before re-entering the Gulf of Mexico.

Examples abound showing how these ocean currents link Gulf beaches with distant places and processes. Some pumice rocks found on Gulf beaches were blown out of volcanoes in the Windward Islands and bobbed at sea for months or years before making their landfall. Buoyant, stone-hard seabeans from Central American vines plop into rivers, float to the Caribbean, and eventually strand on Gulf coasts. Blisterpod fruits, palm nuts, and porcupine seeds exhaled by the Amazon rainforest are swept into ocean currents and arrive in the Gulf after a long Caribbean journey. Waters surrounding the Azores islands in the eastern Atlantic host nursery-age, current-riding loggerhead sea turtles that first scampered to the surf as hatchlings off Florida Panhandle beaches. Floating mats of netting and bamboo framing with solar tracking units, known as a Fish Aggregation Devices (FADs), drift away from tuna purse-seine fishers off western Africa, cross the Atlantic, and strand on Gulf beaches. On any day, one can find a beach-stranded object that has made an astounding journey.

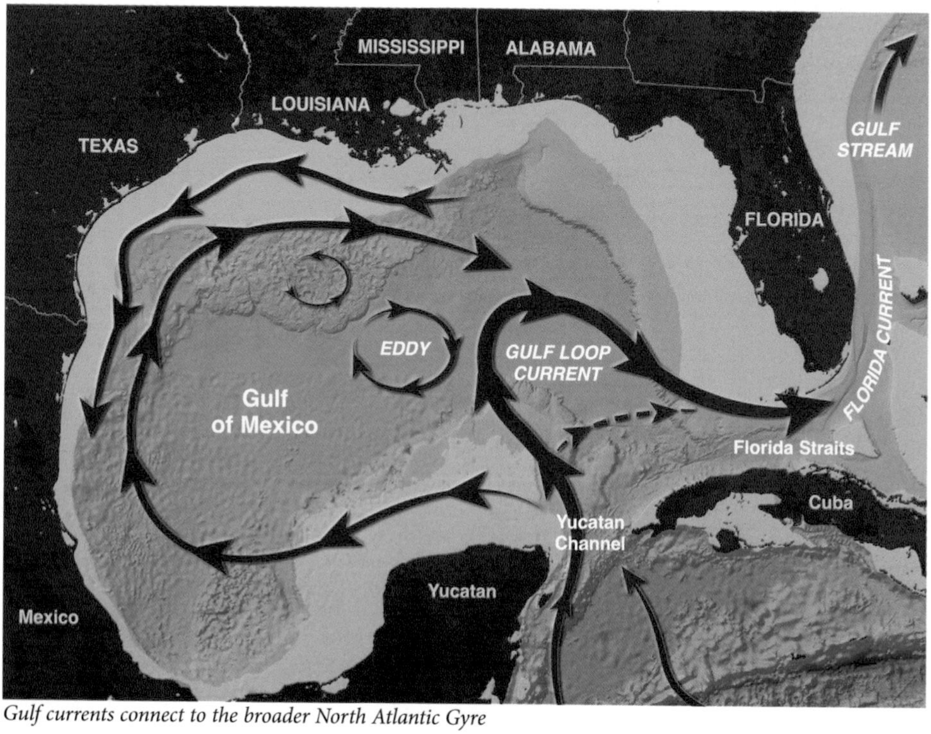

Gulf currents connect to the broader North Atlantic Gyre

Beach Weather

WHAT IS IT? Beach weather, described by temperature, humidity, wind, rain, and sunshine, is often different from inland weather because of a beach's proximity to the sea. In this way, weather and oceanography are intertwined. A beach's climate is its long-term weather patterns. Experiencing atmospheric phenomena from a beach vantage point brings unique perspectives in addition to some hazards.

Flagging seaoats show an afternoon sea breeze

HOW COME? During warm months, inland surface air heats and begins to rise around midday. This rising air is replaced by air rushing in from the Gulf, giving the coast a **sea breeze** that is cooler and more humid than inland air. In winter, the high heat capacity of water, in both the Gulf and coastal lagoons, moderates beach temperatures, making beach areas less prone to freezes. A highly localized form of weather unique to beaches, salt spray, comes from the briny aerosol mist formed during high humidity and rough surf. Sunshine is brighter on the beach; reflection from the sea and sand can intensify the ultraviolet light we receive. By the same token, coastal **sunrises** are more vivid. An open horizon and reflecting sea enhance the visual experience. A person on a wide-open beach during a thunderstorm could become the easiest path to ground for a bolt of **lightning**. Lifeguards in many locations fly **beach warning flags** that identify weather-related hazards.

Texas kite surfers enjoy consistent sea breezes

FOUND: All beaches

SEASONS: Sea breeze and lightning season is late spring through fall. Salt spray is most pronounced on humid, breezy days, late spring through early fall. In winter, beaches have temperatures moderated by surrounding water, but passing polar air masses, called northers, generate strong, cold, north winds, which alter currents and sand movement.

Sunrise at the beach allows a broader view of beauty

Flags announcing moderately high surf and jellyfish

Beaches can be lightning hotspots

DID YOU KNOW? Beach plants include species that are more tropical and salt tolerant than plants found in landward locations. Many plants common in the Caribbean are at home on Gulf beaches. Our changing climate will have conspicuous effects on Gulf beach landscapes.

DANGER: Sunburn increases skin-cancer risk, which kills more than 2,000 Americans each year. Wear a hat, use sunscreen, and enjoy the beach at dawn and dusk. Lightning is a danger due to the electrical exposure of an open beach. Take shelter when thunder is heard. Strong winds often bring dangerous surf and rip currents. Beach warning flags advertise these hazards, but not all beaches have this alert system.

To every thing there is a season. The graphic below shows how some conspicuous beach features vary throughout the year's weather. Prevailing local temperature, humidity, daylength, wind, sea conditions, and erosion/accretion cycles all drive these seasonal occurrences. Cycles of plant flowering, seeding, and dormancy are closely tied to beach weather. Beach infauna (critters living within the sand) move in anticipation of rough surf. And the arrival of drifters, like seabeans, is timed by distant weather and currents.

JAN	FEB	MAR	APR	MAY	JUN	JUL	AUG	SEP	OCT	NOV	DEC
				Salt spray, lightning strikes, waterspouts							
					Hurricanes and tropical storms						
Blue northers										Blue northers	
High erosion, accretion, overwash							High erosion, accretion, overwash				
Narrow berms, distant sandbars				Wide berms, close-in sandbars					Narrow berms		
Infauna offshore				Abundant beach infauna				Infauna migrate offshore			
Plant dormancy									Plant dormancy		
					Flowering						
				Annuals grow							
						Seeds dispersed					
Big surf				Calm surf				Variable surf			

Seasonal changes make each beach visit a unique experience

Tropical Cyclones and Northers

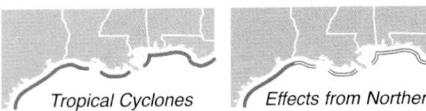

Tropical Cyclones   *Effects from Northers*

WHAT ARE THEY? Tropical cyclones include tropical depressions (sustained winds to 39 mph), tropical storms (up to 74 mph), and **hurricanes** (more than 74 mph). These storms are counterclockwise-spinning low-pressure systems with a spiral arrangement of thunderstorms. They originate from warm seas, as far away as the coast of western Africa. "Northers" are arctic (cold) fronts that collide with warmer Gulf air and bring sudden storms followed by north winds up to 35 mph.

SIZE: Tropical cyclone effects can occupy the entire Gulf region. Northers can sweep from Canada to the Gulf in two days.

HOW COME? These intense storms are fueled by heat energy released from condensing water vapor.

FOUND: Tropical cyclones have affected all Gulf beaches. Northers sweep by all Northern Gulf beaches, but are especially pronounced in Texas, where the locals call these fronts, "blue northers."

SEASONS: Our tropical cyclone season is June through November, with a sharp peak late August through September. Northers occur in late fall and winter.

DID YOU KNOW? Storms shape beaches. In 2005, **Hurricane Katrina** washed over Dauphin Island, Alabama, cutting a mile-wide channel through the island's center. Although occasionally tragic for human residents, assaults by storms set back the advance of land plants, shift sands between land and sea, and foster stunning scenery. Storms make beaches beautiful, wondrous places to visit, but scary places to live.

In 2020, Hurricane Laura spanned the Gulf, TX to FL

A Texas barge stranded by Hurricane Harvey

Homes on Dauphin Is., AL, after Hurricane Katrina

41

Florida marker shows category 1–5 storm surge levels

Hurricane Strikes

Strikes from hurricanes bring about profound beach change. A strike (landfall) occurs where the storm's center moves over the coastline, but storm conditions are likely to extend hundreds of miles on either side. Hurricanes striking Gulf beaches have their most severe effects to the right of their forward movement. Effects include high winds and **storm surge** (elevated sea level). Gulf beaches with gradual offshore slopes have potential surges of 30 ft (9 m). Storm surge was the principal factor in the deadliest natural disaster in US history, the Galveston Hurricane of 1900, which killed 6,000 to 12,000 Texans. One way to show this coastal risk is **hurricane return period**, which is the frequency that hurricanes have struck a given location. A return period of 20 years means that during the previous 100 years, a hurricane struck that location five times.

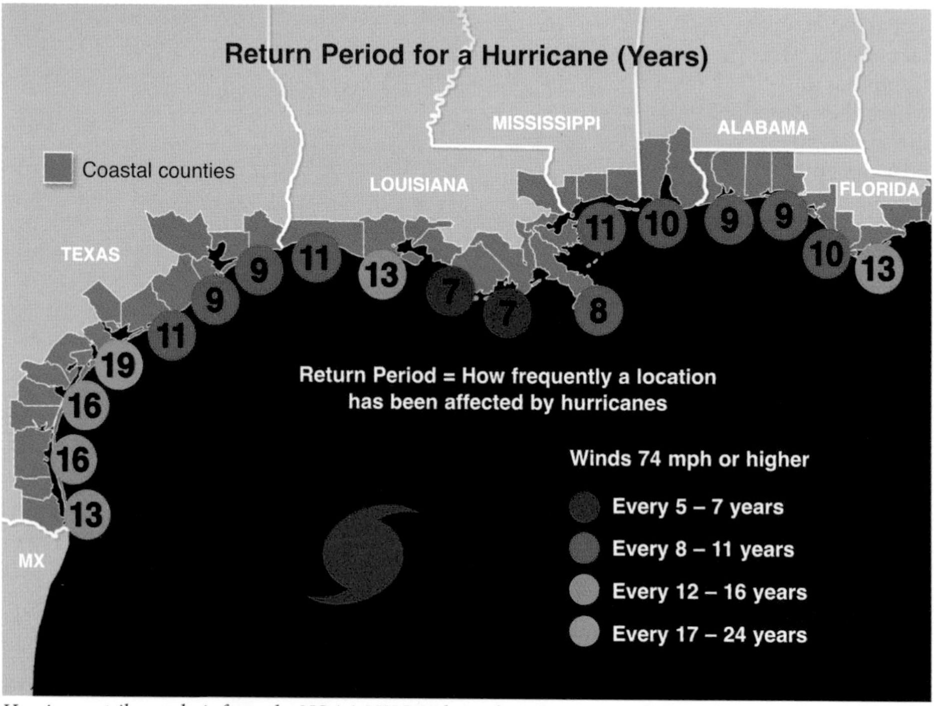

Return Period for a Hurricane (Years)

Coastal counties

MISSISSIPPI

ALABAMA

LOUISIANA

FLORIDA

TEXAS

11 10 9 9

9 11 13

10 13

9

11 7 7

8

19

Return Period = How frequently a location
has been affected by hurricanes

16

16

Winds 74 mph or higher

13

Every 5 – 7 years

MX

Every 8 – 11 years

Every 12 – 16 years

Every 17 – 24 years

Hurricane-strike analysis from the NOAA NHC Risk Analysis Program with data through 2020

The Dead Zone and Red Tide

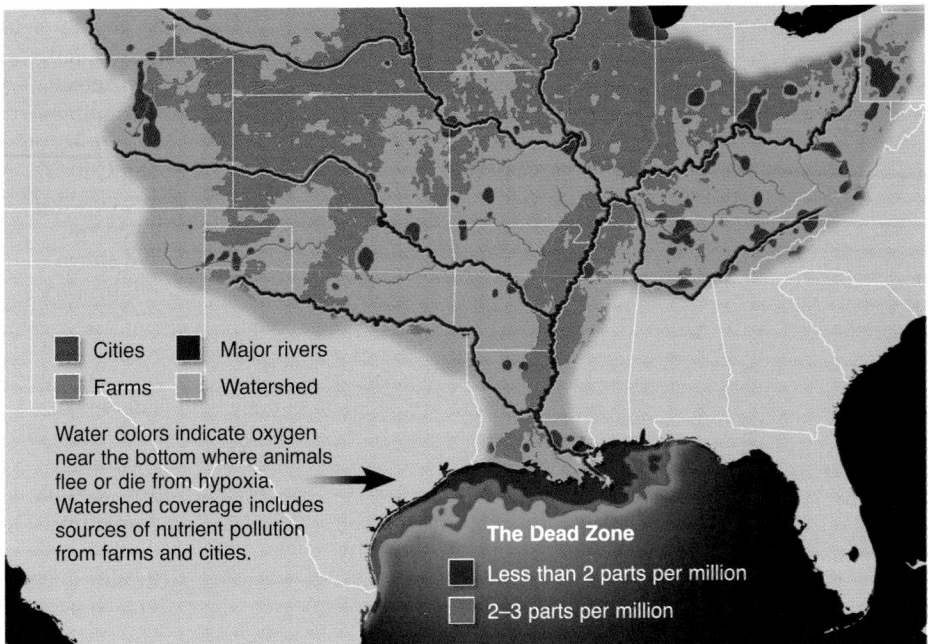

Cities   Major rivers
Farms   Watershed

Water colors indicate oxygen near the bottom where animals flee or die from hypoxia. Watershed coverage includes sources of nutrient pollution from farms and cities.

The Dead Zone
Less than 2 parts per million
2–3 parts per million

The Mississippi watershed and Gulf Dead Zone in 2010. Analysis from the NOAA National Ocean Service

The **dead zone** (hypoxic zone) is a broad, offshore area of low oxygen beneath Gulf surface waters. The phenomenon is caused by nutrients (fertilizer) and sewage effluent from farmlands and cities in the **Mississippi River watershed**. The nutrients flow to the Gulf where they stimulate massive algal growth during warm months. The algae eventually die, sink, and are decomposed by oxygen-consuming bacteria, resulting in oxygen levels near the bottom that are insufficient for most life. These hypoxic waters cause mobile marine animals to flee, immobile animals to die, and bring about reduced reproduction and growth rates. Recently, the summer dead zone has averaged about 6,000 square miles, or the size of Vermont. Researchers have traced most of the harmful nutrients to fertilizer for corn and soybean farming.

Red tides are blooms (population explosions) of toxic dinoflagellates (single-cell, whip-tail algae), most often, *Karenia brevis*. In high concentrations, a toxin made by this alga kills marine life and irritates breathing in humans. Within the range of this book red tides are most common in southern Texas, August through February.

A red-tide fish kill

43

Southern Texas seafoam green

Florida's Emerald Coast was named for its water color

At its clearest, Gulf surf is blue green

Near rivers, Gulf waters can look like chocolate milk

Sky reflection can make any water appear blue

Water Color

WHAT IS IT? From the beach, Gulf waters may appear blue-green, emerald, mint green (seafoam), tea-colored, gray-brown, chocolate milk, or blends of these hues. Apparent water color is influenced by reflection, surface roughness, bottom sands, ambient light, and color from both dissolved and suspended components.

SIZE: Both the algae that tint seawater green, and the suspended clay particles that make waters chalky, are microscopic—smaller than the thickness of spider silk.

HOW COME? Pure shallow seawater is a faint blue, but coastal water has extra stuff in it. Some of this stuff includes algae, which color the water green. When turbidity is low, these algae give shallow waters over white sand an emerald color. During rough surf, suspended silt adds an ashen reflection, giving us "seafoam" hues ranging from mint to gray. The surf just down-drift from a river may receive water stained by tannins from fallen leaves, which give the water a tea-like, red-brown color. Red-brown plus turbidity from silt equals gray-brown water. High turbidity from clays washing down coastal rivers can make surf look like chocolate milk.

FOUND: Gulf waters are the most blue-green in south Texas and the Florida Panhandle, and brownest between central Texas and the Mississippi River.

SEASONS: The bluest waters are in spring and early summer. Rough winter waters tend to be mint green or gray-brown.

DID YOU KNOW? Water absorbs far more yellow and red light than blue light, and seawater contains tiny particles that scatter blue light, which is why the deep sea is so blue. Any water seen at an angle can reflect a blue sky, or a gray one.

Inlets, Passes, and Cuts

WHAT ARE THEY? Inlets, passes, and **cuts** are channels through barrier islands that allow water to flow between the sea and the embayment behind the island. Passes are also the channels between adjacent islands.

SIZE: From 100 ft (30 m) to miles wide. Natural inlets are 3–20 ft (1–6 m) deep, and dredged inlets may be over 40 ft (12 m) deep to accommodate deep-draft ships.

Mobile Pass—a natural channel dredged for ships

HOW COME? These passages are cut by storms or by humans. They are maintained (kept open) by tidal currents or by dredging. Natural inlets can fill with sand when a newly formed inlet into a shared bay "steals" some of the local tidal current. There is a tendency for all inlets to fill in with sand swept down the coast by the longshore current (p. 37). Because of this, inlets that serve boat traffic have a sand-blocking jetty (often made of granite boulders) on either side of their opening to the sea. Some passages to the Gulf are kept open by the flow of major rivers.

FOUND: There are dozens of inlets through and between Texas and northern Gulf beaches. Many others have filled in.

Aransas Pass connects Aransas Bay to the Gulf

SEASONS: All year.

DID YOU KNOW? Dredged inlet jetties block sand transport, and inlet channels allow deep tidal currents that divert nearshore sand away from downdrift beaches. Because of this, inlets are the most important cause of rapid shoreline change (net erosion). Natural inlets have played a critical role in the life history of marine species within coastal estuaries. Many marine species spawn only at inlets.

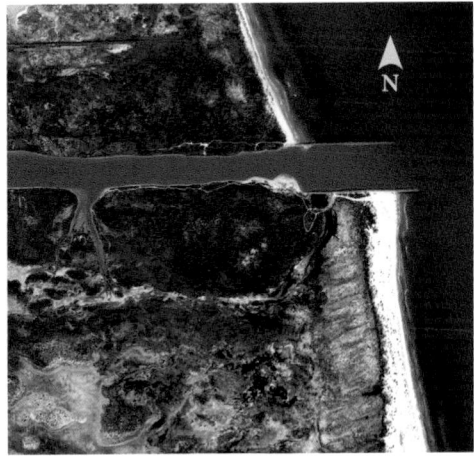

Mansfield Pass steals sand from beaches north

45

Western tip of Dauphin Island, Alabama

Matagorda Island, a wide, Texas barrier island

Barrier Islands

WHAT ARE THEY? These beach-islands are sinuous spits of dry sand parallel to the mainland and separated from it by lagoons, bays, marshes, or open water. Most **barrier islands** have been dry long enough to support diverse plant communities. Some islands have joined the mainland to become a barrier peninsula.

SIZE: Padre Island, Texas, is Earth's longest barrier island at about 113 mi (182 km). Roughly 80 percent of Texas and northern Gulf beaches are on barrier islands, which average several miles long between inlets and passes.

HOW COME? The Gulf's barrier islands were drowned dune ridges or exposed sandbars 8,000–2,000 years ago when sea levels were lower. The islands accreted above the waves as glaciers melted, seas rose, and sediment accumulated. Footholds by dune plants helped sand retention and dune building. Barrier islands unimpeded by coastal development are believed to be increasing in height and migrating landward as our seas continue their rise.

FOUND: Barrier islands rim the entire US Gulf except for the mainland beaches of eastern Texas and western Louisiana, and a small stretch in Florida's Panhandle.

SEASONS: All seasons

DID YOU KNOW? These islands are barriers to wave energy and allow the formation of ecologically diverse lagoons and wetlands behind them. Along the Louisiana coast, rising seas have caused marshes behind some barriers to disappear beneath open waters.

Capes, Spits, and Shoals

Capes

Spits, Shoals

WHAT ARE THEY? Each is a conspicuous, often temporary, pile of sand. **Capes** (cuspate forelands) are triangular landforms that extend seaward. **Sand spits** extend from the downdrift end of barrier islands. **Shoals** are fans of sediment piled up by nearshore currents. Tidal shoals are from back-and-forth tidal flow, which pushes sediment into **ebb shoals** outside the inlet, and into **flood shoals** inside the inlet. River-mouth shoals are called deltas, because of their Δ (delta) shape.

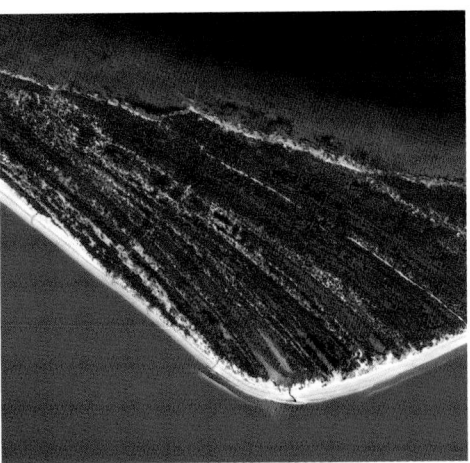

Dune ridges at Cape St. George, Florida Panhandle

SIZE: Up to miles in length and width.

HOW COME? Capes and spits form due to the longshore current (pp. 36, 37) acting in an area of abundant sand. Spits form where this current carries sand to the end of an island. Capes form where longshore currents in opposing directions converge, sculpting sand outward. But these diverted currents likely to have occurred at an ancient river delta, which brought the abundant sand. The sediments of deltas and shoals were carried down rivers from the eroding mainland, and may have been washed in and out of tidal inlets for thousands of years.

Sand spit at the downdrift end of Galveston Island, TX

FOUND: Prominent Gulf capes are Florida's Cape San Blas, Cape St. George, and Alligator Point. Each has associated shoals and is a former river delta.

SEASONS: All year

DID YOU KNOW? Brazos Island is the seaward edge of the old Rio Grande delta in southern Texas. Ancient dune ridge lines on **Cape St. George** mark changes in sea level over the last 4,000 years.

Destin Pass, Florida. Flood shoal (bottom left)

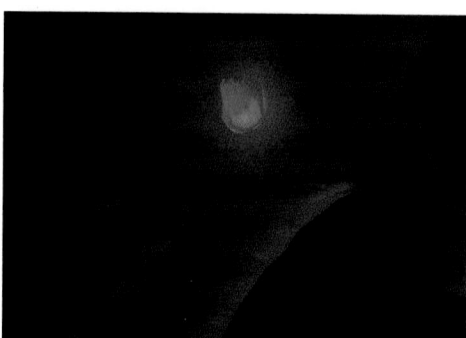

Glowing bioluminescent bacteria on a mole crab shell

South Texas beaches will launch missions to Mars

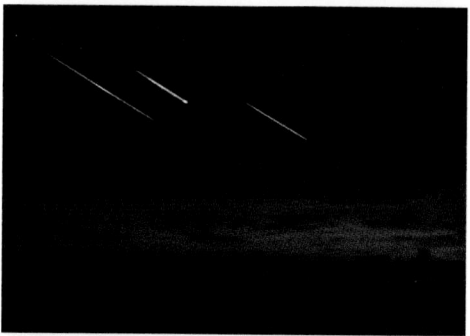

Perseid meteors give a mid-summer night's show

Our Milky Way galaxy

Beaches at Night

Darkest Beaches

WHAT HAPPENS AT NIGHT? Beaches can be splendid places to see nature's nocturnal glow. Nighttime shows include **bioluminescence, meteor showers**, and the **stars** and **planets**.

SIZE: Glowing bacteria and dinoflagellates are microscopic. Luminescent comb jellies are walnut-size. Rockets blast off with boosters that provide millions of pounds of thrust. Brilliant as they are, most meteors are the size of a sand grain. The Universe is infinite.

HOW COME? Bioluminescence on darkened beaches is visible in the surf from dinoflagellates and comb jellies that glow green when disturbed. In the swash zone, parts of mole crabs and other animals have a greenish glow from a coating of luminescent bacteria. Because many beaches are distant from large, lighted cities, they are some of the last convenient vantage points where urban glow has yet to bleach the heavens. On their way to the heavens, specifically Mars, SpaceX's **Starship rockets** launched from Boca Chica in south Texas will be visible for dozens of miles.

FOUND: The best night sights are at beaches preserved as public land.

SEASONS: Consult astronomical charts for seasonal night-sky features. Perseid meteors fly in mid-August, and Leonid meteors can be seen on clear nights in mid-November.

DID YOU KNOW? To protect nesting sea turtles, many beaches enforce light-control ordinances, which also enhance stargazing.

BEACH ANIMALS

AND MICROBES

What are Beach Microbes?

"Microbe" means tiny life. Most microbes are only a single microscopic cell, but many can grow in conspicuous abundance. Some microbes can cause disease, but the vast majority on beaches play hidden-but-critical ecological roles.

Beach sands are habitat for microbes including bacteria, fungi, protozoa, and algae. These organisms interact within a community dependent upon the organic material brought to the beach by incessant ocean waves. Bacteria and fungi break down organic chunks, and protozoans feed on bacteria and fungi, which produce nutrients used by algae that feed the bacteria and fungi … and so on. These cycles are essential to the movement of carbon and nitrogen between sea, land, and atmosphere. Basically, these microbes run the beach, and the oceans, and the world. They are tiny but massive. If you weighed everything alive in the sea, 90 percent of that weight would be microbes.

Microbes that make scary headlines generally come to the beach from someplace else. The most infamous are fecal bacteria and waterborne pathogens, which originate in sewage or farm runoff. These bacteria and viruses can cause disease. Because of public health concerns, government agencies monitor "fecal indicator bacteria" in many recreational areas to compare against federal and state standards and to determine whether waters are safe for swimming. Where bacteria levels frequently exceed standards, biologists investigate potential pollution locations in an effort to eliminate the sources.

Although microbe pathogens give the whole group a bad rap, the preponderance of microbes are more accurately portrayed as the keepers of healthy ecosystems. They are tiny cleaners that rid beaches and waters of waste, and they out-compete pathogens in a way that prevents disease rather than spreads it. Microbes are essential for ocean ecosystems to thrive. Without them, our world would not be livable.

Harmful bacteria resulting from pollution are monitored on busy beaches

Bacteria and Forams

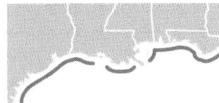

RELATIVES: Bacteria are prokaryotes—simple organisms with a single cell that has no nucleus. **Forams** (Foraminifera) are eukaryotes, which have chromosomes contained within a nucleus. Other eukaryotes include the animals, plants, and fungi that follow this section. Forams are most closely related to other protists such as red algae (pp. 314–315), brown algae (pp. 310–311), diatoms, dinoflagellates, and amoebae.

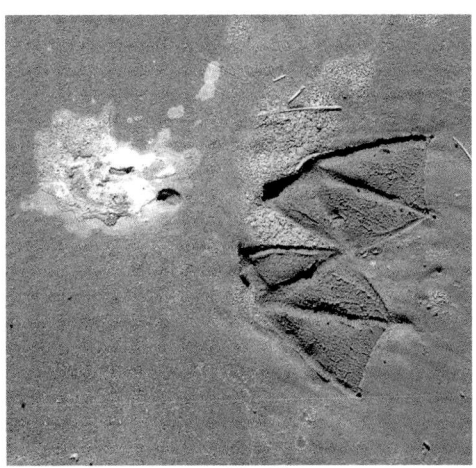

One non-human origin of fecal bacteria

IDENTIFYING FEATURES: Individual bacteria can't be seen, but some densely abundant bacterial colonies stain sands with color. **Purple sulfur bacteria** stain sand purple. Hundreds of foram species inhabit the Gulf. All forams have a single amoeboid cell and a calcium-carbonate test (shell). Most are smaller than a sand grain. Their tests are as elaborate as spirals with divided chambers (like **Ammonia**) or as simple as open tubes or spheres. Forams are abundant, but to appreciate them as the smallest shells in your shell collection, you'll need a microscope.

HABITAT: Bacteria are everywhere. But don't worry, most are friendly. Some even help us with digestion as part of a mutualistic symbiotic relationship. Bad bacteria are generally associated with dead things and poop. To be safest, step around **bird droppings** and wash your hands after examining something stinky. Most forams live on the seabottom in fine sediments.

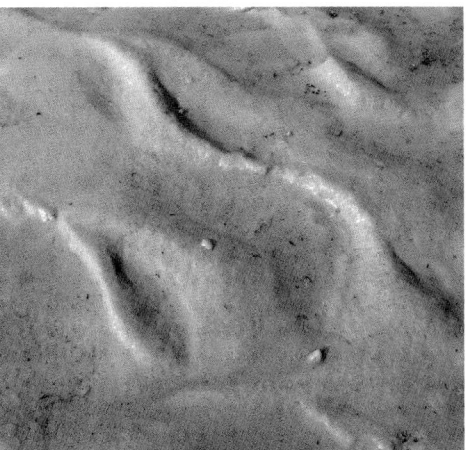

Purple sulfur bacteria in the swash zone

DID YOU KNOW? Purple pigments of sulfur bacteria function in photosynthesis to turn carbon dioxide into carbohydrates using hydrogen sulfide instead of water. Fossil traces of bacteria in rock show they have been on Earth for 3.5 billion years.

Ammonia foram in fine silt. Width of view is 1 mm.

What are Beach Animals?

Animals are multi-celled eukaryotes in the kingdom Animalia. For the most part, animals eat organic stuff, breathe oxygen, move around, and have sexual reproduction. This section highlights animals whose life path includes Gulf beaches. For some animals, this means living near, on, or inside the beach. For others, a beach visit is merely a brief stop during a grand life journey. And for many, the beach is reached only at life's end.

Figuring out what is and is not an animal, or what used to be, can be tricky. Although most folks could place a bird, crab, or fish into a general animal category, some of the lowlier critters can be a puzzle. A wide variety of beach animals seem to be plants, rocks, blobs, trivial specks, or visitors from outer space. Some of these mystery items are what they seem (and are featured in other sections of this book), but others may be animals, colonies of animals, or their lingering parts.

As a rule, live animals twitch when prodded, and dead ones smell worse than rocks or plants. But the sniff test may fail to identify an animal's mineral remnants, and among these there may be impostors. Sun-bleached, brittle, and branching things could be coral or bryozoan animals, but also could be algae. Fibrous tufts could be skeletons from bryozoans or sponges, or algae. Quivering jelly, soft lumps, and gobs of goo are generally animal in nature, and could be tunicates, jellyfish relatives, or a host of other invertebrates.

This section also includes tracks, burrows, and other evidence from some familiar but elusive beach animals. Beach sands can provide an elegant track record of animals that are rare, nocturnal, or shy. Note that we've placed fossils from animals in the *Geology* section and that human traces are in the *Hand of Man* section.

A mass of gooseneck barnacles on freshly stranded driftwood

Swash Meiofauna

RELATIVES: Meiofaunal animals are tiny and diverse. More than a dozen phyla (major animal groups) are known.

IDENTIFYING FEATURES:

Swash meiofauna (MY-o-fawna) are tiny animals (less than 1 mm) within beach sands and include **ostracods** (small crustaceans) (**A**); **nematodes** (**B**); **water bears** (tardigrades, chubby critters with stubby arms and bearlike claws) (**C**); **marine "earthworms"** (Oligochaetes) (**D**); and **bristle worms** (polychaetes) (**E**).

HABITAT: Swash meiofauna live with other larger "infauna" in the sand between the tides. Meiofaunal animals are small enough to move in the spaces between sand grains. This space makes up about a third of beach-sand volume. Punishing extremes of wet and dry require these small animals to be tolerant of dehydration and inundation. This habitat is shared by even tinier bacteria and algae, and with plankton pushed onshore by waves, all of which form the base of the food web in this miniature world.

DID YOU KNOW? Meiofauna have an important place in a living beach. The cleanest beaches have the most diverse array meiofaunal critters. These animals gather detritus, munch bacteria, and hunt each other, wiggling between wave-pounded sand grains and up to several feet deep in the beach. Water bears shrivel dry at low tide but reanimate when waves return. Some of these tardigrades can remain in a state of extreme desiccation (anhydrobiosis) for years, withstanding harsh temperatures and even the vacuum of space.

Meiofaunal invertebrates live between sand grains

White sponge showing its multiple, small oscules

Branching tube sponges have large, terminal oscules

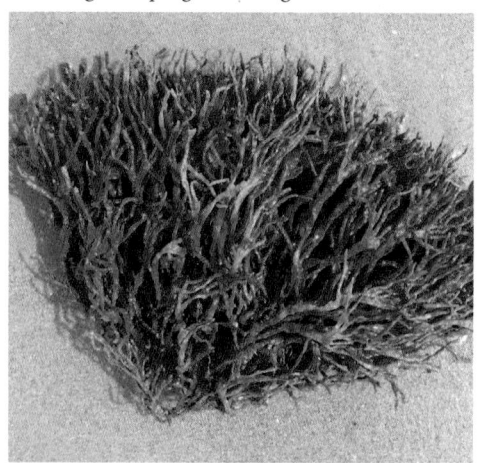

Redbeard sponge

Sponges
(White, Branching Tube, Redbeard)

White, Tube   Redbeard

RELATIVES: Sponges are in the phylum Porifera. Species on the following pages are related to bath sponges (class Demospongiae).

IDENTIFYING FEATURES: Most sponges are difficult to tell apart using only shape and color, but these species have characteristic forms. Their colors vary, especially after stranding. Sponges have **ostia**, which are pores all over the sponge body that let in water and food, and one or more **oscules**, which are larger openings to let water and waste out.

White sponges *(Geodia gibberosa)*, to 24 in (60 cm) across, are firm, dirty-white, lobed masses with clusters of small oscules.

Branching tube sponges *(Aiolochroia crassa)*, to 6 in (15 cm) long, are rubbery tubes with either an irregular surface or one covered with pores (ostia). They have distinct oscules at their terminal end and are yellow in life, but turn dark brown after stranding.

Redbeard sponges *(Clathria prolifera)*, 8 in (20 cm), are bouquets of slim, red, velvety branches that turn to brown after death.

HABITAT: These sponges grow on hard-bottom and in seagrass beds.

DID YOU KNOW? Sponges provide important habitat for many species of fishes and invertebrates. Redbeard sponges are used in labs to study cellular reaggregation: the reassembly of disembodied cells. After being turned to soup in a blender, these sponges' disembodied cells can find themselves and rebuild their correct structure.

Sponges *(Tube, Vase, Chicken Liver, Devil's Fingers)*

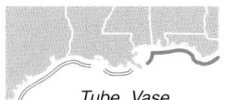
Tube, Vase

Chicken Liver, Devil's Fingers

IDENTIFYING FEATURES:

Tube sponges *(Callyspongia vaginalis)* have individual chimneylike tubes about 2 in (5 cm) wide.

Vase sponges *(Ircinia campana)*, 24 in (60 cm), have a vase-shape and smell really bad after being beached.

Chicken liver sponges *(Chondrilla nucula)*, 4 in (10 cm), look like their name when fresh, and are fibrous and tan when beach-worn.

Orange devil's fingers *(Axinella polycapella)*, 24 in (60 cm), are spongy, finger-thick, branches attached at a thin wrist. Colors are orange, fading to brown after beaching.

Red devil's fingers *(Axinella pomponiae)*, 12 in (30 cm), tend to have tapering branches, often fusing together and with occasional lumps. The sponge is red in life and brown after stranding.

HABITAT: Beached sponges have been torn free from shallow hardbottom or seagrass beds.

DID YOU KNOW? Sponges are simple animals without organs. They've persisted for 600 million years and were probably Earth's first multi-celled animals. These colonial animals grow in place by filtering organic particles from the water. Sponges filter an amazing amount of seawater through their bodies—up to 10,000 times their own volume each day.

Tube sponge (L), vase sponge (R)

Chicken liver sponges in various stages of decay

Orange devil's fingers

Red devil's fingers

55

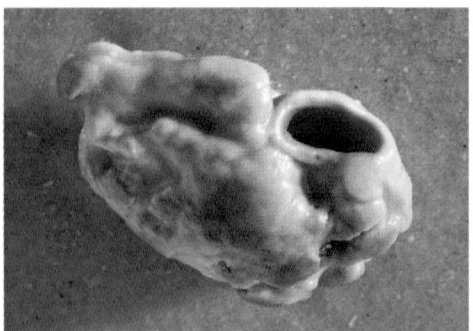

Florida hermit crab sponge (FHCS), orange version

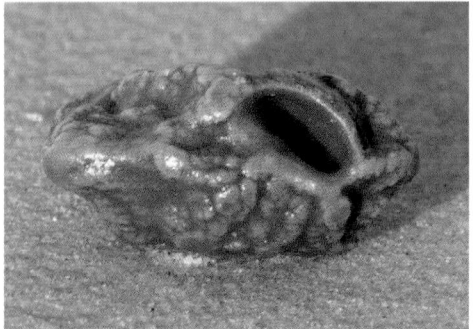

FHCS, green version, showing the distinct hermit hole

Golfball sponge

Sheepswool sponge

Sponges
(Hermit Crab, Golfball, Sheepswool)

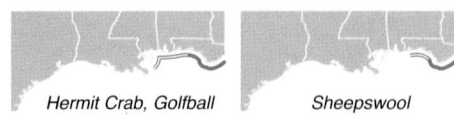

Hermit Crab, Golfball   Sheepswool

IDENTIFYING FEATURES

Florida hermit crab sponges *(Pseudo-spongosorites suberitoides)*, 2.3 in (6 cm), are rubbery masses formerly inhabited by a hermit crab and have a single large oscule opposite a hermit-crab's hole. Colors range from orange to green in life, and tan to turquoise after death.

Golfball sponges *(Tethya* spp.*)*, 2 in (5 cm), are soft spheres with distinct pores. They are yellow or orange in life and turn brown on the beach.

Sheepswool sponges *(Hippospongia lachne)*, 10 in (25 cm), are soft, brown, classic bath-sponge shapes and have abundant holes of varied sizes.

HABITAT: Shallow hardbottom and seagrass beds

DID YOU KNOW? Hermit crab sponges attach to a hermit's seashell when the sponge colony is small and do not keep the hermit from crawling. But as the sponge grows, the hermit has a difficult time dragging it around. If a hermit leaves its sponge-hobbled shell for another, the sponge may be inhabited by a series of larger hermits. Golfball sponges move on their own using rhythmic body contractions at the breakneck speed of 4 mm per day, enough to position them in better habitat. Sheepswool sponges were once the focus of a large commercial sponge-diving effort in Florida. Bath sponges were replaced with synthetic varieties in the 1940s. But sponges are still harvested as marine curios in Apalachicola and Tarpon Springs, Florida.

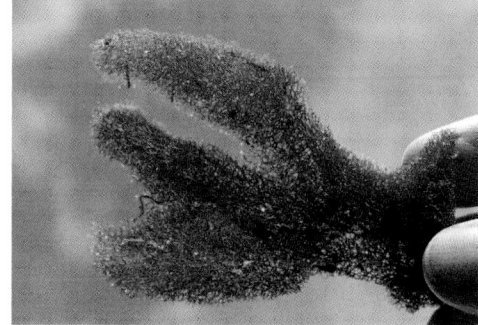

Sponges
(Boring, Crumb-of-bread, Spongin)

Yellow Boring   Crumb of Bread

IDENTIFYING FEATURES

Yellow boring sponges *(Cliona celata)* penetrate and live within shells, rocks, or coral rubble. Beached sponges are fist-size lobes, or are encrustations surrounding the item they washed in with. Exposed living parts are bright yellow, soft, and lumpy. Detached lobes on beaches dry into smooth, woodlike lumps. The shells they invade have telltale perforations (p. 150).

Crumb-of-bread sponges *(Halichondria panicea)*, 10 in (25 cm), are firm but easily torn, lumpy, and occasionally branching sponges, often with short oscular chimneys. They are green-yellow to cream in life and brown or pale after stranding.

Unknown branching sponges, to 24 in (60 cm), are brownish or gray and have lost their identifying features. Sometimes, only the brown fibrous **spongin skeleton** remains.

HABITAT: Shallow coastal waters

DID YOU KNOW? Boring sponges are known to severely weaken concrete pilings that support bridges. These sponges are most common near dense human development where septic tanks leak sewage into nearby waters. The extra nitrates (fertilizer) grow algae on which the sponges thrive. The bath-type sponges on this and the preceding pages are "spongy" because of spongin, the fibrous protein that gives sponges skeletal support. This spongin skeleton is all that remains after the living sponge tissue has decayed.

A freshly beached yellow boring sponge in rock

Crumb-of-bread sponge

Oscular "chimneys" of a crumb-of-bread sponge

A beach-worn spongin skeleton, unknown species

57

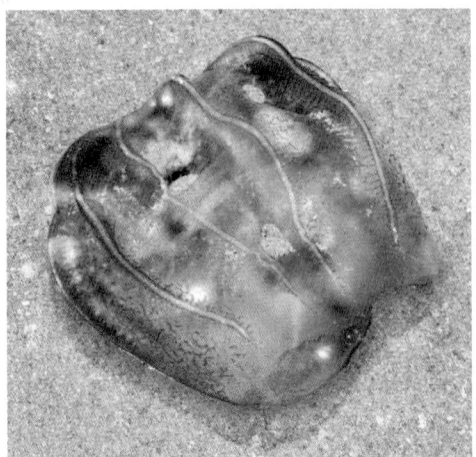

Ovate comb jelly. Cilia glisten in sunlight

Many-ribbed hydromedusa beached, and alive (inset)

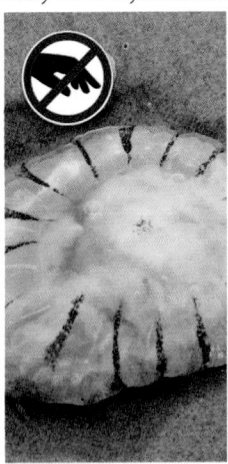
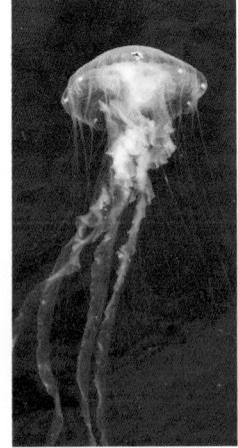

Beached sea nettle (L), pale form in water (R)

Comb Jelly, Hydromedusa, and Sea Nettle

RELATIVES: Comb jellies belong to the phylum Ctenophora. Hydromedusae are cnidarians (phylum Cnidaria) in the class Hydrozoa, and sea nettles are medusa jellies in the cnidarian class Scyphozoa. In addition to their silent "C," both cnidarians and ctenophores have a hydrostatic "skeleton" of jelly called mesoglea.

IDENTIFYING FEATURES:

Ovate comb jellies *(Beroe ovata)*, 4.5 in (11 cm), have pinkish egg-shape bodies bearing eight rows of faint cilia. Sea walnut comb jellies *(Mnemiopsis leidyi)* are a similar size but have two distinct lobes.

Many-ribbed hydromedusae (*Aequorea* spp.), 3.9 in (10 cm), are translucent with 50 or more radiating ribs and thin trailing tentacles. Beachworn specimens look like the thick bottom of a glass Coke bottle.

Sea nettles *(Chrysaora quinquecirrha)*, 10 in (25 cm), have long tentacles. Their bell has rusty radiating stripes or is pale. Got stung? See page 61.

HABITAT: Comb jellies and sea nettles live in coastal waters. Hydromedusae are more common offshore.

DID YOU KNOW? Comb jellies and hydromedusae in the surf at night pulse with brilliant luminescence. Beached jelly animals are often identifiable only as mesoglea, which washes in as lumps of clear, firm nonsticky jelly. Mesoglea is mostly water, with a fibrous net of collagen. Scyphozoan and hydrozoan jellies start as larvae that attach to the bottom and grow as polyps. These produce medusae—the sexually reproducing adult form.

Jellyfishes
(Sea Wasp, Moon, Mushroom Cap)

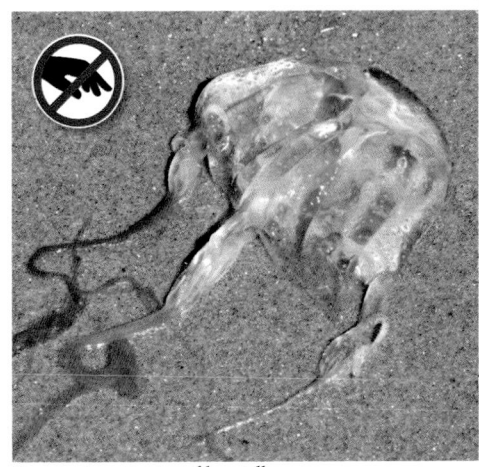

Sea wasp, a species of box jelly

RELATIVES: Jellyfishes are in the phylum Cnidaria, class Scyphozoa.

IDENTIFYING FEATURES:

Sea wasps *(Chiropsalmus quadrumanus)*, 5 in (12 cm), have a cube-shape bell and handlike appendages with streaming tentacles at four corners of the cube. A similar box jelly, *Tamoya haplonema*, has four, thin, flattened appendages.

Moon jellies *(Aurelia aurita)*, 12 in (30 cm), have a saucerlike bell that is clear except for 4–7 lobe-shape gonads arranged in a flower pattern. Color of the gonads ranges violet-pink to yellow. Moon jellies have hundreds of short, marginal tentacles and four, frilly, oral "arms."

Mushroom cap jellies *(Rhopilema verrilli)*, 20 in (51 cm), have a whitish, transparent bell and a limp central cluster of brown-fringed oral arms.

HABITAT: Open ocean and coastal waters

Moon jelly

DID YOU KNOW? On sensitive skin, moon and mushroom cap jellies give a mild, burning sting. In contrast, sea wasps and other box jelly species are famous for inflicting extremely painful, even life-threatening stings. Extensive (numerous) stings result in initial pain followed by cardiac and respiratory threats, and by a rash that may last for months. Swimmers in waters where stinging jellies are common can benefit from divers' "skin suits," and anti-sting lotions like "Safe Sea." Stings come from tentacles, each bearing thousands of nematocysts that fire a tiny venom harpoon. Stung? See page 61.

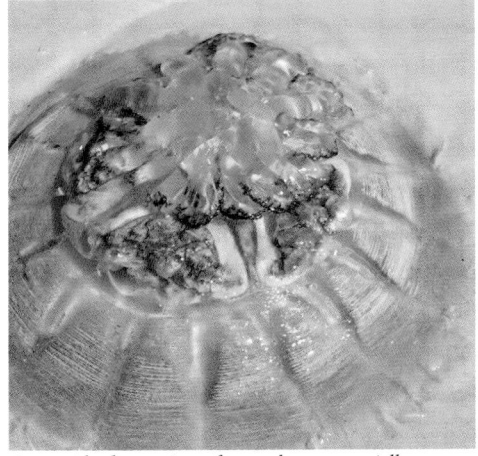

An upside-down view of a mushroom cap jelly

59

Cannonball jelly

Variously beachworn parts of cannonball jellies

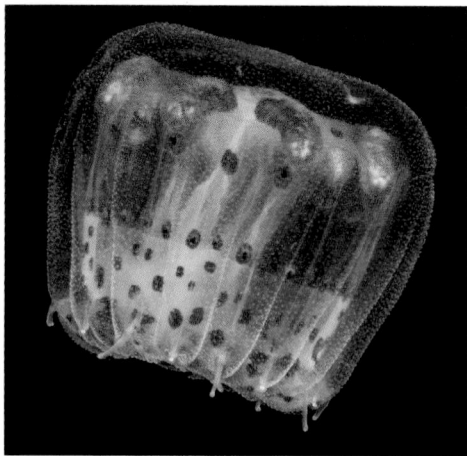

Adult thimble jellyfish

Jellyfishes
(Cannonball, Thimble)

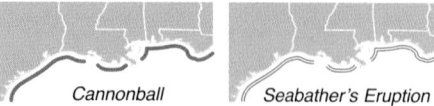

Cannonball   Seabather's Eruption

RELATIVES: Other scyphozoa (jellyfish).

IDENTIFYING FEATURES

Cannonball (Cabbagehead) jellies *(Stomo-lophus meleagris)*, 8 in (20 cm), have a thick, firm, domelike bell that is yellow in life with brown around the margin. Beached specimens often lose the bell, leaving only the stocky, fluted cylinder at the animal's center. The fluted folds are the animal's mouths, which end in short, protruding oral arms.

Seabather's eruption, inaccurately called "sea lice," results from contact with the planula larvae of the **thimble jellyfish** *(Linuche unguiculata)*. The larvae look like specks of finely ground pepper. Although they are too tiny to notice, their stings are not. Exposure results in a rash where bathing suit and skin meet. The adult jelly reaches an inch long (2.5 cm) and looks like a brownish cap with tiny tentacles. Adult thimble jellies also sting, but are typically less numerous than the tiny larvae.

HABITAT: Open ocean and coastal waters

DID YOU KNOW? Cannonball jellies are a common food item in Asia. Their crunchy texture makes them suitable for "sea salads." Coastal abundance of these jellies in the Gulf increases in spring and summer as the animals migrate in from offshore waters. Thimble jellyfish larvae sting when they are trapped against the skin. Swimmers can reduce this irritation by showering with bathing suits off after surf swimming. The peak in seabather's eruption is April–July. Blooms (population spikes) of jellies are linked to fertilizer runoff and overfishing.

Portuguese Man-o-War

RELATIVES: These are hydralike animals, class Hydrozoa, in the phylum Cnidaria. They are siphonophores, which are distantly related to blue buttons and by-the-wind sailors (p. 62).

IDENTIFYING FEATURES:

Portuguese man-o-war (*Physalia physalis*) look like a blue-tinged balloon with a pink-crested sail and trailing tentacles. The float-sail (pneumatophore) is 1–10 in (2.5–25 cm) long. Tentacles can be 6 ft (2 m) on the beach and reach 150 ft (46 m) when fully extended at sea.

Portuguese man-o-war. Arrow shows tentacles

HABITAT: The wide-open sea, except at the end of their voyage when they are found in the wave wash and wrack line.

CAUTION! Their tentacles sting and cling. If stung by any jelly animal, remove tentacles without rubbing. Vinegar deactivates the stinging nematocysts, and a credit card can be used to shave off the tentacles. Don't use fresh water, which causes more nematocysts to fire their tiny venom harpoons. Immediate immersion of a stung area in hot water helps ease initial pain. After that, an ice pack can relieve soreness and swelling. Get medical attention for dizziness or breathing difficulties.

A man-o-war float in beach wrack

DID YOU KNOW? These animals are only remotely related to jellyfish. Apparent individuals are actually colonies of many polyps (balloon, stinging, feeding, and breeding). Right- and left-sailing forms of man-o-war travel in different directions. Their paralyzing tentacles capture small fish and shrimp for food. Beachings are most common December through May. April and May are peak months for abundance and individual size.

A right-sailing form. Front (oral) end in foreground

61

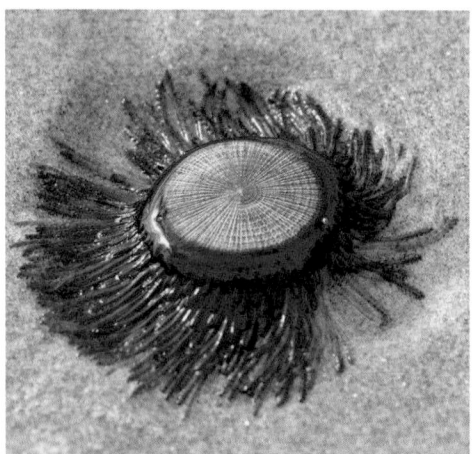

Blue button

By-the-wind sailor

Newly beached and older (inset) by-the-wind sailors

Blue Button and By-the-Wind Sailor

RELATIVES: These are hydralike animals, class Hydrozoa, in the phylum Cnidaria. They are chondrophorines, distantly related to siphonophores like the Portuguese man-o-war.

IDENTIFYING FEATURES:

Blue buttons *(Porpita porpita),* 1 in (2.5 cm), have a small disclike float surrounded by blue-green tentacles.

By-the-wind sailors *(Velella velella),* 2 in (5 cm), have an oval float bearing a crest-like sail and deep blue tentacles beneath.

HABITAT: These animals float on the surface of the wide-open sea, except when beached. Their cellophane-like "floats" linger in the wrack for weeks.

DID YOU KNOW? Neither species is a jellyfish. Like the Portuguese man-o-war, they are colonies of many individual animals. Blue buttons and by-the-wind sailors feed on small animals captured by their tentacles (which are actually individual animals called zooids). The tentacles bear stinging nematocysts that are not potent enough to effect humans, except in sensitive areas like the eyes. By-the-wind sailors have mirror-image forms that are either right-sailing or left-sailing. Because of these directional tendencies, beaching events usually involve only one form. Both of these open-sea colonial animals are eaten by blue glaucus sea slugs (p. 105), purple sea snails (p. 87), and young pelagic sea turtles.

Hydroids
(Snail Fur, Christmas Tree, Tubular)

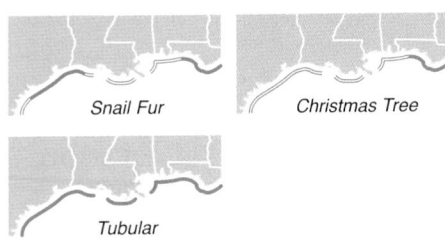

Snail Fur   Christmas Tree

Tubular

RELATIVES: These are athecate hydroids, order Anthoathecata, class Hydrozoa, phylum Cnidaria.

IDENTIFYING FEATURES:

Snail fur *(Hydractinia echinata)* forms a fuzzy, tan coating on marine snail shells. The "fur" mat is composed of stiff spines that project between the individual hydroid polyps in the colony.

Christmas tree hydroid colonies *(Pennaria disticha)*, 12 in (30 cm), are plumes with many alternating, wire-thin, dark branches growing on objects formerly on the sea bottom. Whitish polyps on branches disappear after beach exposure.

Pink-mouth tubular hydroid colonies *(Ectopleura crocea)*, 6 in (15 cm), look like fist-size tufts of coarse, kinky blonde hair. In life, each stiff "hair" is actually a tube with an individual polyp that is pink and white with a ring of long tentacles.

HABITAT: Snail fur grows mostly on shells inhabited by hermit crabs. The other two hydroid colonies grow in shallow waters on sunken objects like rocks, wood, and debris.

DID YOU KNOW? These hydroids are colonies of individual polyps that share a continuous gastrovascular system. If two snail fur hydroid larvae start colonies on the same shell, they recognize whether they share a specific gene. If they don't, the growing colonies fight it out by shooting stinging nematocysts at each other where they meet on the shell.

Snail fur hydroids covering a sharkeye snail

Christmas tree hydroids on a beached crab trap

An exfoliated patch of pink-mouth tubular hydroids

63

Red stick-hydroids

Clytia *pelagic sargassum hydroids on sargassum algae*

Sertularia *sargassum hydroids on sargassum*

Hydroids *(Red Stick-, Sargassum)*

RELATIVES: These are colonial hydroids in the class Hydrozoa. Red-stick hydroids are athecate hydroids related to snail fur, Christmas tree, and tubular hydroids. Sargassum hydroids include both athecate and thecate (order Thecata) varieties.

IDENTIFYING FEATURES:

Red stick-hydroids *(Eudendrium carneum)*, 5 in (12 cm), have irregular tree-like branches bearing pink-orange polyps.

Pelagic sargassum hydroids (orders Anthoathecata and Thecata), 1/4 in (6 mm), add a sparse tan fuzz to drifting items within the open-ocean sargassum community. They are small colonies with single or branched stems bearing polyps, and are connected over the sargassum surface by a network of tubes. More than 50 species are known. Those shown are thecate, meaning that each polyp lives in a protective sheath. **Clytia hydroids** *(Clytia* spp.) look like cups on single stems. **Sertularia hydroids** *(Sertularia* spp.), look like plumes with alternating branches of tiny, nested chevrons.

HABITAT: Red stick-hydroids grow on bottom rubble. Pelagic sargassum hydroids grow on ocean-drifting objects of all kinds, especially sargassum algae (p. 310), which may have most of the plant's surface covered by the hydroid's rootlike stolons that connect the hydroid colony.

DID YOU KNOW? Sargassum hydroids feed on plankton and are themselves food for young sea turtles, juvenile fishes, and sargassum nudibranchs (sea slugs, p. 103).

Sea Pansy

RELATIVES: Sea pansies are with corals and anemones in the phylum Cnidaria and class Anthozoa, and are with octocorals in the order Pennatulacea.

IDENTIFYING FEATURES:

Gulf sea pansies *(Renilla muelleri),* 2 in (5 cm), look like a purplish, thick, leaflike pad on a short purple stalk called a peduncle. **Feeding polyps** show the eight-fold symmetry characteristic of all octocorals.

HABITAT: Sea pansies live on intertidal sands, especially on current-swept flats near inlets and passes.

DID YOU KNOW? Sea pansies are a collection of polyps, the largest of which is the extendible peduncle that anchors the colony in the sand. Anemone-like feeding polyps cover the top of the purplish pad (rachis), which has other specialized polyps for deflating at low tide and inflating when flooded at high tide. These animals have some amazing behaviors to avoid predation by fishes, crabs, and sea slugs. When disturbed, sea pansies flash a green bioluminescent wave over their surface to dazzle would-be predators. The light comes from photocytes with a fluorescent protein, which is currently being used as a tool in biochemical and medical research. Tough calcium carbonate sclerites provide additional defense, as well as a skeleton to keep the colony's leaflike form. As a last measure, the sea pansy may quickly retract its peduncle and allow surf currents to sweep it away. Sea pansies feed on particle-size plants and animals, which are ingested by the feeding polyps and shared with the rest of the colony.

Gulf sea pansy, bottom, showing stubby peduncle

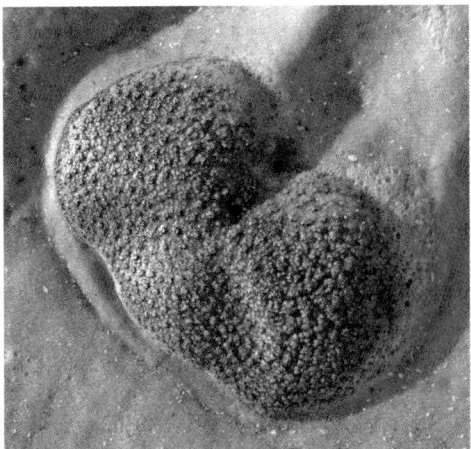

Gulf sea pansy, top, showing polyp surface

Closeup of sea pansy feeding polyps

65

A tangled bundle of straight sea whips

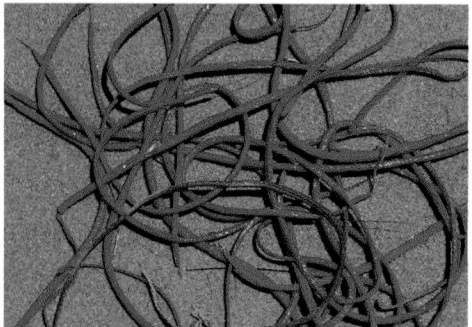

Straight sea whips

Sea whips may have purple or yellow coenenchyme

Partially buried axial skeletons of stranded sea whips

Soft Corals *(Straight Sea Whip)*

RELATIVES: Soft corals (class Anthozoa) are octocorals in the order Gorgonacea. Gorgonians lack the rigid skeleton of hard (stony) corals. Like the sea pansies, polyps of these octocorals have eight tentacles, each with its own body-cavity partition (mesentery).

IDENTIFYING FEATURES: Soft corals are colonies with a tough-but-flexible **axial skeleton** covered by an external tissue layer called **coenenchyme**, which includes the colonies polyps, supportive spicules, and tubes connecting the polyps' gastrovascular cavities.

Straight sea whips *(Leptogorgia setacea)*, 3 ft (0.9 m) in length and 1/8 in (3 mm) thickness, are gorgonian colonies that beach as stringy, limber, unbranched rods. The colonies are mostly bright yellow, but are occasionally whitish, lavender, or purple. A single colony may be anchored to a shell, or be unattached. They often strand en masse in tangled clumps. On the beach, the colorful coenenchyme wears away leaving a wirelike axial skeleton.

HABITAT: Shallow coastal waters including tidal creeks and bays

DID YOU KNOW? Unlike stony corals, soft corals do not need symbiotic algae (zooxanthellae) to survive, so they can live in more turbid areas without the light required for algal photosynthesis. Soft coral colonies are flexible but tough because of the hornlike protein (gorgonin) in their axial skeletons.

Soft Corals
(Colorful Sea Whip, Regal Sea Fan)

Colorful Sea Whip

Regal Sea Fan

RELATIVES: These are octocorals in the order Gorgonacea.

IDENTIFYING FEATURES:

Colorful sea whips *(Leptogorgia virgulata)* reach 24 in (60 cm) in length. Branches average 3/16 in (4 mm) wide, but are thicker at the colony's distinct base. Color of the outer coenenchyme tissue may be purple, red, orange, or yellow. Polyps are white and alternate in rows along two sides of each branch. Like other gorgonian colonies, these sea whips may beach as only their tough, dark, **axial skeleton**, which looks like a naked, miniature tree.

Regal sea fans *(Leptogorgia hebes)*, 9 in (23 cm), grow as densely branched fans with a distinct base. They are most commonly orange, but also can be red, dark yellow, or purple.

HABITAT: Shallow coastal hardbottom and rubble

DID YOU KNOW? Color variation in sea whips is genetic, and many areas may have more than one color. The colors are from pigments fixed within the colony's supportive, calcium carbonate spicules. Sea whips can live over 15 years. During that time they may form complex interspecies relationships. Although sea whips produce chemicals that inhibit predator munching and attachment by fouling organisms, some animals are not deterred. These include the one-tooth simnia (p. 80), which feeds on sea whip tissue and develops the color of what it eats, and the sea whip barnacle (p. 158), which attaches as a larva and becomes enveloped by the living tissue of the colony.

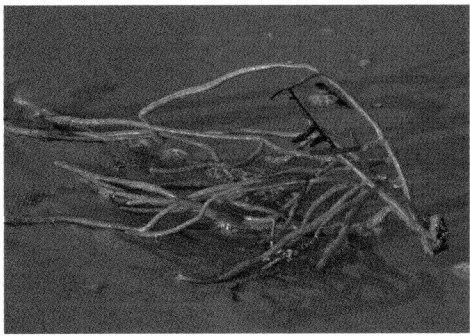

Colorful sea whip, yellow version

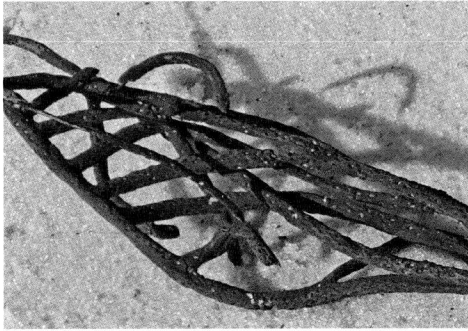

A purple, colorful sea whip

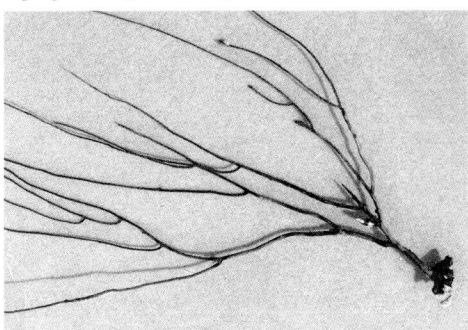

Beached axial skeleton of a colorful sea whip

Regal sea fan

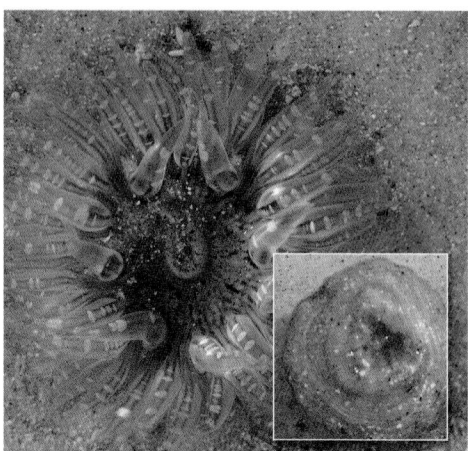

Gray warty anemone exposed, retracted (inset)

Onion anemone retracted, exposed (inset)

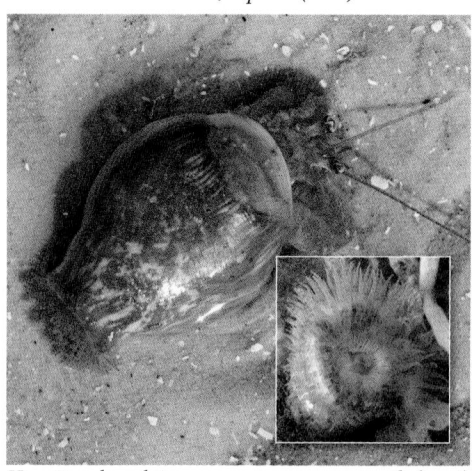

Hermit crab and its anemone, anemone mouth (inset)

Sea Anemones

RELATIVES: Sea anemones are related to corals and sea pansies, class Anthozoa.

IDENTIFYING FEATURES:

Gray warty anemones *(Anthopleura krebsi),* 4 in (10 cm), appear from the sandy bottom in protected intertidal areas as "flowers" made up of about 100 stubby tentacles that are translucent, striped, and flecked with white. When uncovered, these anemones pucker up into a warty gray ball.

Onion anemones *(Paranthus rapiformis),* 3 in (7.5 cm), have short, thin tentacles and anchor in the sand by a small disc at the base. When eroded from the sand they assume the shape and appearance of a cocktail onion.

Hermit crab anemones *(Calliactis tricolor),* 2 in (5 cm), have a fringe of fuzzy tentacles and a column that is tinted purple and yellow with basal dark spots.

HABITAT: Sandy bottom near beaches. Hermit crab anemones live on bottom debris and on gastropod shells occupied by hermit crabs.

DID YOU KNOW? Hermit crabs and their anemones have a cooperative agreement to trade travel for stinging protection. Hermit crabs are known to take their anemones with them when they upgrade to larger shells. The anemones defend themselves by extruding stinging, orange or white filaments, called acontia. When not used in defense, the filaments line the gut of the anemone and help to subdue swallowed prey. Each of these anemones can reproduce by dividing lengthwise, producing two anemones.

Stony Corals
(Brain, Ivory Bush, Northern Cup)

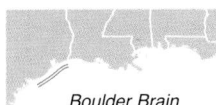

Boulder Brain

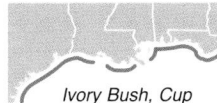

Ivory Bush, Cup

RELATIVES: Stony corals (class Anthozoa, subclass Hexacorallia, order Scleractinia) are more closely related to anemones than to soft corals.

IDENTIFYING FEATURES: These are hard lumps or branches with distinct texture from scattered **corallites** with radiating, dividing walls called **septa**, which are the skeletal cups formed by the individual coral polyps. Pieces are bone white, or grayish if sediment stained.

Boulder brain coral *(Colpophyllia natans)* forms domes up to 6 ft (1.8 m), but beach specimens are merely broken chunks. The surface is a maze of interlinking valleys containing the corallites.

Ivory bush coral *(Oculina diffusa)*, 12 in (30 cm), has short, crooked, pencil-thick branches with separated, raised corallites, each with 12 primary septa.

Northern cup (star) coral *(Astrangia poculata)*, 2 in (5 cm), is a lone ball or encrusting lump with crowded, deep, 0.2-in (5-mm) corallites, each with 25 to 35 septa. This coral is similar to an extinct species, *Septastrea marylandica* found as fossils, but northern cup coral has twice the septa per corallite.

HABITAT: Northern cup and ivory bush corals grow in shallow coastal waters. Brain coral found on Gulf beaches may be remains of old, storm-rolled, boulders from reefs like the Flower Garden Banks.

DID YOU KNOW? Northern cup and ivory bush corals can live in turbid coastal waters because they don't require the symbiotic, photosynthetic, zooxanthellae algae that provide food for other stony corals.

A beached chunk of boulder brain coral

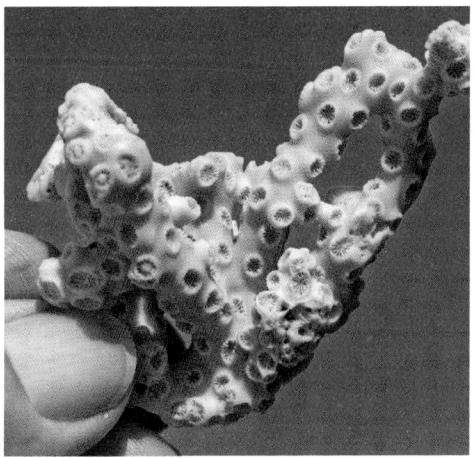

Ivory bush coral

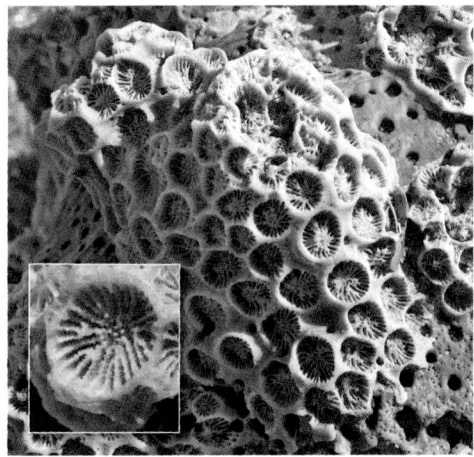

Northern cup coral, corallite with septa (inset)

69

Shelled Mollusk Anatomy

Seashells are the protective or supportive skeletons of mollusks (phylum Mollusca). The most common shells are from snails (with one coiled shell) and bivalves (with two hinged shells). Snails are gastropods, as are sea slugs and sea hares, which have an internal shell or none at all. Other shelled mollusks include tusk shells and some squids.

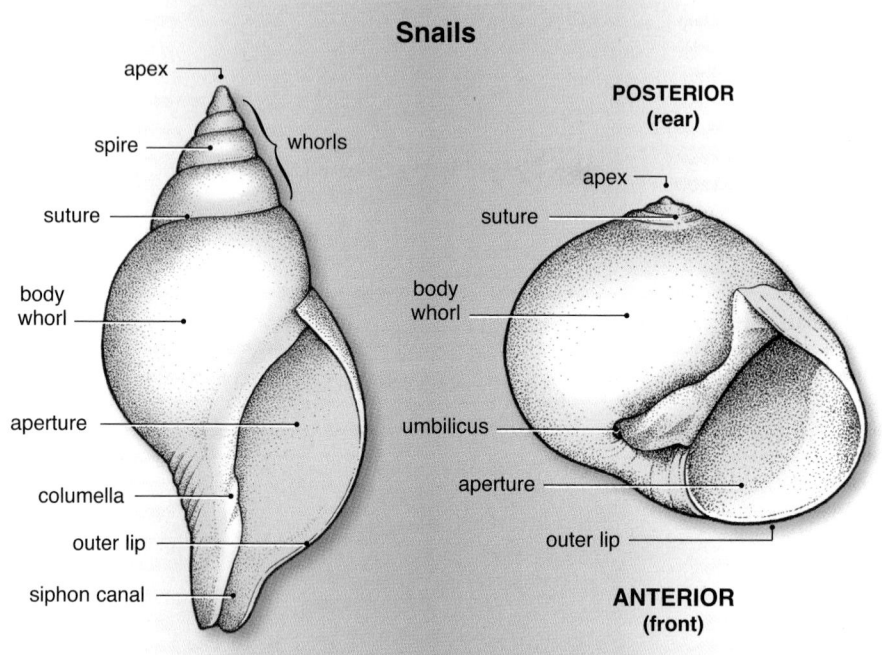

Snails

apex
spire
suture
body whorl
aperture
columella
outer lip
siphon canal
whorls

POSTERIOR (rear)

apex
suture
body whorl
umbilicus
aperture
outer lip

ANTERIOR (front)

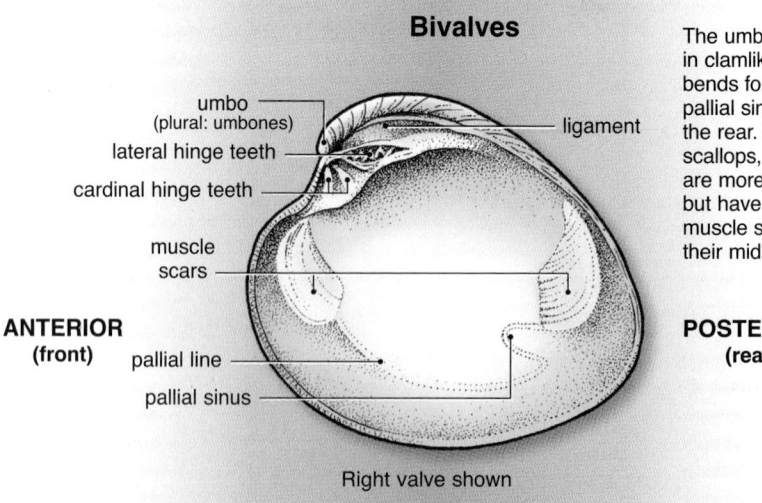

Bivalves

umbo (plural: umbones)
lateral hinge teeth
cardinal hinge teeth
muscle scars
ligament

The umbo is dorsal, and in clamlike bivalves bends forward. The pallial sinus is toward the rear. Mussels, scallops, and oysters are more symmetrical, but have their single muscle scar to rear of their midline.

ANTERIOR (front)
pallial line
pallial sinus

POSTERIOR (rear)

Right valve shown

Chiton, Limpets, and Turban

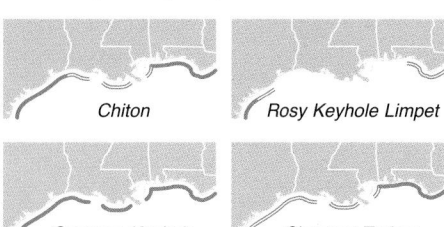

RELATIVES: Chitons belong to the mollusk class Polyplacophora. Limpets (family Fissurellidae) and turbans (Turbinidae) are mollusks in the class Gastropoda.

IDENTIFYING FEATURES:

Mesh-pitted chitons *(Ischnochiton papillosus)*, like all chitons, have eight connected dorsal plates surrounded by a fleshy girdle. This species has plates with tiny beads in angled rows, and is green, pale, purplish, or is mottled with these colors.

Rosy keyhole limpets *(Fissurella rosea)*, like all limpets, are shaped like oval volcanoes with a top opening offset from center. This species has a broad oval top hole and pink or purple rays.

Cayenne keyhole limpets *(Diodora cayenensis)* have a keyhole opening offset from center, and irregular radial ribs crossed by concentric ridges.

Chestnut turbans *(Turbo castanea)* have a coiled snail shape with rounded whorls like a beaded turban. Their mottled color is as their name suggests. The shell's aperture is circular, as is the thick, calcified **operculum** (aperture door), which collectors call a cat's eye or Shiva's eye shell.

HABITAT: Shallow rocks and seagrass

DID YOU KNOW? These distantly related mollusks feed on algae they scrape from rocks with their radula, a rasplike structure with tiny teeth. A chestnut turban's operculum reveals the snail's growth without shape change, in the form of a "golden" (Fibonacci) spiral pattern.

Mesh-pitted chiton, max 0.7 in (1.8 cm), on penshell

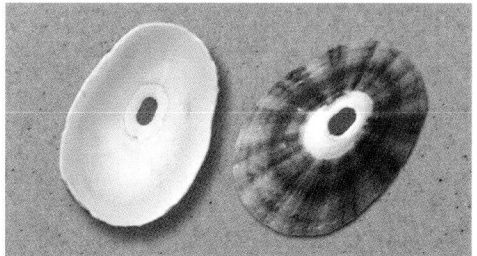

Rosy keyhole limpet, max 1.5 in (3.8 cm)

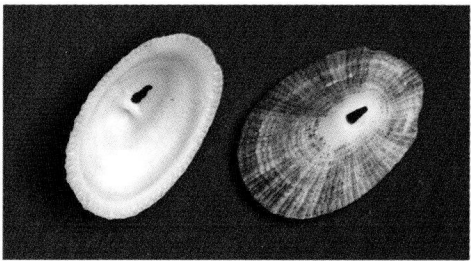

Cayenne keyhole limpet, max 2 in (5 cm)

Chestnut turban, max 1.5 in (4 cm)

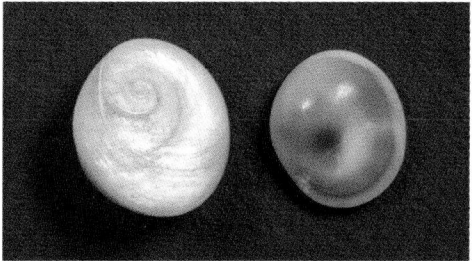

Chestnut turban opercula, max 0.7 in (1.8 cm)

71

Cat. #17969

Olive nerite, max 0.5 in (1.3 cm)

Virgin nerite, max 0.5 in (1.3 cm)

Antillean nerite, max 1 in (2.5 cm)

Four-tooth nerite, max 1 in (2.5 cm)

Nerites

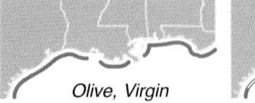

Olive, Virgin

Antillean, Four-tooth

RELATIVES: Other gastropods in the family Neritidae.

IDENTIFYING FEATURES: Nerites have thick shells with rounded spires and D-shape apertures that look like a toddler's smile.

Olive nerites *(Vitta usnea)* have smooth whorls that are brownish green with dark suture lines.

Virgin nerites *(Vitta virginea)* are glossy and variably patterned with waves and swooshes.

Antillean nerites *(Nerita fulgurans)* have spiral ridges separated by light sutures, and two central teeth on the inside lip.

Four-tooth nerites *(Nerita versicolor)* have spiral ridges with blurred markings of black, greenish-white, and maroon, and have four large teeth on the inside lip.

HABITAT: Olive and virgin nerites live in shallow estuarine waters and seagrass beds. Antillean and four-tooth nerites live on and under rocks in intertidal areas.

DID YOU KNOW? Virgin nerites migrate up rivers, where these freshwater forms show darker shells than their vivid counterparts in saline lagoons. All nerites have a thick, calcified operculum (aperture door) that keeps them protected from desiccating heat at low tide. These shells are named for the handsome Greek sea god *Nerites,* who refused an invitation from Aphrodite to join the Gods of Olympus and as punishment was transformed into a shellfish. He's still pretty good looking, right?

Topsnail, Tegula, Button Snail, and Hornsnail

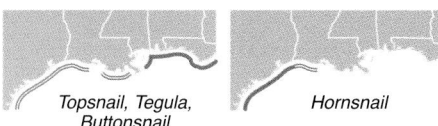

Topsnail, Tegula, Buttonsnail *Hornsnail*

Scuptured topsnail, max 1 in (2.5 cm)

RELATIVES: Topsnails and tegulas are in the family Trochidae. Button snails (family Modulidae) and hornsnails (Potamididae) are only distantly related.

IDENTIFYING FEATURES:

Sculptured topsnails *(Calliostoma euglyptum)* have a dark-tipped apex and rounded whorls forming a wide cone.

Smooth (silky) Atlantic tegulas *(Tegula fasciata)* are mottled pink-gray or brown with white, and have a round-top spire with a turban shape.

Button snails *(Modulus modulus)* have a pale, gray- or brown-streaked, ridge-sculptured body whorl and a low spire.

Smooth Atlantic tegula, max 0.8 in (2 cm)

Plicate hornsnails (*Cerithideopsis pliculosa*), max 1.3 in (3.3 cm), are sharply conical with a thickened aperture. Their color is brownish with cream mid-whorl lines and **varices.**

HABITAT: Topsnails live on shallow hardbottom. Smooth Atlantic tegulas and button snails prefer seagrass beds. Hornsnails are abundant on lagoon mudflats.

DID YOU KNOW? These snails graze on algae, detritus, and small bottom animals, and are most commonly beached near inlets. Varices (singular, varix) are thickened axial ridges spaced at intervals around the whorls, and are formed by thickening of the outer aperture lip during a resting stage in the growth of the shell. This means that hornsnails, and other snails with varices, have periodic growth spurts and resting phases.

Button snail, max 0.5 in (1.3 cm)

Plicate hornsnails have light lips and varices

73

Dark cerith, max 1.5 in (3.5 cm)

Stocky cerith, max 1 in (2.5 cm)

Variable cerith, max 0.7 in (1.8 cm)

Fly-speck cerith, max 1 in (2.5 cm)

Ceriths

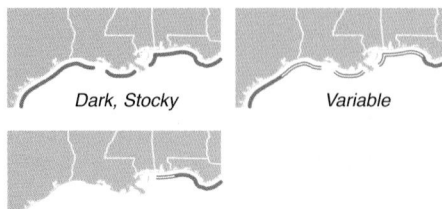

Dark, Stocky

Variable

Fly-speck

RELATIVES: Ceriths are in the family Cerithiidae and are distantly related to hornsnails, worm shells, and wormsnails.

IDENTIFYING FEATURES: All are coarsely sculptured. Their colors are typically yellow to brown, but many beached shells are bleached white.

Dark ceriths *(Cerithium atratum)* have 18–20 beaded ridges per whorl and occasional larger lumps. Their beached shells are light to dark, and are often spiraled with brown and white.

Stocky ceriths *(Cerithium litteratum)* are compact, and compared to dark ceriths, have fewer but larger beads over their rough-sculptured whorls.

Variable (dwarf Atlantic) ceriths *(Cerithium lutosum)* are brown with a light apex and a thick aperture lip.

Fly-speck ceriths *(Cerithium muscarum)* have 9–11 ridges per whorl, which are crossed by spiral lines. New shells are "fly-specked" with spiral rows of brown dots.

HABITAT: All live on subtidal sandy bottom and seagrass.

DID YOU KNOW? Ceriths feed on algae and detritus. The dark cerith is the most common beached cerith shell on most Gulf beaches, but the variable cerith is often abundant near Texas inlets. Empty cerith shells are popular homes for small hermit crabs.

Turretsnail and Wormsnails

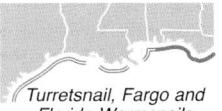 

Turretsnail, Fargo and Florida Wormsnails   *Corroding Wormsnail*

RELATIVES: Turrets and wormsnails are in the family Turritellidae, and are distantly related to ceriths.

IDENTIFYING FEATURES:

Boring turretsnails *(Turritella acropora)* have pale, sharp spirals with rounder whorls than the similarly shaped auger shells. Turretsnail apertures have no siphon canal.

Fargo wormsnails *(Vermicularia fargoi)* as young snails are coiled and have three spiral cords leading to a squarish aperture. Older Fargo wormsnails abandon the coil theme after reaching about an inch (2.5 cm) long and begin growing freestyle. Colors range pale to gray-brown.

Florida wormsnails *(Vermicularia knorrii)* grow wormlike after reaching about 1/2 in (1.3 cm) and except for growth lines have smooth whorls. Colors range pale to reddish brown, but the spiral tip is most often translucent white.

Corroding wormsnails *(Dendropoma corrodens)* have apertures to 0.2 in (6 mm), and are lumpy white outside, shiny brown inside. They attach as individuals or in groups to rocks and shells.

HABITAT: Boring turrets live in shallow sandy bottom. Wormsnails live amid sponges on sandy hardbottom.

DID YOU KNOW? Wormsnail juveniles can move around, but larger snails stay in place, feeding on suspended plankton and detritus. Florida wormsnails often gather as juveniles and literally tie themselves together in knots as they grow.

Boring turretsnail, max 1.5 in (3.8 cm)

Adult Fargo wormsnail, max 3 in (7.2 cm)

Florida wormsnail

Corroding wormsnail colony, individual (inset)

75

Live sargassum snail on a sargassum algal leaf

Mangrove periwinkle, max 1 in (2.5 cm)

Cloudy periwinkles, max 1 in (2.5 cm)

Cloudy periwinkles on a wave-washed tree trunk

Sargassum Snail and Periwinkles
(Mangrove, Cloudy)

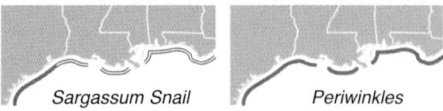

Sargassum Snail   Periwinkles

RELATIVES: Sargassum snails (family Litiopidae) are distantly related to ceriths. Periwinkles are in the family Littorinidae.

IDENTIFYING FEATURES:

Sargassum snails (*Litiopa melanostoma*), max 0.2 in (0.6 cm), are conelike and smooth except for clear axial and spiral lines in the topmost whorls. Colors range pale golden to brown, typically matching the sargassum seaweed (p. 310) the snails are found on.

Mangrove periwinkles (*Littoraria angulifera*) have thin shells, sharp aperture lips, and a groove in their lower columella and inner aperture, which is circular and has the outer-shell color pattern.

Cloudy periwinkles (*Littoraria nebulosa*) are shaped like mangrove periwinkles, but have a D-shape aperture and a rust-color interior. Often, their light body whorl contrasts with a darker patterned spire.

HABITAT: Sargassum snails live only at the surface of the open sea on the algae for which they are named. Periwinkles live just above the tide. Mangrove periwinkles live on mangrove shorelines, and cloudy periwinkles live on stumps and rocks near wave splash.

DID YOU KNOW? The best way to find sargassum snails is to patiently peer into clumps of freshly beached sargassum. Periwinkles feed out of the water on algae that grows on plants and rocks, and their beached shells are most common near inlets. These snails are an important link in the food chain between estuarine plants and dozens of crab, fish, and bird species.

Periwinkles *(Marsh, Interrupted)*

RELATIVES: Periwinkles are in the family Littorinidae.

IDENTIFYING FEATURES:

Marsh periwinkles *(Littoraria irrorata)* have flattened whorls, thick aperture lips, and are patterned with dashed streaks of tan, or purple-brown on their spiral ridges.

Interrupted (placid) periwinkles *(Echinolittorina placida,* formerly thought to be *E. interrupta)* have rounded whorls and are variably patterned with white, and with diagonal, wavy, interrupted, purple-brown or rusty lines. In many populations, these lines match the color of a dark band spiralling around the base of each whorl.

HABITAT: These periwinkles live just above the tide. Marsh periwinkles live on marsh reeds, and interrupted periwinkles live on rocks of jetties and groynes (p. 392).

DID YOU KNOW? Marsh periwinkles are a favorite food of diamondback terrapins (p. 204). The natural range of the interrupted periwinkle was only as far north as southernmost Gulf, where sparse rock outcrops emerge on the Yucatan Peninsula, Mexico. But in the last century, humans have constructed artificial habitats throughout the Gulf that the snails seem to like, in the form of sea walls, jetties, and groins. For some reason, interrupted periwinkles living on jetties with sandstone boulders tend to grow larger than those living on granite.

Marsh periwinkle, max 1 in (2.5 cm)

Live marsh periwinkle on Spartina *stem*

Interrupted periwinkles, max 0.5 in (1.3 cm)

Live interrupted periwinkle on a sandstone jetty

77

Hawk-wing conch, max 4 in (10 cm)

Florida fighting conch, max 4 in (10 cm)

Fighting conchs vary greatly in color

Juvenile hawk wing (left) and fighting (right) conchs

Conchs

Hawk-wing

Florida Fighting

RELATIVES: These are true conchs in the family Strombidae.

IDENTIFYING FEATURES:

Hawk-wing conchs *(Lobatus raninus)* as adults have coarse spiral cords on the body whorl and a rear projection to their widely flared aperture lip. Their colors are mottled brown and white. **Juveniles** 2 in (5.0 cm), have smooth whorls with only faint knobs.

Florida fighting conchs *(Strombus alatus)* have thick shells with smooth, blunt-knobbed whorls. Colors vary from pale yellow to chestnut-brown with occasional light spots and zigzags. **Juveniles** 2 in (5.0 cm), have whorls with distinct spiral cords and shoulder knobs.

HABITAT: Hawk-wing conchs in the northern Gulf live on offshore reefs and are occasionally washed ashore following hurricanes. Florida fighting conchs live in sandy shallows, including the swash zone of low-energy beaches.

DID YOU KNOW? These conchs (pronounced "konks") feed on algae and detritus. Fighting conchs get their name from occasional bouts between rival males. They are spry for snails. A beached fighting conch can quickly flip itself using its foot and sickle-shape operculum. To escape, the snail can also stretch out, dig in its operculum "claw," and leap down the beach. Fighting conchs are being farmed experimentally as an edible alternative to the rarer and slower-maturing queen conch *(Aliger gigas).*

Slippersnails

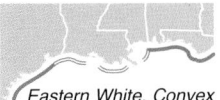 

Eastern White, Convex   *Atlantic*

RELATIVES: Slippersnails are gastropods together in the family Calyptraeidae.

IDENTIFYING FEATURES: Slippersnails are shoe-shape snails, with a conspicuous ventral shelf.

Eastern white slippersnails *(Crepidula depressa)*, max 1 in (2.5 cm), are white, thin, and flattened with a small pointed apex. The inner shelf is flat in convex shells and is convex in recurved shells.

Atlantic slippersnails *(C. fornicata)*, max 2.5 in (6.5 cm), have a coiled apex bent to one side, a smooth exterior, and a shelf over half the aperture.

Convex slippersnails *(C. convexa)* are lumpy and brownish, mottled or pale outside, and shiny brown inside. Their shelf extends over one third of the aperture.

HABITAT: Slippersnails live in shallow waters on rocks and other shells. White slippersnails prefer to be inside other shells.

DID YOU KNOW? Slippersnails grow where they settle as tiny "spat" and have shell shapes that conform to their location. Atlantic slippersnails are famous for growing in **stacks.** The bottom snail in a stack began life as a male and switched to female. A snail arriving to grow on the bottom female remains male until another snail settles upon it. Each arriving young snail assumes a male's role until another snail arrives, a process that can continue to the height of 10 or more slippersnails. The stacks do function in reproduction, but the snail's species name may innocently refer to its curved shape. *Fornix* is Latin for arch.

Eastern white slippersnail, middle shows inner shelf

Atlantic slippersnail, right shows inner shelf

Arching stacks of live Atlantic slippersnails on a clam

Convex slippersnail, max 0.5 in (1.3 cm)

79

Atlantic deer cowrie, max 5 in (13 cm)

Measled cowrie, max 4.5 in (11 cm)

One-tooth simnia, max 0.75 in (2 cm)

A live, beached, one-tooth simnia on a sea whip

Cowries and Simnia

Deer Cowrie   Measled Cowrie

Simnia

RELATIVES: Cowries (family Cypraeidae) are distantly related to siminias (family Ovulidae).

IDENTIFYING FEATURES:

Atlantic deer cowries *(Macrocypraea cervus)* are glossy, egg-shape shells with a body-length, grinning aperture. Their colors are chocolate-brown with solid white spots, or hazy brown with light bands.

Measled cowries *(Macrocypraea zebra)* are similar to deer cowries, but are more elongate (less domed) and have dark centers to their side spots.

One-tooth (sea-whip) simnias *(Simnialena uniplicata)* are shiny, spindle-shape shells with an aperture stretching between pointed ends. Colors are yellow (most common), dark (sediment-stained), cream, or purplish.

HABITAT: Deer and measled cowries live on shallow hardbottom and offshore reefs. Only rarely do unbroken shells make it ashore. Simnias live and feed on soft corals near shore. Look through clumps of these beached gorgonians to find stranded simnias.

DID YOU KNOW? The Atlantic deer cowrie is one of the largest of the world's 190 cowrie species. In life, the glossy shell of a cowrie is completely covered by a grayish frilly mantle with hundreds of soft papillae. Simnias feed on variously colored sea whip gorgonians (p. 66), and often take on the color of their prey's sclerites.

Moonsnails *(Shark Eye, Gould's)*

RELATIVES: Moonsnails share the family Naticidae with baby's ears.

IDENTIFYING FEATURES: Moonsnails have a large body whorl, gaping aperture, smooth rounded spire, and a deep umbilicus partially filled with a thick callus.

Shark eye moonsnail

Shark eye moonsnails *(Neverita duplicata),* max 3 in (7.6 cm), have a smooth, dome-like shell with a low spire. In many shells, an azure band on the lower whorls spirals inward to form a blue "eye." The color of the eye may also be purple, chestnut, or orange. The umbilicus is almost completely covered by a brown callus, which contrasts with the pale lower shell. The upper shell is tan, pinkish, brown-gray, blue-gray, or faded. The snail's thin **operculum** (aperture covering) is translucent amber.

Gould's moonsnail

Gould's moonsnails *(N. delessertiana),* max 2.7 in (7 cm), are similar to shark eyes, but have a more conical spire. Other differences are in the umbilicus, which has a ridged edge to the umbilical wall and is only two-thirds covered by the callus. Shell color is tan, gray, or brownish, with a spiral "eye" less contrasting than in *N. duplicata.*

HABITAT: Sandy shallows out to moderate depths. Live shark eye snails are common in the swash zone during low tide off beaches with silty sands and protective shoals. The amber **opercula** from dead snails persist in the beach's wrack line.

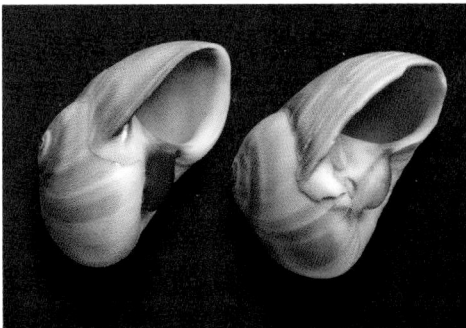

Profiles of shark eye (L) and Gould's (R) moonsnails

DID YOU KNOW? Moonsnails are predators that specialize in preying on other mollusks, especially clam species. The predators find their prey on plowing patrols through surf-zone sands. Unlucky clams are enveloped by the snail's foot, while an acidic secretion softens the spot

Shark eye moonsnails have amber opercula

81

Live shark eye moonsnail in the swash

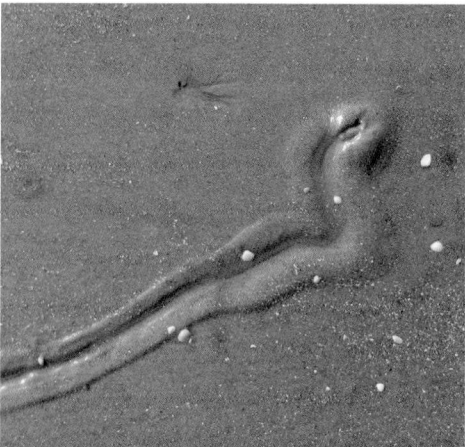

A shark eye makes tracks at low tide

Shark eye "sand collar" eggs, max 4 in (10 cm)

on the shell where a tooth-studded tongue (radula) rasps a beveled hole (p. 150). This hole allows a visit from the snail's proboscis, which injects enzymes to digest the clam's adductor muscles. With no muscles to hold it closed, the clam opens, allowing the moonsnail to complete its meal of clam chowder. A moonsnail's favored diet includes surfclams (p. 141) and coquina clams (p. 136). Juvenile snails eat small clams, and larger adults eat large clams, each at a rate of almost a clam a day.

When plow-prowling for clams, moonsnails detect their prey by "smelling" for telltale clam proteins. The clams are also able to smell the predatory snails and may flee to the sand's surface during a slow-motion attack. On the prowl, most of the moonsnail's body is out of its shell and inflated with seawater. When picked up by a curious beachcomber, the snail must squirt out this water before it can withdraw into its shell and close its operculum behind it.

Moonsnails breed in the surf zone by cementing their eggs with sand into a curled, gelatinous ribbon that cures into a rubbery **sand collar**. A circular opening atop the sand collar is where the snail's aperture was positioned as the collar formed. The collar is a study in hydrodynamics, being just the right shape to remain upright in the surf. A close examination reveals each collar to contain thousands of transparent pockets. These pockets are the minute eggs, embedded within a single-layered matrix of sand grains cemented in gelatin. The collar disintegrates when eggs hatch, so whole collars found in the swash zone probably contain developing little snails. Other moonsnails make similar egg collars.

Moonsnail *(Miniature),* Baby's Ear, and Scotch Bonnet

RELATIVES: Moonsnails and baby's ears share the family Naticidae. Scotch bonnets are in the family Cassidae with cowrie-helmets and helmet shells, and are distantly related to tuns and figsnails.

IDENTIFYING FEATURES:

Miniature moonsnails *(Tectonatica pusilla)* are glossy, ovate-shape (not flattened) moonsnails with an umbilicus almost completely filled by a callus. The shells are patterned with pale, blurry, reddish-brown or gray blotches and a light band on the upper whorls.

White baby's ears *(Sinum perspectivum)* are like a flattened moonsnail with an expansive aperture. The body whorl is sculptured with broad spiral grooves. Shells are dull white or stained. **Live animals** have a white body enveloping their shell, and look like a poached egg.

Scotch bonnets *(Semicassis granulata)* have a thin shell with a pointed spire and a thick, toothed, aperture lip. Axial grooves cross spiral cords, giving the shell a beaded feel. Colors range from white, to cream with dark squares.

HABITAT: Sandy shallows and intertidal flats

DID YOU KNOW? Baby's ear snails have an enormous foot that cannot be withdrawn into their shell. They prey on buried bivalves by chasing them down beneath soft sediments. The baby's ear subdues the bivalve with its muscular foot, drills through the prey's shell with a rasping radula, and inserts a proboscis to feed.

Miniature moonsnail, max 0.3 in (0.8 cm)

White baby's ear, max 2 in (5 cm)

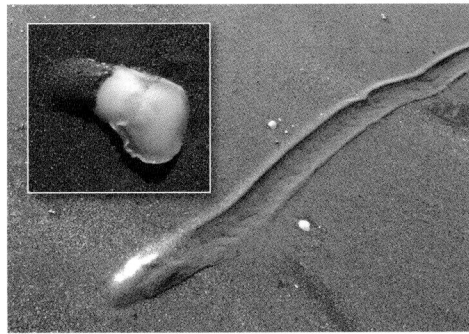

A white baby's ear (uncovered, inset) moves in wet sand

Scotch bonnet, max 4 in (10 cm)

83

Reticulate cowrie-helmet, max 3 in (7.6 cm)

Giant tun shell, max 10 in (25 cm)

Giant hairy triton, max 3.5 in (9 cm)

Cowrie-helmet, Giant Tun, and **Triton**

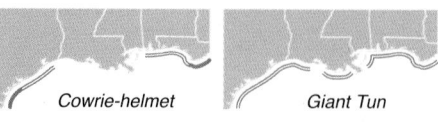

Cowrie-helmet

Giant Tun

Hairy Triton

RELATIVES: Cowrie-helmets share the family Cassidae with Scotch bonnets and helmet shells, and are distantly related to tun shells (family Tonnidae), tritons (Ranellidae), and distorsios.

IDENTIFYING FEATURES:

Reticulate cowrie-helmets (*Cypraecassis testiculus*) are dense, egg-shape shells with smooth spiral grooves and growth lines, and a rounded spire. They are chestnut to salmon with darker, blurry squares.

Giant tun shells (*Tonna galea*) are almost spherical in shape with a wide aperture, prominent spiral ridges, and a cream or brown color. Most beach finds are pieces.

Giant hairy tritons (*Monoplex parthenopeus*) are sculpted with spiral bands and have a thick, wavy outer lip.

HABITAT: Cowrie-helmets and giant hairy tritons live on rocky reefs. Giant tuns are most common offshore.

DID YOU KNOW? Tun shells feed on other mollusks, sea cucumbers, and fishes by engulfing their prey within a large expandable proboscis. "Hairy" tritons are named for the coat of frilly **periostracum** that covers the living snail (and still covers the unworn shell in the bottom image). Several other triton species rarely strand on Gulf beaches.

Distorsio, Figsnail, Miniature Cerith, and Wentletrap *(Angulate)*

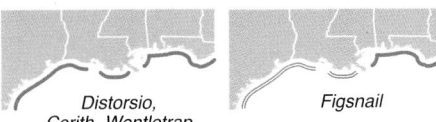

Distorsio, Cerith, Wentletrap

Figsnail

Cat. #16794

Atlantic distorsio, max 3 in (7.6 cm)

RELATIVES: Distorsios (family Personidae) are closer to tun shells than to figsnails (Ficidae). Miniature ceriths are in the family Cerithiopsidae with triphoras, and are not directly related to ceriths. Wentletraps are in the family Epitoniidae, distantly related to purple sea snails.

IDENTIFYING FEATURES:

Atlantic distorsios *(Distorsio clathrata)* have a latticed sculpture, thick aperture lip, long siphon canal, and whorls at angles presenting a distorted look. Colors are dull tan with a glossy white aperture.

Atlantic figsnails *(Ficus papyratia)* are delicately tapered at the front, have a low spire, and are sculptured with fine spiral ridges. Their colors range from cream to tan, sometimes with faint brown dots.

Atlantic figsnail, max 5 in (13 cm)

Adam's miniature ceriths *(Seila adamsii)* have sharply conical shells with spiral cords like screw threads. Color is typically a uniform yellow-brown.

Angulate wentletraps *(Epitonium angulatum),* are glossy white with the body whorl circled by 9–10 bladelike ribs, which are angled at the whorl shoulders.

HABITAT: Distorsios live on rocky reefs, and figsnails live in sandy shallows. Adam's miniature ceriths live under rocks in coastal waters. Wentletraps live in sandy areas to moderate depths.

Adam's miniature cerith, max 0.5 in (1.2 cm)

DID YOU KNOW? Angulate wentletraps chew chunks off living anemones by soothing them with a purple anesthetic. "Wentletrap" is Dutch for a winding staircase. The winding ribs probably offer protection against predatory snails that bore holes through shells of their prey.

Angulate wentletrap, max 1 in (2.5 cm)

85

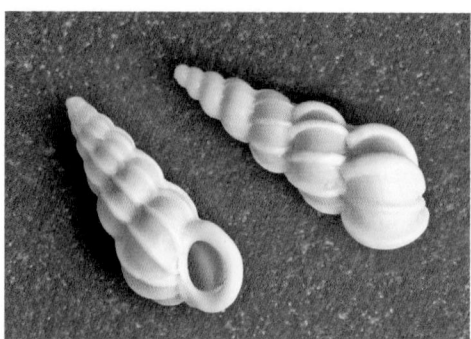

Humphrey's wentletrap, max 1 in (2.5 cm)

Many-ribbed wentletrap, max 0.5 in (1.3 cm)

Brown-banded wentletrap, max 0.8 in (2 cm)

Mitchell's wentletrap, max 2.5 in (6.3 cm)

Wentletraps *(Humphrey's, Many-ribbed, Brown-banded, Mitchell's)*

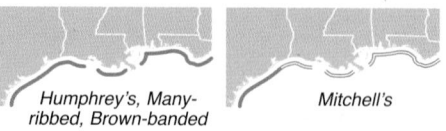

Humphrey's, Many-ribbed, Brown-banded   Mitchell's

RELATIVES: Wentletraps are in the family Epitoniidae, distantly related to purple sea snails.

IDENTIFYING FEATURES: These shells have a rounded, thick-lipped aperture, distinct ribs, and a porcelain sheen.

Humphrey's wentletraps *(Epitonium humphreysii)* are glossy white with 8–9, thick, rounded ribs on the body whorl.

Many-ribbed wentletraps *(Epitonium multistriatum)* are glossy white with about 19 low ribs on the body whorl.

Brown-banded wentletraps *(Gyroscala rupicola)* have rounded ribs of varying strengths, and spiral bands of white, tan, and brown.

Mitchell's wentletraps *(Amaea mitchelli)* have slightly raised ribs that cross less distinct spiral cords and an aperture lip that is thinner than in other wentletraps. Colors are cream with a thick brown band in the center of each whorl.

HABITAT: Wentletraps live in sandy areas to moderate depths. Mitchell's wentletraps are found farthest offshore and are the rarest of these beach finds.

DID YOU KNOW? Wentletraps have family relatives comprising about 56 species in the Gulf. Most live in southern Gulf waters or are found offshore, out as far as the dark, icy-cold reefs under two miles of water. There and in Gulf shallows, this group of snails specializes in eating coral polyps and anemones.

Purple Sea Snails

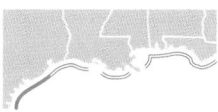

RELATIVES: Purple sea snails are in the family Janthinidae, distantly related to wentletraps.

IDENTIFYING FEATURES: Also known as "storm snails," these gastropods have fragile, violet shells. Free-floating live snails produce a translucent, **bubble raft** arcing from their aperture.

Common purple sea snails (*Janthina janthina*) have a low spire and D-shape aperture. Their top whorls are pale and their base is violet.

Globe purple sea snails (*Janthina globosa*) are violet, with their body whorl slightly paler. The shell has rounded whorls with a pointed spire. An wide furrow in the body whorl leads to an indented outer aperture. A similar species, the pale purple sea snail (*J. pallida*), has a more rounded aperture than *J. globosa*, no furrow and aperture indentation, and is only faintly purple.

HABITAT: Purple sea snails live adrift on the open ocean. Unbroken snails can be found in freshly beached wrack.

DID YOU KNOW? The shell-tinting of purple sea snails blends in with the color of deep ocean waters, presumably hiding them from the birds and young sea turtles that would love to eat them. These snails raft on, and nibble on, their floating hydrozoan prey, which includes Portuguese man-o-war, by-the-wind sailors, and blue buttons (p. 62). A purple dye secreted by the snails paralyzes their prey, allowing them to eat unretracted tentacles. When not sailing upon their prey, purple sea snails construct a mucous-bubble raft for buoyancy, allowing them to float to their next meal.

Common purple sea snail, max 1.5 in (3.5 cm)

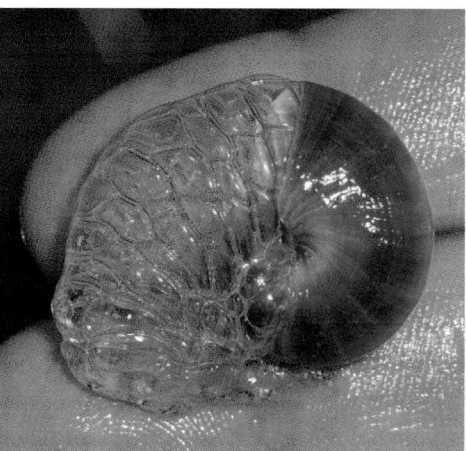

Common purple sea snail with bubble raft

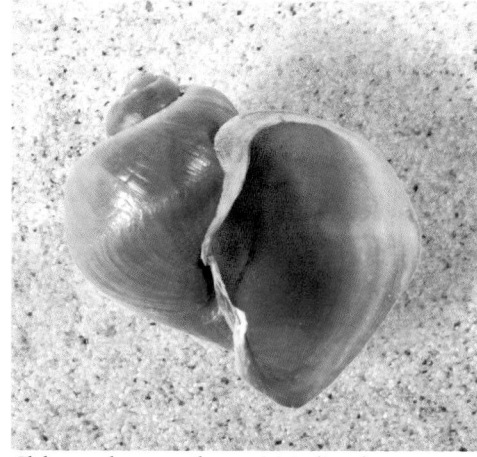

Globe purple sea snail, max 0.8 in (2 cm)

Brown-line niso, max 0.3 in (0.7 cm)

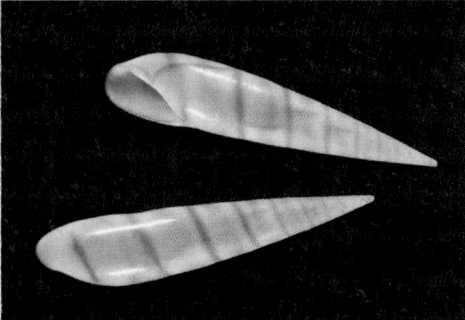

Two-band eulima, max 0.5 in (1.3 cm)

Apple murex, max 4.5 in (12 cm)

Communal apple murex egg mass, max 19 in (50 cm)

Niso, Eulima, and Murex *(Apple)*

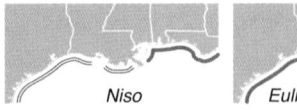
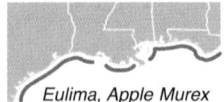

Niso   *Eulima, Apple Murex*

RELATIVES: Nisos and eulimas share the family Eulimidae. Not directly related, murices share the family Muricidae with rocksnails and drills.

IDENTIFYING FEATURES:

Brown-line nisos *(Niso aeglees)* have glossy shells with an angled body whorl and a thin brown line on each suture.

Two-band eulimas *(Eulima bifasciata)* have slender, glossy shells with a translucent apex and elongate aperture. Brownish bands lie just below the suture, and fainter bands are mid-whorl. The similarly shaped Jamaica eulima *(Melanella jamaicensis)* has a glossy white base and spire tip.

Apple murices *(Phyllonotus pomum)* are cream with brown bands and have three lumpy varices per whorl. Their inner aperture lip has a thin, flared margin and a dark blotch opposite the siphon canal.

HABITAT: These snails live in sandy areas to moderate depths. Eulimas and nisos are closely associated with echinoderms.

DID YOU KNOW? The Eulimidae are a family of parasites. Most feed by chewing through the skin of sea cucumbers or sea urchins, and sucking their blood. Some species live inside the bodies of these echinoderms. Apple murex **egg masses** belong to multiple females that take part in group spawning events where egg capsules are combined within a common mass. These communal egg masses can accomodate the spawn of up to 100 females and be dozens of times larger than an individual murex snail.

Murices *(Giant Eastern, Lace)* and **Oyster Drill**

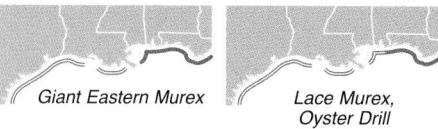

Giant Eastern Murex

Lace Murex, Oyster Drill

RELATIVES: Murices share the family Muricidae with drills and rocksnails.

IDENTIFYING FEATURES: Snails in the murex family have highly sculptured shells, often with varices (thick axial ridges).

Giant eastern murices *(Hexaplex fulvescens)* are a turnip-shape with a rounded aperture and tubular siphon canal. The body whorl is sculptured by about eight axial ridges, each bearing hollow spikes. Beached shells are white to gray and may have only worn knobs instead of spikes. Reddish-brown smudges or dashes on the spiral cords may persist even in old shells. The persistent **operculum** is thick, oval, and golden brown.

Lace murices *(Chicoreus florifer)* are distinct in having varices liuned with hollow, scoop-shape spines, and in having a simple, circular, unmarked aperture.

Atlantic oyster drills *(Urosalpinx cinerea)* have rounded shoulders and 9–12 rugged axial ridges per whorl. The aperture is oval with an open siphon canal. Colors range between yellow, orange, gray, and white, occasionally with brown streaks.

HABITAT: Atlantic oyster drills frequent shallow oyster reefs. Giant eastern and lace murices live in coastal waters out to moderate depths.

DID YOU KNOW? These snails are predators of bivalves. A venom used to subdue their prey is the source of the ancient garment dye, tyrian purple. The snails use this secretion to sedate their prey and as an antimicrobial lining to protect their egg masses.

Giant eastern murex, max 7 in (18 cm), and operculum

Lace murex, max 3.2 in (8 cm)

Atlantic oyster drill, max 1.5 in (3.8 cm)

Living Florida rocksnails on a jetty boulder

Hay's rocksnail, max 4.5 in (11.4 cm)

Similar-size Hay's (L) and Florida (R) rocksnails

Fresh and empty (inset) rocksnail egg capsules

Rocksnails

RELATIVES: Rocksnails are in the murex family, Muricidae.

IDENTIFYING FEATURES:

Florida rocksnails *(Stramonita haemastoma)*, 3 in (8 cm), have shells with spiral cords, a rounded body whorl, and wide apertures having a thick, inner-toothed outer lip. They are grayish to brownish and frequently show red-brown spots. Aperture color ranges white to orange.

Hay's rocksnails *(Stramonita canaliculata)* have prominent shoulder knobs where axial ridges cross two bands of coarse spiral cords. Whorls are angled, and separated by a channeled suture. The inner aperture lip has a smooth, thick, parietal shield.

HABITAT: Rocksnails are common on jetties and other intertidal rocks.

DID YOU KNOW? Rocksnails prey on barnacles, oysters, and other mollusks. Their variable color and shape may come from hybrids between the species, which confuses identification of individual shells. In the spring, female rocksnails lay fertile eggs within vase-shape **capsules** that are attached to a shell or rock by a short stem in clusters of 20–86. Each capsules contains about 2,000 tiny eggs. Capsule colors are creamy yellow when first deposited, brown when eggs mature, and purple when empty. New rocksnail veliger larvae drift for 2–3 months, and may travel hundreds of miles before settling in shallow habitat as a young snail.

Cantharus Shells

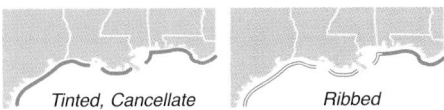

Tinted, Cancellate   Ribbed

RELATIVES: Canthari (singular cantharus) are in the family Pisaniidae, distantly related to dovesnails, spindle shells, whelks, and nassas.

IDENTIFYING FEATURES:

Tinted canthari *(Gemophos tinctus)* have shells with distinct spiral cords that intersect with low axial ridges to form weak bumps at the whorl shoulders. Their outer lip is toothed and the columella is glossy. Colors are cream to bluish-gray, typically with streaks and smudges of brown.

Cancellate canthari *(Solenosteira cancellaria)* have spiral cords crossing axial cords, which give most of the shell a beaded look. The outer aperture lip is scalloped and the siphon canal is slightly upturned. Colors are yellowish to red-brown with white streaks.

Ribbed canthari *(Hesperisternia multangulus)* also called false drills, have large axial ridges, which are sharply angled at the whorl shoulders and that cross numerous spiral cords. Colors range pale orange to dark brown.

HABITAT: Shallow sand, rubble, and seagrass out to moderate depths

DID YOU KNOW? Females deposit clusters of spiked, vase-shape **egg capsules** in spring. Cantharus snails prey on worms, barnacles, and other attached invertebrates. Canthari get their name from the cantharus, sacred cup of Bacchus, Roman god of wine.

Tinted cantharus, max 1.2 in (3 cm)

Cancellate cantharus, max 1.4 in (3.6 cm)

Ribbed cantharus, max 1.2 in (3 cm)

Cancellate cantharus egg capsules, each about 2 mm

91

Lunar dovesnail, max 0.25 in (0.6 cm)

Fat dovesnail, max 0.25 in (0.6 cm)

Well-ribbed dovesnail, max 0.9 in (2.3 cm)

Greedy dovesnail, max 0.8 in (2 cm)

Semiplicate dovesnail, max 0.6 in (1.5 cm)

Dovesnails

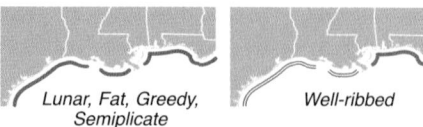

Lunar, Fat, Greedy,
Semiplicate

Well-ribbed

RELATIVES: Dovesnails are in the family Columbellidae and are distantly related to canthari, spindle shells, whelks, and nassas.

IDENTIFYING FEATURES: Dovesnails have tubby shells the size of a pencil eraser with pointed spires, short siphon canals, and toothed aperture lips.

Lunar dovesnails *(Astyris lunata)* are smooth, and orange-brown with dark wavy lines or pale spiral bands.

Fat dovesnails *(Parvanachis obesa)* have plump shells with distinct vertical ribs and fine revolving lines. The outer lip has a few small teeth inside. Colors are dull gray to white with yellow spiral bands.

Well-ribbed dovesnails *(Cotonopsis lafresnayi)* have straight-sided whorls that telescope into the spire. All whorls have spiral cords and prominent axial ribs. Colors are tan to brown with indistinct markings.

Greedy dovesnails *(Costoanachis avara)* are similar to well-ribbed dovesnails, but have a lower spire and distinct white spots on the body whorl.

Semiplicate dovesnails *(Costoanachis semiplicata)* resemble well-ribbed and greedy dovesnails, but have a white background with abundant red-brown splotches, faint spiral cords, and axial ribs only on the body whorl.

HABITAT: Seagrass or rubble out to moderate depths

DID YOU KNOW? These dovesnails graze on algae and detritus. About 46 dovesnail species are known in the Gulf, from coastal lagoons to offshore reefs.

Tulip Shells

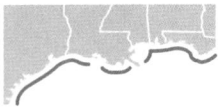

RELATIVES: Tulip snails share the family Fascilariidae with spindle shells.

IDENTIFYING FEATURES: Tulip snails are shaped like a pointed spindles with rounded curves and a stemlike siphon canal.

Eastern banded tulips (*Cinctura hunteria*) are cream to light gray with orange or gray splotches and distinctly fine, continuous, spiral lines of reddish brown. In an orange form, these markings are all shades of orange. The whorls are smooth.

True tulips (*Fasciolaria tulipa*) are similar to banded tulips but have darker brown (or orange) splotches and interrupted, blurry, closer-set spiral lines. Their whorls also differ in having fine ridges below each suture.

HABITAT: Banded and true tulips live on sand in water less than 100 ft (30 m).

DID YOU KNOW? True tulips prey on banded tulips, as well as on other midsize gastropods. Tulip snails crawl into shallow waters during the winter to find firm places where they attach their clustered **egg capsules**, which look like tiny bouquets. Young, miniature tulip snails emerge from holes at the flat end of each frilly capsule. The capsules are formed of a tough, fingernail-like protein. If a beach-found capsule cluster rattles, it's likely to contain tiny tulip shells. Several occupy each capsule. True tulip egg capsules are flattened cones with frilly outer rims. Eastern banded tulips have similar egg capsules with less frilly outer rims.

Eastern banded tulip, max 4 in (10 cm)

True tulip, max 5 in (13 cm)

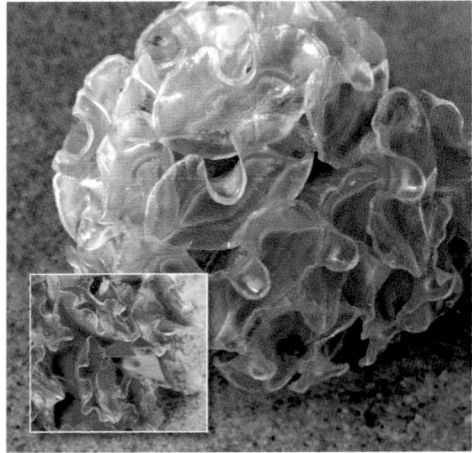

Eastern banded and true tulip (inset) egg capsules

93

Horse conch, max 19 in (48 cm), and egg mass

A horse conch's operculum (A) and orange body (B)

Juvenile horse conch, 2 in (5 cm)

Florida Horse Conch

RELATIVES: Spindle shells like the horse conch share the family Fasciolariidae with tulip snails.

IDENTIFYING FEATURES:

Florida horse conchs (*Triplofusus giganteus,* original name, *Fasciolaria gigantea*) are very large as adults. They have a whitish spire and are often covered with brown, flaky periostracum. Beach-worn adult shells are white with a glossy tan interior. Living horse conchs have an orange-red body and a thick **operculum**.

A **juvenile horse conch** is the same shape as the adult, but is a more uniform, peach-gold color and significantly less worn.

HABITAT: Horse conchs live on sandy bottom in waters as shallow as the low-tide line and out to depths as great as 650 ft (200 m).

DID YOU KNOW? Horse conch **egg masses** comprise dozens of capsules, each shaped like a flattened, ribbed cone, and attached at its base within a twisted cluster. Capsules with neat, round holes in their almond-shape end show escape apertures where young snails have left. The horse conch has the distinction of being Florida's state shell, and the largest snail in North America. Being hefty allows horse conchs to prey on big gastropods like lightning whelks, which can make up almost a third of the horse conch's diet. Large horse conchs may be over a decade old.

Crown Conch

RELATIVES: Crown conchs are actually whelks in the family Melongenidae, and are distantly related to murices, canthari, dovesnails, spindle shells, and nassas.

IDENTIFYING FEATURES:

Crown conchs (*Melongena corona*) have a large aperture leading into a wide siphon canal. Their shell is sculptured by axial ribs along the whorl shoulders that commonly project as hollow spikes, but may be only tiny nubs. Shoulder-spiked shells have similar spines at their base. Juvenile crown conchs have no spikes. Colors range from spiral bands of blue-gray or brown in smaller shells, to tan or bone white in larger shells, which are commonly fouled with algae.

HABITAT: Crown conchs live on shallow sand, mud, and oyster reef (p. 14), typically less than chest deep.

DID YOU KNOW? Crown conchs are predators that specialize in eating other mollusks, including scallops, oysters, clams, and other gastropods. But they are opportunistic when hungry, occasionally feeding on carrion and detritus. Spring through summer, female crown conchs deposit ribbonlike rows of up to 20 **egg capsules**, attaching them to a variety of intertidal objects. Each capsule contains a dozen or more eggs, which hatch in about 20–28 days. Although the crown conch's taste for eastern oysters (p. 116) has prompted some to implicate the snail in declines of commercial oyster production, studies show that significant effects from this predation are doubtful.

Crown conch, max 5 in (13 cm), juvenile, upper right

Crown conch with freshly deposited egg capsules

Live crown conch prowling the intertidal zone

95

Lightning whelks, max 16 in (40 cm)

Live lightning whelks have black bodies

A long coil of lightning whelk egg capsules

Lightning Whelk

RELATIVES: These snails share the family Melongenidae with pear and crown conch whelks.

IDENTIFYING FEATURES:

Lightning whelks (*Sinistrofulgur sinistrum*) have a large body whorl with distinct shoulders and a wide aperture tapering into their long siphon canal. They may have a dozen or more knobs at the shoulders of their body whorl, which spirals to their left (spire is rearward). This **left-handed aperture** separates them from most other marine snails. Colors are cream to gray with younger shells showing brown, lightning-bolt, axial streaks. Living snails have a black body, which also separates them from other whelks. Texas shells tend to have more numerous shoulder knobs (spines), and many Florida shells have smooth shoulders.

HABITAT: Sandy shallows and seagrass including intertidal areas

DID YOU KNOW? Left-handed coiling may help this whelk survive predation from stone crabs, which have their right claw as a "crusher claw." Lightning whelks produce **egg masses** containing dozens of discs attached by a common string. These discs differ from those of other whelks by their edge flanges, as if the discs were fabricated by a poor-fitting mold. For thousands of years, lightning whelks have been used by native Americans, who fashioned the shells into myriad artifacts interred in graves hundreds of miles from the Gulf Coast. The lightning whelk is the state shell of Texas.

Pear Whelks

RELATIVES: These snails share the family Melongenidae with knobbed, lightning, channeled, and crown conch whelks.

IDENTIFYING FEATURES:

Pear (fig) whelks *(Fulguropsis spirata)* are pear-shape snails with a large aperture leading into a long siphon canal. Colors of unworn shells are cream with brown, wavy, axial streaks. A channel along the suture disappears in earlier whorls. Pear whelks have no projections on the edges of their body whorl, and only small projections on the spire whorls.

Shouldered pear whelks *(Busycotypus plagosus)* are similar to pear whelks, but have distinct shoulders on the upper body whorl bearing numerous nodules.

HABITAT: Estuaries and other sandy shallows

DID YOU KNOW? Pear whelks follow the whelk practice of preying on clams. But unlike their larger cousin, the lightning whelk, pear whelks have too dainty an aperture lip to chip their way into a clammed-up bivalve. Instead of chipping, pear whelks use their shell lip as a wedge that, little-by-little, forces a clam's valves open enough to allow the whelk access for its proboscis through the narrow slit. Because pear whelks are smaller than the other whelk species, so are its **egg masses,** although they have the similar theme of disc-shape capsules on a string. In pear whelks, the discs are shaped like inflated coins and are fringed with pointed edge projections.

Pear whelk, max 5.5 in (14 cm)

Shouldered pear whelk, max 4 in (10 cm)

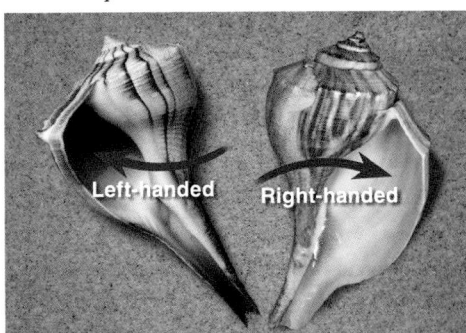

Left-handed Right-handed

Lightning whelk (L), shouldered pear whelk (R)

Pear whelk egg discs. Young occupants (inset)

97

Sharp nassa, max 0.5 in (1.3 cm)

Bruised nassa, max 0.8 in (2 cm)

Florida cone, max 2 in (5 cm)

Alphabet cone, max 3 in (7.5 cm)

Nassas and Cone Shells

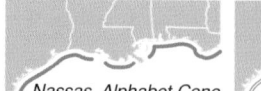

Nassas, Alphabet Cone

Florida Cone

RELATIVES: Nassas are in the family Nassariidae and are distantly related to whelks. Cone shells (Conidae) are distantly related to augers.

IDENTIFYING FEATURES:

Sharp nassas (*Nassarius acutus*) have glossy, oval shells with conical spires. Their whorls have pointed beads connected by brown spiral lines.

Bruised nassas (*Phrontis vibex*) have beaded axial ribs, a broadly conical spire, and dark-lined sutures. Their inner aperture has a wide, glossy callus, which in darker shells bears a purple "bruise." Juveniles, less than 1/4 in (0.6 cm), have no callus.

Florida cones (*Conus anabathrum*) have a pointy, medium-high spire (for a cone shell) and a body whorl with orange or yellow streaks and splotches on a lighter background. Shells often have a mid-body-whorl white band.

Alphabet cones (*Conus spurius*) are variable. Most have a low, concave spire with a distinct shoulder. Colors are white with spiral rows of orange dashes and checks. A few may be uniformly pale, or have markings that range from faint to dark brown.

HABITAT: Nassas are in shallow seagrass, sand, or muddy areas of inshore bays out to moderate depths. These cones are most common in deeper rubble and reef.

DID YOU KNOW? In Latin, *nassa* means "wicker basket." Nassas that smell something dead will emerge with other nassas to swarm the carrion food source. Cone shells have a modified radula tooth that looks and functions like a harpoon. They use this barbed, needle-like weapon to inject venom into the worms they eat.

Augers

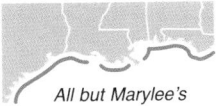 

All but Marylee's   *Marylee's*

RELATIVES: Augers (family Terebridae) are distantly related to cone shells.

IDENTIFYING FEATURES: Augers are glossy, sharp cones with smooth ribs and short, distinct siphon canals.

Common American (eastern) augers *(Neoterebra dislocata)* are white, gray and white, or pale orange with beaded spiral bands between axially ribbed whorls.

Gray Atlantic augers *(Hastula cinerea)* and Sallé's augers *(H. salleana)* are purple- or brown-gray with darker banded, short axial ribs atop each smooth whorl. Only slight variations in sculpture and color separate these two forms across the Gulf.

Marylee's augers *(Hastula maryleeae)* are mostly brown and cream with an inflated band of short axial ribs atop each smooth whorl.

HABITAT: All live in sandy shallows.

DID YOU KNOW? These augers feed on worms, often hunting with a long stride and a quick pace, nearly one "footstep" per second. The snails lunge when they find a worm above the sand and subdue it with a stab from their venomous, radular tooth. If pushed up the beach by a wave, gray Atlantic and Sallé's augers use their foot like a drogue (sea anchor) to catch the water of the next receding wave so they are swept back down the beach. These auger species also have an active mating behavior. Summer mating swarms are in the style portrayed by Burt Lancaster and Deborah Kerr in *From Here to Eternity,* with embracing pairs rolling in the swash zone.

Common American auger, max 2.4 in (6 cm)

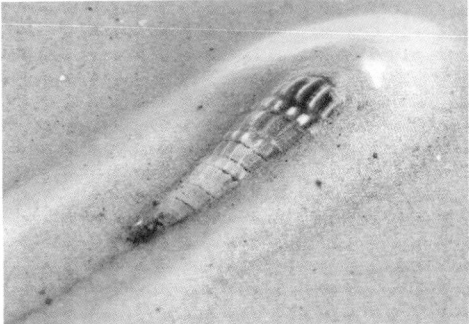

Live common American auger at low tide

Gray Atlantic auger, max 1.5 in (3.8 cm)

Cat. #05365

Marylee's auger, max 1 in (2.5 cm)

Minute dwarf olive, max 0.4 in (1.0 cm)

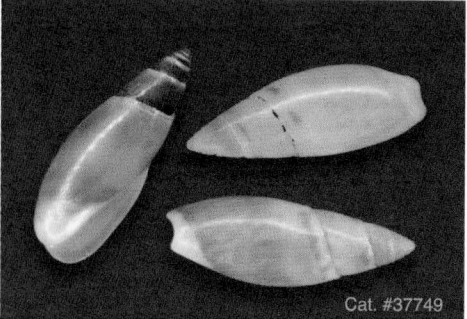

Cat. #37749

Whitened dwarf olive, max 0.5 (1.3 cm)

Variable dwarf olive, max 0.6 in (1.5 cm)

Trail at low tide from a variable dwarf olive

Dwarf Olive Shells

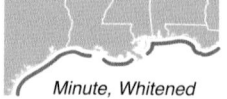
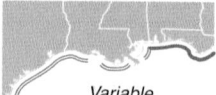

Minute, Whitened   Variable

RELATIVES: Dwarf olives are with olive shells in the family Olividae.

IDENTIFYING FEATURES: Dwarf olives have glossy, olive-shape shells with triangular apertures no more than half the total shell length and a wide, open siphon canal.

Minute dwarf olives (*Olivella minuta*) are medium bodied, with length about 2.3 times their width, and are patterned by purplish wavy lines blurred against a background of cream and yellow. The whorl suture is deeply grooved.

Whitened dwarf olives (*Olivella dealbata*) are moderately narrow, having a length about 2.5 times the width, with an elevated, non-telescoping, flat-sided spire and an indistinctly grooved suture.

Variable dwarf olives (*Olivella mutica*) are tubby, being only twice as long as wide. They are variably patterned with thick spiral bands of gray or brownish, and tend to have their darkest colors at the center body whorl. The whorl suture is deeply grooved.

HABITAT: Intertidal sands and coastal shallows

DID YOU KNOW? These gastropods are tiny, occasionally abundant, surf-zone hunters of small bivalves, crustaceans, worms, forams, and diatoms. Unlike their olive-shell cousins, dwarf olives swallow their prey whole. In the swash zone during low tide on fine-grained beaches, tiny **trails** and faint movement give away the paths of foraging dwarf olives.

Olive Shells and Junonia

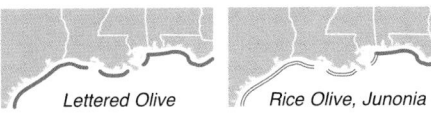

Lettered Olive   Rice Olive, Junonia

RELATIVES: Olive shells are with dwarf olives in the family Olividae. Junonias are volutes (family Volutidae), distantly related to marginellas and nutmegs.

IDENTIFYING FEATURES:

Lettered olives *(Oliva sayana)* have a thick, glossy shell with a small pointed spire about 1/9th of total shell length. Unfaded shells are colored with overlapping, slightly blurry, brown zigzags that are darkest just below the suture and as part of two, broad, spiral bands.

Rice olives *(Olivella floralia)* are narrow, having a length about three times the width, with an elevated, non-telescoping, flat-sided spire and indented sutures. The apex is often dark gray, and the body whorl is typically rusty with white spiral bands top and bottom. But many shells fade from this pattern or are all white.

Junonias *(Scaphella junonia)* have unmistakably flamboyant shells with squared, chestnut spots on a background of pinkish-ivory. They have a wide aperture and rounded whorls that spin into a distinct spire with a rounded apex.

HABITAT: These olives live in sandy shallows. Junonias live on offshore reefs.

DID YOU KNOW? An olive's glossy shell is covered in life by its body mantle and large foot. The foot allows them to burrow easily through sand. **Live lettered olives** are commonly seen at low tide as they prowl in search of coquina clams and other mollusks. Prey are wrapped by the olive's foot, paralyzed with salivary secretions, and eaten alive in bites from the predator's radula-tipped proboscis. The junonia is the state shell of Alabama. They are rare there, but arrive after hurricanes.

Lettered olive, max 2.7 in (7 cm)

A live lettered olive with raised siphon, at low tide

Rice olive, max 0.3 in (0.9 cm)

Junonia, max 3.5 in (9 cm)

101

Atlantic marginella, max 0.5 in (1.3 cm)

Common nutmeg, max 1.7 in (4.5 cm)

Common sundial, max 2.5 in (6.4 cm)

Variable color patterns of common sundials

Marginella, Nutmeg, and Sundial

RELATIVES: Marginellas (family Marginellidae) and nutmegs (Cancellariidae) are distantly related to junonias. Sundials are in the family Architectonicidae, remotely related to bubble shells and some sea slugs.

IDENTIFYING FEATURES:

Atlantic marginellas (*Prunum apicinum*) have a glossy, egg-shape shell with a low spire and a thick outer aperture lip extending up past the body whorl. Colors range white, through pale warm colors, to gray.

Common nutmegs (*Cancellaria reticulata*) have egg-shape shells with a cross-hatched texture and whorls indented at the sutures. The inner lip of the aperture has two white folds on the columella. Shell colors vary, from tan with blurry brown streaks to cream-white.

Common sundials (*Architectonica nobilis*) are distinct in having a deep umbilicus and a spire like a flattened cone.

HABITAT: Sandy shallows. Marginellas also frequent bays with seagrass.

DID YOU KNOW? Marginellas get their name from their wide margin (aperture lip). They feed mostly as scavengers, emerging from the sand en masse and swarming any dead fish or crab they detect. Common nutmegs feed on soft-bodied animals buried in the sand. Sundials spend their days buried spire-down in the sand and emerge at night to feed on sea pansies (p. 65). Sundials themselves often fall prey to polychaete fireworms.

Striate Bubble, Barrel-bubble, and **Sargassum Nudibranch**

Bubble, Barrel-bubble   Nudibranch

RELATIVES: All are gastropods. Bubble shells (family Bullidae) are distantly related to channeled barrel-bubbles (Tornatinidae) and more remotely related to nudibranchs and glaucus sea slugs (opisthobranch gastropods in the order Nudibranchia).

Striate bubble shell, max 1 in (2.5 cm)

IDENTIFYING FEATURES:

Striate bubbles *(Bulla striata)* have a fragile, smooth, mottled-brown, egg-shape shell with a sunken apex and an aperture longer than the body whorl.

Channeled barrel-bubbles *(Acteocina canaliculata)* are whitish with an aperture along three-fourths of the shell length. The spire is low and often has a protruding nipple-like protoconch. The similar Candé's barrel-bubble *(A. candei)* is straight-sided with a more conical spire.

Channeled barrel-bubble, max 0.2 in (6 mm)

Sargassum nudibranchs *(Scyllaea pelagica)*, 2.5 in (6.3 cm), are the color of fresh sargassum algae (p. 310) and have rhinophores shaped like the plant's leaves. On the beach, they may look like an amber blob, but placed in seawater, they will show their full form.

HABITAT: Striate bubbles and channeled barrel-bubbles live in sandy shallows and seagrass. Sargassum nudibranchs live adrift on floating sargassum algae in the open sea.

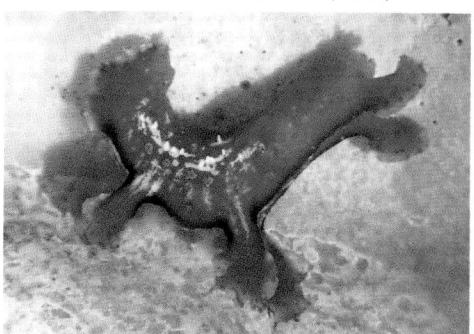

Sargassum nudibranch in water held in a seashell

DID YOU KNOW? Sargassum nudibranchs graze on sargassum hydroids (p. 64) and swallow the sargassum's tiny air bladders to keep from sinking should they loose their grip on the algae.

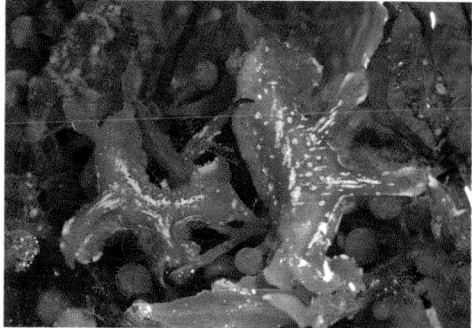

Can you find the two nudibranchs in this sargassum?

103

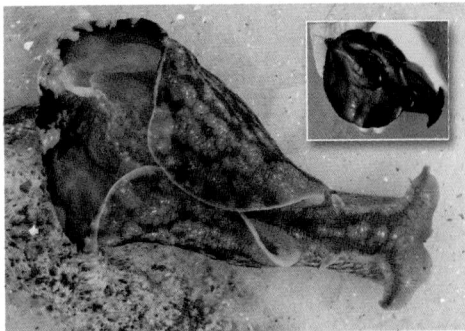

Mottled sea hare. Sooty sea hare (inset)

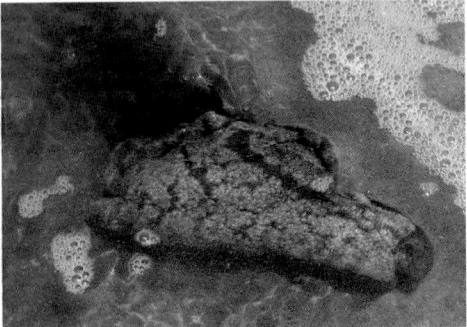

A mottled sea hare, inking in defense

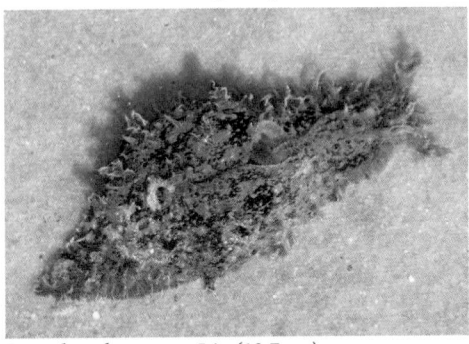

Ragged sea hare, max 5 in (12.7 cm)

Straight-needle pteropods, max 0.4 in (1 cm)

Sea Hares and Pteropod

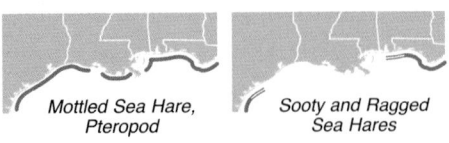

Mottled Sea Hare, Pteropod *Sooty and Ragged Sea Hares*

RELATIVES: Sea hares are opisthobranch gastropods with internal shells in the order Aplysiida, remotely related to pteropods, order Thecosomata.

IDENTIFYING FEATURES:

Stranded **mottled sea hares** (*Aplysia fasciata*), max 10 in (25 cm), are slippery, writhing blobs with green or brown mottling. The blunt head has a rabbitlike face with a fleshy moustache (oral tentacles) and two beady eyes below two soft rhinophores. In water, they swim by undulating winglike parapodia. **Sooty sea hares** (*A. morio*) are similar but dark, and spotted sea hares (*A. dactylomela*) are yellow-green with dark rings.

Ragged sea hares (*Bursatella leachii*) are covered by fleshy, branching papillae that give the animal a frazzled, algal appearance. To camouflage further, they are splotched with olive, yellow, brown, black, and other seabottom tones.

Straight-needle pteropods (*Creseis acicula*) have glassy, needle-shape shells that occasionally beach in mass numbers. This tiny, shelled, sea slug has paired winglike flaps for swimming.

HABITAT: These sea hares live in shallow seagrass. Pteropods inhabit the open sea.

DID YOU KNOW? Sea hares beach after population booms and when rough seas sweep them from the shallows. They are among few animals that feed on toxic cyanobacteria. When stressed, sea hares release **purple ink.** Adults have both male and female working parts and perform as each in mating chains. Pteropods feed by trapping plankton in a mucous web.

Blue Glaucus and Falselimpet

Glaucus

Falselimpet

RELATIVES: Glaucus sea slugs are opisthobranch gastropods in the order Nudibranchia. Falselimpets are pulmonate gastropods (having a pallial lung instead of gills) in the family Siphonariidae, not directly related to keyhole limpets.

IDENTIFYING FEATURES:

Blue glaucus sea slugs *(Glaucus atlanticus)* have tapered bodies and paired, hand-like cerata. From above, they are deep blue on a silvery blue background. Below they are silver gray.

Striped falselimpets *(Siphonaria pectinata)* are shaped like oval-base volcanoes, but with no top hole. Their color is cream with numerous brown radial stripes.

HABITAT: Blue glaucus sea slugs inhabit the surface of the open ocean. Falselimpets live on intertidal rocks.

DID YOU KNOW? The blue glaucus floats upside down (foot up), buoyed by an air bubble in its stomach. Their cerata bear defensive, stinging cells taken in by feeding on the tentacles of oceanic hydroids. Falselimpets invaded the Gulf from the Mediterranean in the 19th century, possibly as hitchhikers on ships. These uncoiled snails feed on algae they scrape from rocks with their radula. Falselimpets breathe air and slowly move up and down rocks to stay wet but not submerged. Their resting spot above the tide is a rock scar that matches the outline of the resident falselimpet. In spring, females deposit gelatinous ribbons containing several hundred **eggs**.

Blue glaucus on fingertip, max 2 in (5 cm)

Bottom (foot) side of a blue glaucus placed in water

Striped falselimpet, max 1 in (2.5 cm)

A falselimpet deposits her eggs (L) among barnacles

105

Stranded land snails in river wrack

Texas liptooth, max 0.4 in (1.1 cm)

Southern flatcoil, max 0.6 in (1.5 cm)

Rockpile liptooth, max 0.4 in (1.1 cm)

Lowland pillsnail, max 0.3 in (0.9 cm)

Land Snails

Texas Liptooth

Flatcoil,
Rockpile Liptooth

Pillsnail

RELATIVES: These land snails are gastropods in the order Stylommatophora (pulmonate snails and slugs) and in the family Polygyridae.

IDENTIFYING FEATURES: Land snails have thinner shells than most marine snails and are less colorful. They arrive on the beach where rivers deposit debris. Liptooths and pillsnails are wider than tall with a flared (reflected) lip and 1–3 aperture teeth.

Texas liptooth snails *(Linisa texasiana)* have a distinct umbilicus, a V–shape parietal (inner aperture) tooth, and two outer-aperture teeth.

Southern flatcoils *(Polygyra cereolus)* have a pronounced umbilicus and a single inner-aperture (parietal) tooth. Whorls have a ropelike sculpture and are bone white (orange brown in life). The similar Florida Flatcoil *(Polygyra septemvolva)*, viewed from the bottom, has a deeper umbilicus and tighter appearing whorls.

Rockpile liptooth snails *(Daedalochila auriformis)* have a large parietal tooth and two outer-lip teeth that make their aperture look like a wrinkled kiss.

Lowland pillsnails *(Euchemotrema leai)* have a single, wide parietal tooth across most of the aperture.

HABITAT: Wet dunes, backyards, and river banks

DID YOU KNOW? Southern flatcoils are the most abundant land snail in the coastal Gulf region.

Land Snails

Texas Oval

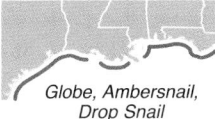
Globe, Ambersnail, Drop Snail

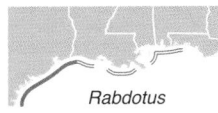
Rabdotus

Texas oval landsnail, max 0.9 in (2.3 cm)

RELATIVES: Oval landsnails and globe snails (family Polygyridae), rabdotus (Bulimulidae), and ambersnails (Succineidae) are in the order Stylommatophora. Globular drop snails are prosobranch gastropods (Helicinidae), more related to nerites (p. 72) than to other land snails.

IDENTIFYING FEATURES:

Texas oval landsnails *(Patera roemeri)* are round, flat, smooth, and lack an umbilicus. Their aperture has a single, low parietal tooth that is sometimes lacking.

White-lip globe, max 1 in (2.5 cm)

White-lip globe snails *(Mesodon thyroidus)* have round, shells with a pronounced spire and no umbilicus, the aperture has one low, ridgelike parietal tooth.

White-washed rabdotus *(Rabdotus dealbatus)* have a conical shell with 5–6 whorls. Colors are tan with faint axial stripes.

White-washed rabdotus, max 0.7 in (1.8 cm)

Ambersnails/chalksnails *(Succinea* spp.) have elongate, thin, fragile shells with 3–4 whorls and a large aperture. Colors are pale with faint growth lines (amber in life).

Globular drop snails *(Helicina orbiculata)* have smooth, nearly spherical shells with no umbilicus, and are yellow or brown with lighter sutures. The thick basal lip bends at an angle before joining the columella.

Spotted ambersnail, max 0.7 in (1.8 cm)

HABITAT: Wet dunes, backyards, and river banks. Drop snails are semi-aquatic.

DID YOU KNOW? Land snails crawl using waves of muscular foot contractions upon secreted mucus, which acts as both temporary glue and lubricant.

Globular drop snail, max 0.4 in (1.1 cm)

Atlantic nutclam, max 0.4 in (1.0 cm)

Pointed nutclam, max 0.4 in (1.0 cm)

Cat. #06521

Concentric nutclam, max 0.7 in (1.8 cm)

Red-brown ark, max 1.3 in (3.3 cm)

Nutclams and Ark *(Red-brown)*

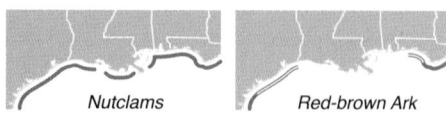

Nutclams   Red-brown Ark

RELATIVES: Nutclams (family Nuculidae) and pointed nutclams (Nuculanidae) are primitive bivalves not directly related to ark shells (Arcidae), which are more closely allied with mussels, scallops, pen shells, and oysters.

IDENTIFYING FEATURES: Nutclams all have the hinge of their valves lined with interlocking, comblike teeth. Pointed nutclams have similar hinge teeth, and have pointed shells. Arks have thick shells with forward umbones and distinct ribs.

Atlantic nutclams *(Nucula proxima)* have whitish, smooth, triangular shells.

Pointed nutclams *(Nuculana acuta)* are pointed at the posterior end, which has a distinct ridge on the dorsal side. Their sculpture has distinct concentric lines.

Concentric nutclams *(Nuculana concentrica)* have whitish to yellowish, smooth, posterior-pointed shells.

Red-brown arks *(Barbatia cancellaria)* are finely beaded and brown except for a light streak from the umbo down.

HABITAT: These nutclams live within muddy-sandy sediment out to moderate depths. Red-brown arks live at similar depths in calcareous rubble.

DID YOU KNOW? Nutclams feed by gleaning the freshly settled layer of organic material (detritus) from the sea floor. Their adaptations for this "deposit feeding" include tentacle-like proboscis-palps, which extend outside the shell and draw in edible particles by ciliary action. This food is conveyed to labial palps inside the shell, which sort out the grit and move food particles to the mouth.

Arks *(White Bearded, Delicate, Baughman's, Cut-ribbed)*

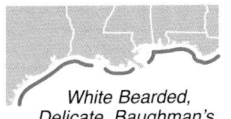

White Bearded, Delicate, Baughman's

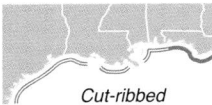

Cut-ribbed

RELATIVES: Ark shells (Arcidae) are distantly related to mussels, scallops, pen shells, and oysters.

IDENTIFYING FEATURES:

White bearded arks *(Barbatia candida)* are rounded in front, angled in back, and have crossing ribs and growth lines that give the shells a beaded look. Their color is white beneath a brown periostracum.

Delicate arks *(Barbatia tenera)* have yellowish, oval shells with tiny riblets between more prominent radial ribs.

Baughman's arks *(Anadara secernenda)* have white shells with fewer than 30 prominent ribs and a long, straight hinge bearing numerous comblike teeth.

Cut-ribbed arks *(Anadara secticostata)* are similar to Baughman's arks but have a more angled posterior. Their anterior ribs (toward umbo) have a central groove.

HABITAT: White bearded and delicate arks attach with byssal threads to rocky substrates. Baughman's and cut-ribbed arks live within sand. All occur from nearshore shallows out to moderate depths.

DID YOU KNOW? These ark shells in life are covered in brown, fuzzy periostracum (shell skin), which is worn away after beaching. Byssal threads, used by some arks to anchor themselves, are strong, silky fibers composed of proteins secreted by their byssus gland within the ark's foot. Although the threads keep an ark from drifting, the anchored ark can still move by lengthening and shortening this tether.

White bearded ark, max 3 in (7.6 cm)

Delicate ark, max 1.4 in (3.6 cm)

Baughman's ark, max 2 in (5.1 cm)

Cut-ribbed ark, max 4.5 in (11.5 cm)

109

Incongruous ark, max 3 in (7.5 cm)

Chemnitz's ark, max 2.8 in (7.1 cm)

Transverse ark, max 1.4 in (3.6 cm)

Blood ark, max 3 in (7.5 cm)

Arks *(Incongruous, Chemnitz's, Transverse, Blood)*

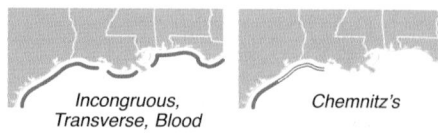

Incongruous, Transverse, Blood   *Chemnitz's*

RELATIVES: Ark shells are in the family Arcidae and are distantly related to mussels, scallops, pen shells, and oysters.

IDENTIFYING FEATURES: These are thick shells with a prominent umbo. Fresh shells may retain the brown, fuzzy periostracum they had in life.

Incongruous arks *(Anadara brasiliana)* are thin-shelled for an ark. Their roughly 25 ribs have distinct dash-shape beads.

Chemnitz's (triangular) arks *(Anadara chemnitzii)* are similar to incongruous arks, but their radial ribs are thinner, and the furrows between ribs are wider.

Transverse arks *(Anadara transversa)* have an elongate oval shell with a relatively straight hinge-line spanned by mostly vertical teeth. The 30–35 ribs are beaded on the left valve and smooth on the right valve.

Blood arks *(Lunarca ovalis)* have a thick, oval shell with 26–35 heavy radial ribs. A slightly arched hinge-line bears all of the most conspicuous hinge teeth, which are angled backward and behind the umbo.

HABITAT: All live in sandy or muddy shallows out to moderate depths.

DID YOU KNOW? The blood ark gets its name from the red, hemoglobin pigment throughout its tissues, giving its flesh an orange tint. This hemoglobin, the same pigment as in our own blood, efficiently transports oxygen from the ark's gills to its body tissues and helps these bivalves live in oxygen-poor sediments. Most arks have robust shells that wear down slowly in the surf, making these species some of the most common shell bits in patches of shell hash (p. 26).

Arks *(Ponderous, Mossy, Turkey Wing)* and **Giant Bittersweet**

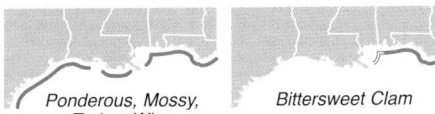

Ponderous, Mossy, Turkey Wing

Bittersweet Clam

Mossy ark, max 2.5 in (6.3 cm)

RELATIVES: Mossy and turkey wing arks (family Arcidae) are related to ponderous arks (Noetiidae), but only distantly allied with bittersweets (family Glycymerididae).

IDENTIFYING FEATURES:

Mossy arks *(Arca imbricata)* have a straight hinge line, nearly as long as the shell, below a broad, triangular ligament area. In life, the shell's numerous beaded ribs and mostly chestnut brown color are hidden by a frilly (mossy) periostracum.

Turkey wings (zebra arks) *(Arca zebra)* have elongate shells with rough ribs, less beaded than in mossy arks. The hinge is as long as the shell, which is striped by nested, red-brown Vs or Ws at the umbo that turn into oblique lines or zigzags.

Turkey wing, max 3.6 in (9 cm)

Ponderous arks *(Noetia ponderosa)* differ from Arcidae arks by having their hinge ligament only at the umbo. In this species, the ligament area is grooved, and often remains dark. They have a very thick, triangular shell with 27–31 wide, flat ribs.

Giant bittersweets *(Glycymeris americana)*, like others in their family, have thick, round shells with curved hinge-lines bearing prominent teeth on either side of the umbo. Giant bittersweets have roughly 50 flattened ribs and are glossy cream with concentric, blurry necklaces of tan or rust.

Ponderous ark, max 2.8 in (7 cm)

HABITAT: Mossy and zebra arks live byssally anchored to shallow-water rubble. Ponderous arks and bittersweets live in nearshore sands.

DID YOU KNOW? Mossy and zebra ark shells have a mid-shell gape where their byssal anchor protrudes between valves.

Giant bittersweet, max 4 in (10 cm)

111

Scorched mussel, max 1.5 in (4 cm)

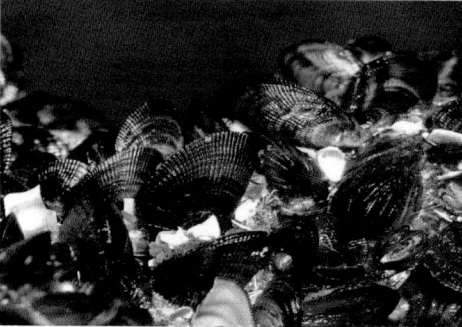

A mass of scorched mussels growing on a jetty

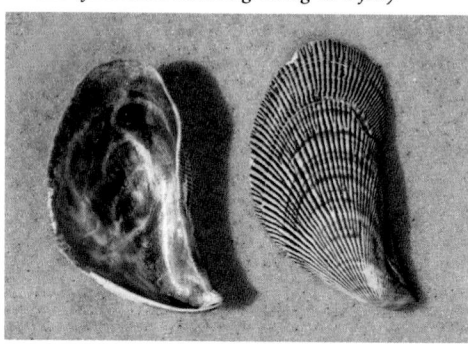

Hooked mussel, max 2 in (5 cm)

Green mussel, max 3.5 in (9 cm)

Mussels *(Scorched, Green, Hooked)*

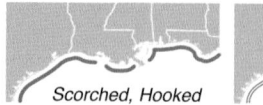

Scorched, Hooked

Green

RELATIVES: Mussels (family Mytilidae) are remotely related to arks, oysters, and penshells.

IDENTIFYING FEATURES:

Scorched mussels *(Brachidontes exustus)* have shells with radiating ribs and two or three hinge teeth under the umbo.

Hooked mussels *(Ischadium recurvum)* have a curved, triangular shell with radial ribs that branch toward the end opposite the umbo.

Green mussels *(Perna viridis)* are smooth and green, especially at their ventral edge. The similar brown mussel *(Perna perna)* is brown outside and pale purple inside.

HABITAT: Shallow-water rubble. Scorched mussels are common within the intertidal crevices of jetty rocks.

DID YOU KNOW? Both green and brown mussels are alien invaders from Asia. Mussels are filter feeders, and although they live attached, they don't always stay in one place. Juveniles use their byssal-thread anchors like climbing ropes, attaching and pulling themselves forward in successive moves. Larger adult mussels can regenerate threads and re-attach if they become dislodged. The byssal threads are as strong as they are stretchy, retaining tensile strength even when pulled to more than half again their length. Relative to size, a mussel's anchoring byssus is five times the strength of our own tendons.

Mussels
(Mahogany Date, Horse, Ribbed)

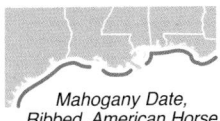

Mahogany Date, Ribbed, American Horse

Southern Horse

RELATIVES: Mussels (family Mytilidae)

IDENTIFYING FEATURES:

Mahogany date mussels *(Leiosolenus bisulcatus)* are elongate with a diagonal furrow that runs from the umbo to the ventral shell margin. Because they are delicate, whole shells are only found by cracking their rock living space.

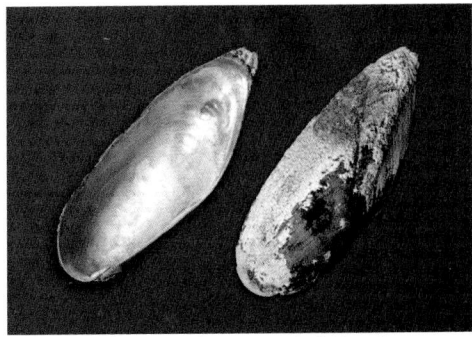

Mahogany date mussel, max 1.7 in (4.3 cm)

Horse mussels *(Modiolus* spp.) have smooth, inflated shells and an umbo below their upper end. Two similar species occur. Southern horse mussels *(M. squamosus)* reach 2.5 in (6.5 cm), have less-inflated umbones, and are whitish or purple after beach wear. American horse mussels *(M. americanus)* reach 4 in (10 cm), have bulbous umbones, and are bright red through their golden periostracum.

A mahogany date mussel revealed in a fractured rock

Southern ribbed mussels *(Geukensia granosissima)* have shells with radiating ribs and no hinge teeth. Shells without the brown periostracum are yellowed gray with occasional purple tinges.

HABITAT: Mahogany date mussels bore into limestone, shells, and hard coral. Horse mussels anchor to soft corals. Ribbed mussels attach to intertidal rocks and saltmarsh plants.

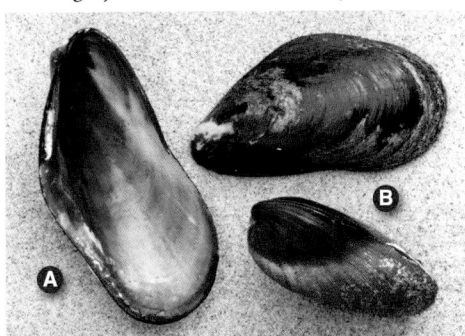

Southern (A) and American (B) horse mussels

DID YOU KNOW? Mahogany date mussels bore into rocks using eroding chemicals secreted by pallial glands at the edge of their mantle. The chemicals are not acids, but are a calcium-binding mucus (mucoprotein). The byssal threads that anchor all mussels are made of tough but elastic proteins secreted by a gland within the mussel's foot. The threads have the tensile strength of nylon fishing line.

Southern ribbed mussel, max 3 in (7.6 cm)

113

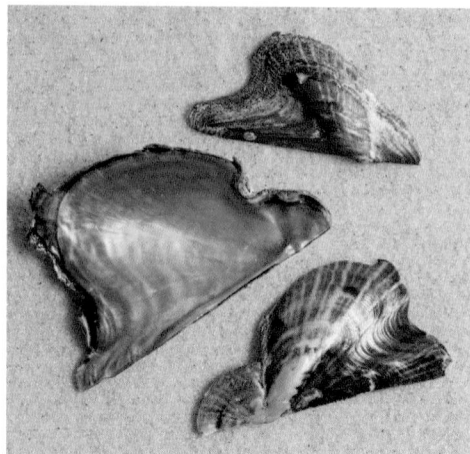

Atlantic wing oyster, max 3.5 in (9 cm)

Atlantic pearl oyster, max 3.5 in (9 cm)

Byssal threads anchoring an Atlantic pearl oyster

Oysters *(Atlantic Wing, Altantic Pearl)*

RELATIVES: Wing and pearl oysters, family Pteriidae, are related to flat tree-oysters.

IDENTIFYING FEATURES:

Atlantic wing oysters *(Pteria colymbus)* have valves with a triangular front wing near the umbo and a long rear wing extending past the rest of the shell.

Atlantic pearl oysters *(Pinctada imbricata)* have valves with short, triangular, front and rear wings, and have a scaly, fringelike periostracum.

HABITAT: These loosely leashed bivalves live attached by their byssal threads. Atlantic wing oysters attach to offshore soft corals and to man-made debris, such as trap ropes and buoys. Atlantic pearl oysters are more common in rocky areas from the shoreline out to moderate depths.

DID YOU KNOW? These oysters are commonly found after storms wash in the items to which they are anchored. Atlantic pearl oysters produce pearls, although rarely. In the southern Caribbean, the oysters were harvested to near depletion by the Spanish in the late 1500s. Prior to that, the value of pearls shipped from the Americas was greater than the region's mined silver and gold. Pearls are formed inside the oyster as defense against a threatening irritant, such as a parasite. An oyster's mantle seals off the irritation in a pearl sac by depositing layers of aragonite (a form of calcium carbonate) within a matrix of complex proteins called conchiolin. The matrix is called nacre, and is the same as the layers of smooth, iridescent, mother-of-pearl lining the interior of the oyster's shell.

Oysters *(Flat Tree-, Frond, Crested)*

Flat Tree-

Frond

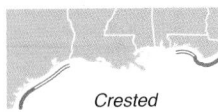

Crested

Flat tree-oyster, max 4 in (10 cm)

RELATIVES: Flat tree oysters (family Isognomonidae) are related to frond and crested oysters, which share the family Ostreidae with eastern oysters.

IDENTIFYING FEATURES:

Flat tree-oysters *(Isognomon alatus)* have both valves equally flat, and a straight hinge with 8–12 distinct grooves.

Frond oysters *(Dendostrea frons)* have yellow- or purple-colored oval shells with strong radial ridges ending in interlocking scalloped margins. Those attached by fingerlike shell projections to the branches of soft corals tend to have the most elongate shell shape.

Frond oysters on sea whips, max 2.7 in (7 cm)

Crested (horse) oysters *(Ostrea equestris)* have oval, lumpy shells with tiny **hinge teeth** in the right (upper, unattached) valve. Unworn shells have a rayed, flaky periostracum. The left valve is flattened and fused to a rock or debris.

HABITAT: Flat tree- and crested oysters attach to shallow, hard substrates like jetty rocks, wood, sunken trash, and other shells. Frond oysters attach flexible objects like soft corals, trap ropes, and floats.

Attached valves of frond oysters on a trap float

DID YOU KNOW? Frond oysters start out as males, grow to become females, and as such are caring parents. The mother frond oyster broods her offspring within the inhalant chamber of her mantle cavity. She releases her young as tiny veliger larvae, which are better able to survive than if they were broadcast as unfertilized eggs.

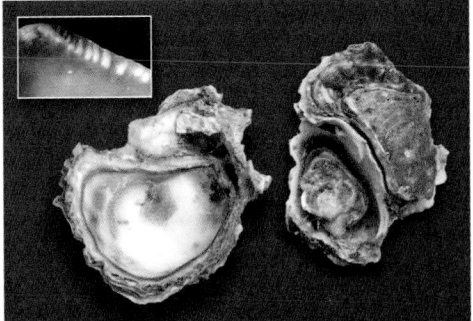

Crested oyster, max 2 in (5 cm). Hinge teeth (inset)

115

Eastern oyster. Left shell shows purple muscle scar

A pile of highly variable eastern oyster shells

A cemented bouquet from an oyster reef

Oyster *(Eastern)*

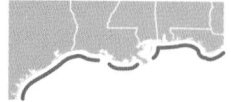

RELATIVES: Eastern oysters are allied with crested and frond oysters in the family Ostreidae.

IDENTIFYING FEATURES:

Eastern oysters *(Crassostrea virginica),* max 6 in (15 cm), have shells that vary from distorted-oval to clown-shoe shapes. Their exterior is lumpy or shingled with brownish purple marks on white, and the inner surface is smooth white with a **purple muscle scar.** The lower, attached valve (the left one) is cupped, and the upper (right) valve tends to be flattened, but often bent. Oyster shells occur in **bountiful piles** and in **cemented clusters.**

HABITAT: These oysters prefer shallow brackish waters and live attached to rocks, debris, and other oysters.

DID YOU KNOW? Eastern oysters have provided value to humans living along the Gulf Coast for many thousands of years. Oyster shell middens, some covering an area greater than a football field and up to 30 ft (9 m) tall, record the time that Native Americans spent eating these mollusks and piling up their remains. The mounds formerly occurred from the middle Texas coast through Florida, marking human habitation at sites continually occupied by people over a span of 3,500 years. Older shell middens are now sunken by rising seas and shifting sediments. The eastern oyster is the state shell of Mississippi, and is the origin of the state gemstone of Louisiana—"LaPearlite," which is a cabochon (rounded, rather than faceted, gemstone), cut and polished from an oyster shell.

Penshells

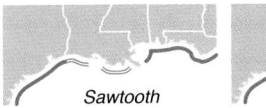
Sawtooth

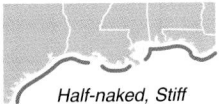
Half-naked, Stiff

RELATIVES: Penshells (family Pinnidae) are distantly related to arks, mussels, and oysters.

IDENTIFYING FEATURES: Penshells have thin, amber-brown, rough-sculptured, fanlike valves with a pearly iridescent (or cloudy) nacre inside.

Sawtooth penshells *(Atrina serrata)* have about 30 radiating ribs bearing hundreds of short, hollow prickles.

Half-naked penshells *(Atrina seminuda)* have about 15 radiating ribs bearing a few to dozens of long tubular spines. Their posterior (fan end) muscle scar is completely within their pearly (or cloudy) nacreous area.

Stiff penshells *(Atrina rigida)* have 15–20 radiating ribs and tubular spines similar to half-naked penshells but are darker, broader, and have their **posterior muscle scar** outside the shiny nacre.

HABITAT: Penshells live in colonies with individuals buried in soft sediment out to 20 ft (6 m) deep.

DID YOU KNOW? Penshells anchor themselves with golden-colored byssal threads, which lead from their pointed (front) end to a small bit of rubble beneath the sand. Like most bivalves, they are filter feeders. Many living penshells have pale, soft-bodied pen shrimp or pea crabs (p. 168) that live within their mantle cavity. The crabs and shrimp feed on items filtered from the water by their penshell host.

Sawtooth penshell, max 9 in (23 cm)

Half-naked penshell, max 9 in (23 cm)

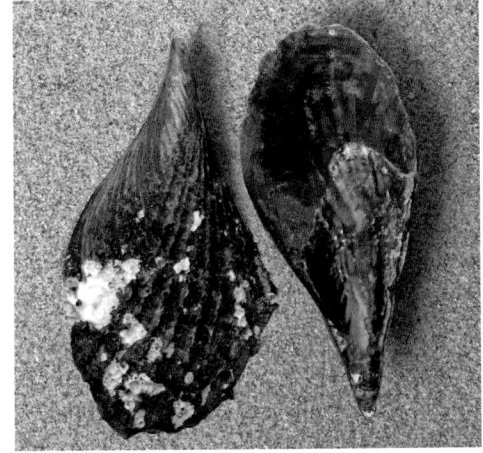
Stiff penshell, max 11 in (28 cm), arrow shows scar

Atlantic bay scallops from Texas (A) and Florida (B)

Atlantic calico scallop, max 2.7 in (7 cm)

Ribs of bay scallop (A) and calico scallop (B)

Scallops *(Atlantic Bay, Calico)*

RELATIVES: Scallops (family Pectinidae) are distantly related to kittenpaws and thorny oysters.

IDENTIFYING FEATURES: Scallop shells have unequal, winglike projections (ears) near the umbo.

Atlantic bay scallops *(Argopecten irradians)* have radiating ribs that are **squared in comparison** to calico scallops. Shell color may be white, gray-brown, or orange. The lower (right) valve is more inflated and mostly white. Two subspecies occur in the Gulf. *A. irradians amplicostatus* occurs in Texas and has 17–18 widely spaced ribs. *A. irradians concentricus* is found east of the Mississippi Delta and has 19–21 narrowly spaced ribs.

Atlantic calico scallops *(Argopecten gibbus)* have shells with 19–21 **rounded ribs**. Shell colors vary through white, yellow, orange, red, purple, and gray, generally with splotches of dark on light. Their ears are often worn.

HABITAT: Bay scallops live in shallow seagrass. Calico scallops live on sand bottom offshore.

DID YOU KNOW? Bay and calico scallops, like many in their family, can swim in short bursts by squirting jets of water through vents near their hinge. Scallops draw water into their mantle cavity by opening their valves, then rapidly clap shut to push water back through vents, which jets the animal forward. A scallop may clap and jet a dozen times to escape a predator such as a gastropod, squid, octopus, sea star, or crab. These bivalves are also unique in having a way to visually detect threats with their numerous, small eyes. Both of these species are harvested by humans for food.

Scallops *(Round Rib, Rough, Lion's Paw)*

Round-rib

Rough

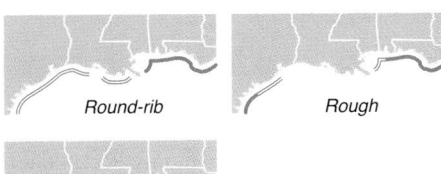

Lion's Paw

Round-rib scallop upper valves. Both valves (inset)

RELATIVES: These scallops share the family Pectinidae with Atlantic bay and calico scallops.

IDENTIFYING FEATURES:

Round-rib scallops *(Euvola raveneli)*, max 2 in (5 cm), have a mostly white, domelike lower valve and an upper valve that looks as if it were melted flat. Exterior colors for the flat valve are reddish, orange or purple-brown, with lighter rays. The interior valve is white with burgundy trim. Ribs are rounded and smooth with wide gaps.

Rough scallops *(Aequipecten muscosus)* have about 19 ribs that are roughened by tiny spoon-shape prickles. Beach-worn shells are less prickly. Most rough scallops are solid-colored lemon, peach, or tangerine, but some are mottled with plum.

Northern lion's paws *(Nodipecten nodosus)* have thick, flattened shells with 7–8 large, roughly ridged ribs bearing occasional knuckles. Their outer shell color is commonly orange or brick red, but may range from pale to purple.

HABITAT: Round-rib and rough scallops live on sand bottom out to moderate depths. Lion's paws live on offshore reefs.

DID YOU KNOW? Round-rib scallops lie with their flat upper valve flush with the sandy bottom. If disturbed by a potential predator, they launch from the bottom by clapping their valves to create jets of propulsion. Like some other scallops, they use an array of tiny eyes to sense danger and steer their escape.

Rough scallop, max 1.5 in (4 cm)

Lion's-paw, max 6 in (15.2 cm)

Atlantic kittenpaw, max 1.2 in (3 cm)

Atlantic kittenpaws have variable shapes and colors

Atlantic thorny oyster, max 5.1 in (13 cm)

Antillean lima, max 1.1 in (2.8 cm)

Kittenpaw, Thorny Oyster, and **Lima**

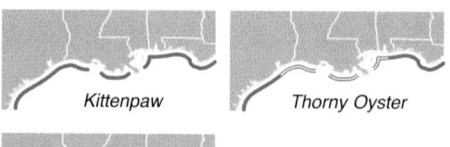

Kittenpaw

Thorny Oyster

Lima

RELATIVES: Kittenpaws (family Plicatulidae) are related to thorny oysters (Spondylidae) but are only remote kin to limas (family Limidae).

IDENTIFYING FEATURES:

Atlantic kittenpaws *(Plicatula gibbosa)* have thick, tough, flattened shells with 6–10 digitlike ribs. They are pale except for their tabby-orange ribs, which are marked with numerous, red-brown lines.

Atlantic thorny oysters *(Spondylus tenuis)* have thick, circular, lumpy valves with occasional thorns (long in unworn shells). The hinge on the cup-shape lower valve has two large cardinal teeth separated by a split, and the upper valve hinge has two corresponding sockets. Colors are commonly orange and brick red. A rarer, deep-water thorny oyster, *S. americanus,* is usually white.

Antillean limas *(Limaria pellucida)* have thin white shells with no hinge teeth, a single blunt ear, and fine riblets.

HABITAT: Atlantic kittenpaws live attached to rocks out to moderate depths. Atlantic thorny oysters are cemented to offshore reefs, where Antillean limas occupy crevices.

DID YOU KNOW? Upper valves of kittenpaws and thorny oysters are most common because the lower (right) valve often remains attached where the animal lived. The flat attachment site on the right valve retains an impression of the shell or rock on which it grew.

120

Jingle Shell and Pandora

RELATIVES: Jingle shells (family Anomiidae) are related to scallops, and only indirectly related to pandoras (Pandoridae).

IDENTIFYING FEATURES:

Common jingle shells *(Anomia simplex)*, max 2 in (5 cm), are round and translucent with no obvious umbo or hinge site. Nearly all beached jingle shells are the unattached, upper (left) valve. Valve sets (top image) show the **right (lower) valve** with a center hole. The hole marks the place where a thick, calcified byssus attached the bivalve to a rock or other hard substrate. Colors include silver-gray, white, yellow, and orange. Black shells have been stained by sulfurous sediments.

Threeline pandoras *(Pandora trilineata)* have whitish, smooth, flat shells with arching margins. The bluntly pointed hind end allows the bivalve's siphon to exit its closed valves.

HABITAT: Common jingle shells live attached to the bottom in shallows out to moderate depths. Threeline pandoras live in muddy bay sediments.

DID YOU KNOW? Jingle shells can be crushed into **glittery flakes**. The glitter bits show how the shell is constructed. Flakes represent hundreds of layers of foliated (sheet) calcite (a form of calcium carbonate), laid down in laths that alternate in orientation and are adhered within a protein matrix. Overlapping sheets add strength and inhibit crack propagation. It's an example of benefits from composite materials. Calcite is strong but brittle, and the binding protein has tension strength but is flexible. Together in an organized matrix, the materials make a light, resilient, protective (and pretty) shell.

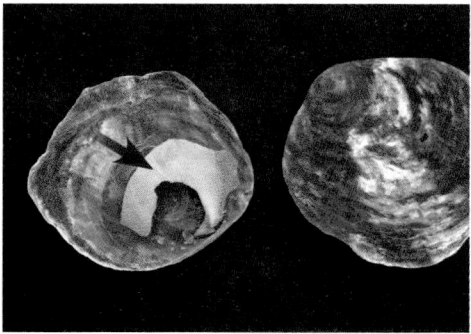
Common jingle shell. Arrow shows right valve

Common jingle shells vary widely in color

A jingle shell crushed into glittery flakes

Threeline pandora, max 1 in (2.5 cm)

121

Three-tooth carditid, max 0.25 in (6 mm)

Broad-ribbed cardita, max 2.5 in (6.3 cm)

Unequal spoonclam, max 1 in (2.5 cm)

Buttercup lucine, max 2.5 in (6.4 cm)

Carditids, Spoonclam,
and **Lucine** *(Buttercup)*

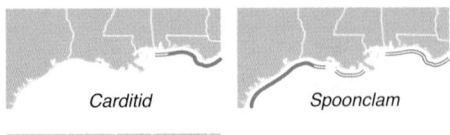

Carditid   Spoonclam

Buttercup Lucine

RELATIVES: Carditids are in a unique group of bivalves (Archiheterodonta) in the family Carditidae, not directly related to spoonclams (Periplomatidae) and lucines (Lucinidae).

IDENTIFYING FEATURES:

Three-tooth carditids *(Pleuromeris tridentata)* have small, thick triangular shells with beaded, radial ribs. Colors are cream and rusty pink.

Broad-ribbed carditas *(Cardites floridanus)* have thick valves with about 15 strong, beaded ribs. They are most often white with bands of chestnut.

Unequal spoonclams *(Periploma margaritaceum)* have faintly lined, opaque-white shells. The left valve is larger (more inflated) than the right. Inside, there is a conspicuous spoon-shape chondrophore under the umbo.

Buttercup lucines *(Anodontia alba)* have inflated shells with a forward flare in the shell that protrudes more than the umbo. The outer shell is dull white with fine growth lines, and the inner shell is glossy yellow or cream.

HABITAT: All live in sandy shallows out to moderate depths.

DID YOU KNOW? Spoonclams have no hinge teeth, but do have a well developed chondrophore—a lobe to which the internal shell ligament is attached. This ligament springs bivalves open when they relax their strong adductor muscles.

Lucines *(Chalky, Thick Buttercup, Pennsylvania, Florida)*

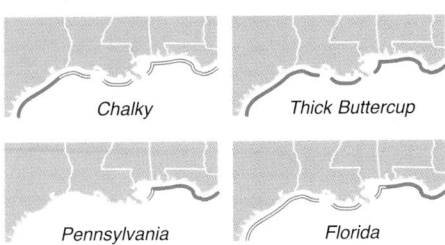

Chalky

Thick Buttercup

Pennsylvania

Florida

Chalky lucine, max 4.3 in (11 cm). Arrow shows scar

RELATIVES: Lucines (family Lucinidae) are distantly related to jewelboxes, cockles, and other clamlike bivalves.

IDENTIFYING FEATURES: Lucines have circular shells with a furrow radiating from a central umbo.

Chalky lucines *(Pegophysema schrammi)* are similar to buttercup lucines, but have a chalky interior and an angled anterior muscle scar separated from the pallial line.

Thick buttercup lucines *(Phacoides pectinatus)* are thick-shelled and compressed, with coarse growth lines. Colors are whitish to pale yellow.

Thick buttercup, max 2.7 in (6.7 cm)

Pennsylvania lucines *(Lucina pensylvanica)* have thick, glossy, white valves with a deep furrow behind the umbo.

Florida lucines *(Stewartia floridana)* have compressed shells with a squared-off posterior end above the furrow. Colors are white to cream.

HABITAT: All live in muddy sand.

DID YOU KNOW? Lucines filter detritus from water drawn into a mucus-lined tube maintained by their long foot. But they also benefit nutritionally from a symbiotic relationship with sulfide-oxidizing bacteria that live within their specialized gill filaments. The bacteria help the lucine by fixing carbon (from CO_2) into organic compounds that feed the lucine during lean times. Lucines are named for Lucina, an aspect of the Roman goddess Juno who represented light and childbirth.

Pennsylvania lucine, max 2 in (5 cm)

Florida lucine, max 1.5 in (3.8 cm)

Cross-hatched lucine, max 1 in (2.5 cm)

Tiger lucina, max 3.5 in (8.9 cm)

Atlantic diplodon, max 0.7 in (1.8 cm)

Corrugate jewelboxes, max 1 in (2.5 cm)

Lucines *(Cross-hatched, Tiger Lucina),* **Diplodon,** and **Jewelbox** *(Corrugate)*

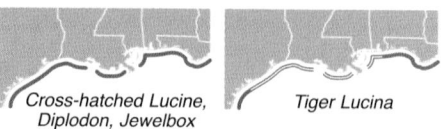

Cross-hatched Lucine, Diplodon, Jewelbox   *Tiger Lucina*

RELATIVES: Lucines (family Lucinidae) are distantly related to diplodons (Ungulinidae) and jewelboxes (Chamidae).

IDENTIFYING FEATURES:

Cross-hatched lucines *(Divalinga quadri-sulcata)* are moderately inflated with relatively thin valves sculptured by numerous, olique, parallel lines that make the shell appear covered with fingerprints. Beached shells are glossy white, chalky, or ivory.

Tiger lucinas *(Codakia orbicularis)* are compressed with thick, yellow-white valves sculptured with fine riblets and concentric growth ridges. Their lunule is a small deep pit beneath the pointed umbo. Insides of the youngest shells are yellow, rimmed with pink.

Atlantic diplodons *(Diplodonta punctata)* are moderately inflated with fine, concentric growth lines, smoother near the umbo. Diplodons have no pallial sinus line. Colors are cream to white.

Corrugate jewelboxes *(Chama congregata)* are highly variable with a rough exterior and are red, orange, or purple outside, reddish, purplish or pale, inside.

HABITAT: These lucines, and diplodon, live in muddy sand out to moderate depths. Corrugate jewelboxes live with their left valve cemented to reef substrates.

DID YOU KNOW? When corrugate jewelboxes are larvae, tiny enough to fit through the eye of a small sewing needle, they attach themselves to rock, shell, or coral with a flexible byssus. They soon become cemented for life and gain adult shell characteristics when they are still smaller than a pinhead.

Jewelboxes *(Leafy, Florida Spiny)* and **Marsh Clam**

Jewelboxes   Marsh Clam

RELATIVES: Jewelboxes (family Chamidae) are distantly related to marsh clams (Cyrenidae) and other clamlike bivalves.

IDENTIFYING FEATURES:

Leafy jewelboxes *(Chama macerophylla)* have thick, oval shells covered in numerous scaly ridges. Beach-worn shells are lumpy, but new shells may have long, hollow scales. They are generally yellow or chalky, and occasionally orange or lavender.

Leafy jewelbox, max 3.1 in (8 cm)

Florida spiny jewelboxes *(Arcinella cornuta)* are shaped like tubby commas bearing about eight radiating ridges with hollow spines (or knobs, if beach-worn). They are white with a pinkish interior.

Southern marsh clams *(Polymesoda floridana)* have rounded triangular shells with a pointed end. Their sculpture is smooth with irregular growth lines, and colors range through cream, browns, pinks, and purples. Light and dark colors may alternate in rays or in growth bands. Fresh shells have a dirty brown periostracum.

Florida spiny jewelbox, max 2.5 in (6.3 cm)

HABITAT: Leafy jewelboxes live with their left valve cemented to reef substrates. Florida spiny jewelboxes have a similar start, but detach when young to grow free within sandy rubble. Southern marsh clams live in bay sediments and have a high tolerance for salinity changes.

DID YOU KNOW? The ornate spines of the Florida spiny jewelbox seem important for deterring attacks from gastropod predators. Marks from predation (bore holes) show that gastropods avoid the spikes, forcing them to attempt boring into the thickest part of the jewelbox shell.

Southern marsh clams, max 1 in (2.5 cm)

125

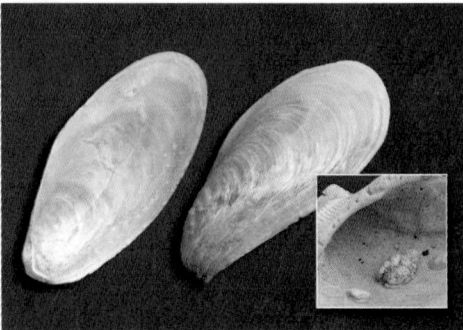

Chimney clam, max 0.7 in (1.8 cm), dwelling (inset)

Yellow pricklycockle, max 2.5 in (6.4 cm)

Florida pricklycockle, max 2.7 in (6.9 cm)

Atlantic strawberry-cockle, max 2 in (5.1 cm)

Chimney Clam, Pricklycockles, and Strawberry Cockle

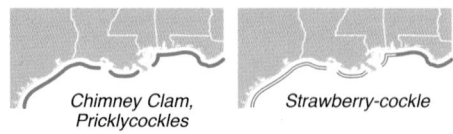

Chimney Clam, Pricklycockles

Strawberry-cockle

RELATIVES: Chimney clams (family Gastrochaenidae) are distantly related to cockles (Cardiidae) and clamlike bivalves.

IDENTIFYING FEATURES:

Atlantic chimney clams (*Rocellaria stimpsoni*) bore into limestone and shells, where they form bottle-shape dwellings of calcium carbonate that look like roughened blisters. Their remnant holes are common (p. 150). The shell is the same shape as its holes and has a large gape.

All **cockle shells** are oval, inflated, and have a large umbo with one central tooth and socket.

Yellow pricklycockles (*Dallocardia muricata*) have about 35 ribs roughened by small scales. They are tinged yellow inside and out, and may tend toward peach with occasional red-brown streaks.

Florida pricklycockles (*Trachycardium egmontianum*) have about 30 pronounced ribs covered by strong scales (in unworn shells) and end in a hind margin that is deeply serrated. Their external color is cream with tan or purple-brown splotches. Valves inside are salmon and/or purple.

Atlantic strawberry-cockles (*Americardia media*) are cream with red-brown specks and have numerous flattened ribs that feel like sandpaper. An angled ridge runs across the longest part of the shell.

HABITAT: Chimney clams live inside rocks and shells out to moderate depths. These cockles live in sandy shallows.

DID YOU KNOW? Well-developed ribs and scales may deter boring by gastropod predators and keep these pricklycockles anchored in the sand.

Giant Cockle and Egg Cockles
(Common, Painted)

RELATIVES: Egg cockles are allied with other cockles in the family Cardiidae.

IDENTIFYING FEATURES:

Atlantic giant (heart) cockles *(Dinocardium robustum)* are cream with brown or tan marks along their shell ribs, which are rounded and bumpy on one side (front), flattened and smooth on the other (rear).

Common egg cockles *(Laevicardium serratum)* are smooth with only faint riblets. Their valves are an oblique oval shape, with ridges along the inner margin. They are glossy white or yellow with occasional rosy tinges. The hinge is often purplish.

Painted egg cockles *(Laevicardium pictum)* are smooth outside, compressed for a cockle, and have a triangular shape. They are cream with blurry zigzags and spatters of brown or yellow-orange.

HABITAT: These cockles live in sandy or muddy shallows.

DID YOU KNOW? The family name of cockles, Cardiidae, refers to their heart shape when viewed head-on. Giant cockles found on Gulf beaches have more pronounced color patterns than Atlantic shells and were formerly separated as a subspecies called Van Hyning's cockle. Many cockles escape predators by leaping from the sea bottom. In a leap, they extend and flick their long, muscular foot. The jump propels the shell backwards, and often occurs in a sequence of several bounces that ends with a twist. This sticks a landing with the cockle's foot in an ideal position to immediately burrow beneath the sand. Some egg cockles can swim during their jumps by rapidly repeating the jumping movements in mid-water.

Atlantic giant cockle, max 5.2 in (13.2 cm)

A living Atlantic giant cockle exposed on a Gulf beach

Common egg cockle, max 3 in (7.6 cm)

Painted egg cockle, max 1 in (2.5 cm)

127

Yellow egg cockle, max 1 in (2.5 cm)

Texas Venus, max 3 in (7.6)

Disc dosinia, max 3 in (7.6 cm)

Elegant dosinia, max 3 in (7.6 cm)

Egg Cockle *(Yellow)* and Venus Clams *(Texas, Dosinias)*

RELATIVES: Egg cockles are allied with other cockles in the family Cardiidae, and are distantly related to Venus clams (Veneridae) and other clamlike bivalves.

IDENTIFYING FEATURES:

Yellow egg cockles *(Laevicardium mortoni)* are smooth, and almost evenly rounded with a central umbo. They are colored by relatively distinct rows of brown, purple, or orange zigzags.

Texas Venus clams *(Agriopoma texasianum)* have oval, smooth, moderately inflated shells with fine concentric growth lines. An S-shape posterior cardinal tooth under the umbo in each valve separates this clam from similar species.

Disc dosinias *(Dosinia discus)* have ivory-colored, circular shells with sharp, forward-pointing umbones. Fine concentric lines leave the exterior relatively smooth.

Elegant dosinias *(Dosinia concentrica)* are similar to disc dosinias, but their flattened concentric ridges are broad, easily seen, and readily felt.

HABITAT: Yellow egg cockles prefer shallow inlet areas and lagoons. The Texas Venus and these dosinias live in muddy sand and off protected beaches.

DID YOU KNOW? Because dosinias have a strong internal hinge ligament, they are commonly found with their open valves still connected. This ligament springs a bivalve open when its valves are not drawn together by its adductor muscles. The dark brown hinge ligament is made of an elastic, fibrous protein. Soaking a partially opened dosinia in water will allow you to test this valve-spring mechanism without breaking the ligament.

Venus Clams
(Pointed, Cross-barred, Sunray)

RELATIVES: Venus clams are allied within the family Veneridae

IDENTIFYING FEATURES:

Pointed Venus clams *(Anomalocardia cuneimeris)* have glossy shells with rounded concentric ridges and a pointed posterior end. Their colors vary from dirty white to shades of blue, green, gray, and brown.

Cross-barred Venus clams *(Chione elevata)* have sharp, concentric ridges that cross radial riblets, giving even beach-worn shells a distinct, cross-hatched look. Most are gray-white with occasional darker streaks and a white or purple interior, which is seen externally in highly worn shells. Concentric ridges are extra sharp and elevated in young clams.

Sunray Venus clams *(Macrocallista nimbosa)* have smooth, elongate shells that are purplish-brown with darker, narrow rays streaking from the umbo. The exterior, sunray shell color is from a thin, tough, shiny periostracum. Well worn beach shells may be bone white.

HABITAT: Pointed Venus clams are most common in salty bays. Cross-barred and sunray Venus clams live in sandy bottom near shore.

DID YOU KNOW? Seafood producers have explored commercial aquaculture of sunray Venus clams. A sticking point has been the clam's unreliable spawning. It seems that rather than spawning according to seasonal and temperature cues, which can be imitated in tanks to initiate group spawning, the clams spawn irregularly whenever they are well fed by the phytoplankton they filter from the water.

Pointed Venus clam, max 0.8 in (2 cm)

Cross-barred Venus clam, max 1.3 in (3.3 cm)

Sunray Venus clam, max 6 in (15.2 cm)

A live sunray Venus clam exposed at low tide

Boring choristodon (L), in its stony home (R)

Thin cyclinella, max 1 in (2.5 cm)

Imperial Venus clam, max 1 in (2.5 cm)

Thick-ringed Venus, max 1.3 in (3.3 cm)

Venus Clams *(Choristodon, Cyclinella, Imperial, Thick-ringed)*

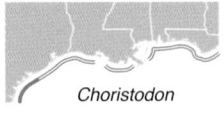

Choristodon

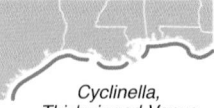

Cyclinella, Thick-ringed Venus

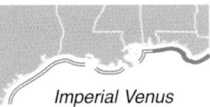

Imperial Venus

RELATIVES: Venus clams are allied within the family Veneridae.

IDENTIFYING FEATURES:

Boring choristodons *(Choristodon robustus)*, max 1 in (2.5 cm), have strong, chalky white shells with irregular radial ribs crossing finer growth lines. They are most often found peaking out from the cavity in the rock or coral chunk where they lived.

Thin cyclinellas *(Cyclinella tenuis)* are flat white with fine but irregular growth lines. They are smaller and have thinner shells than the dosinias.

Imperial Venus clams *(Lirophora varicosa)* have their shells thickened by 5–9 concentric, chunky ribs. The dorsal side of the largest ribs comes to a sharp edge, leaving a sunken channel atop each rib. They are whitish, light gray, or mottled tan with a few blurry rays.

Thick-ringed Venus clams *(Lirophora obliterata)* are similar to the imperial Venus, but have thick ribs that are rounded or slightly flattened in cross-section, without a sharp edge. Colors are cream with rose or light-brown mottling.

HABITAT: Boring choristodons live in rocky reef and rubble. Thin cyclinellas live in sandy shallows. Imperial and thick-ringed Venus clams occur offshore to moderate depths.

DID YOU KNOW? The thick shell rolls of the imperial Venus may help this shallow-burrowing clam avoid predation from drilling gastropods.

Venus Clams *(Gray pygmy, Calico, Lady-in-waiting)* and **False Anglewing**

RELATIVES: Venus clams and false angelwings are in the family Veneridae.

IDENTIFYING FEATURES:

Gray pygmy Venus clams *(Chioneryx grus)* have ribs crossed by growth lines and are cream or gray, often with a purple-brown streak covering the hind end.

Calico clams *(Megapitaria maculata)* have smooth, creamy shells with blurry brown rectangles and smudges.

Lady-in-waiting Venus clams *(Chionopsis intapurpurea)* have strong concentric ridges that are serrated on the hind end. Colors are cream, tan, or gray, often with brownish streaks. Many have a purple smudge inside.

False angelwings *(Petricolaria pholadiformis)* are elongate with large radial ribs that fade toward the hind end. They have a simple hinge margin with three (left valve) or two (right valve) cardinal teeth. Colors are white or clay-stained.

HABITAT: The gray pygmy Venus lives in shallow sand and rubble. Calico and lady-in-waiting clams live in similar habitat farther offshore. False angelwings live in shallow, clay bottom.

DID YOU KNOW? There are over 400 living species of Venus clams. False angelwings are unusual within this group because they have evolved to burrow deep into hard sediment just like the unrelated angelwings and piddocks. They live within clay, dense mud, peat, wood, and even limestone, by boring further into it as they grow. The clam breathes and filters food through two long siphons.

Gray pygmy Venus clam, max 0.4 in (1 cm)

Calico clam, max 3.5 in (8.9 cm)

Lady-in-waiting Venus clam, max 1.6 in (4.1 cm)

False angelwing, max 2 in (5 cm)

Southern quahog, max 5.9 in (15 cm)

Cat. #00664

Texas quahog, max 6 in (15.2 cm)

Growth bands of southern (L) and Texas (R) quahogs

Southern quahogs stained by clay

Quahogs

 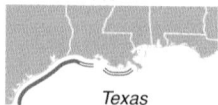

Southern   Texas

RELATIVES: Quahogs (pronounced KO-hogs) and other Venus clams are in the family Veneridae.

IDENTIFYING FEATURES:

Southern quahogs *(Mercenaria campechiensis)* have very thick shells with an inflated, central umbo. Their mid-shell growth lines are clearly visible, the largest of which are as wide as a pencil lead. These concentric ridges (**growth bands**) are prominent at the beaks (umbones). The shell is gray outside with occasional purple zigzags and broad rays, especially in the smallest clams.

Texas quahogs *(Mercenaria texana)* are similar to southern quahogs, but have thin concentric ridges that are eroded into wide, smooth **growth bands** across the middle shell. Another difference is in the **lunule**, which is smaller than in the southern quahog.

HABITAT: These clams live in the muddy sands of shallow bays and lagoons. Southern quahogs prefer more saline waters with fewer fluctuations in salinity.

DID YOU KNOW? The Texas quahog is endemic to the state (it is found no place else). This clam would seem to overlap with the southern quahog, also found in Texas, but the two clams have different living spaces. Texas quahogs favor estuaries into which fresh water occasionally flows, and southern quahogs are minimally tolerant of waters less salty than seawater. Enormous southern quahogs, as wide as your open hand, may be fossil shells from the late Pleistocene (more than 12,000 years ago). Fossils are sometimes dredged up during beach renourishment (p. 393).

Tellins *(Tampa, Texas, Alternate, Taylor's Alternate)*

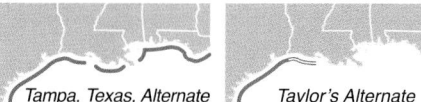

Tampa, Texas, Alternate   Taylor's Alternate

RELATIVES: Tellins are allied within the family Tellinidae and are distantly related to coquina, semele, and tagelus clams.

IDENTIFYING FEATURES: Valves in each of these species have a rounded front end and a tapered rear, which has a recognizable rightward bend (outward in the right valve).

Tampa tellins *(Tampaella tampaensis)* have thin, strong, cream-color shells that are smooth except for extremely fine outer growth lines.

Texas tellins *(Ameritella texana)* have opaque white shells and are steeply sloped behind the hinge.

Alternate tellins *(Eurytellina alternata)* are a pearly yellow-white with numerous concentric grooves between flattened concentric ridges. They may have yellow or pink radiating from the umbo.

Taylor's alternate tellins *(Eurytellina tayloriana)* are similar to alternate tellins, but are colored shades of pink, with darker pink inside.

HABITAT: Texas tellins live in muddy bays. Tampa and alternate tellins prefer sandy shallows. Taylor's alternate tellins live in sand out to moderate depths.

DID YOU KNOW? A tellin's bladelike form and strong foot allow it to burrow rapidly to escape predators. This energetic escape is possible even in oxygen-depleted sediment, thanks to unique oxygen-storing hemoglobin surrounding the tellin's nerves. Tampa and Texas tellins are often abundant in coastal hypersaline lagoons. They represent many local shells that are common where they live but only occasionally make it to the beach.

Tampa tellin, max 1 in (2.5 cm)

Texas tellin, max 0.5 (1.3 cm)

Alternate tellin, max 2.7 in (6.9 cm)

Taylor's alternate tellin, max 2.7 in (6.9 cm)

133

Rose petal tellin, max 1.3 in (3.3 cm)

Iris tellin, max 0.5 in (1.2 cm)

Many-colored dwarf tellin, max 0.5 in (1.2 cm)

Striate tellin, max 1 in (2.5 cm)

Tellins *(Rose Petal, Iris, Many-colored Dwarf, Striate)*

RELATIVES: Tellins and macomas share the family Tellinidae, and are distantly related to coquina, semele, and tagelus clams.

IDENTIFYING FEATURES:

Rose petal tellins *(Eurytellina lineata)* have thick, opaque shells and are rosy white to deep pink.

Iris tellins *(Oudardia iris)* have fragile, translucent shells tinged with pink. Microscopic lines and broad low ridges (called sissulations) at the anterior end run at oblique angles to the shell's fine concentric growth lines.

Many-colored dwarf tellins *(Ameritella versicolor)* are similar to iris tellins, appearing translucent white or rose, but older shells turn a frosty cream color. They also differ from iris tellins in having no sissulations, with growth lines so fine that the shell appears smooth.

Striate tellins *(Serratina aequistriata)* have triangular shells with compact, thin growth lines. The posterior end has two distinct radial ridges in the right valve, and one ridge in the left valve. Colors range from translucent-white to chalky white.

HABITAT: All live in sandy and muddy shallows.

DID YOU KNOW? Tellins lie deep beneath sediment, left valve down, so that their posterior curves upward. This allows their long, mobile, intake siphon to reach the sediment surface where it vacuums in morsels that have settled to the bottom. This way to make a living is called deposit feeding.

Tellin *(White-crested)* and **Macomas**

RELATIVES: Tellins and macomas share the family Tellinidae, and are distantly related to coquina, semele, and tagelus clams.

IDENTIFYING FEATURES:

White-crested tellins *(Tellidora cristata)* have flat, white, solid shells, with a pointed umbo that curves away from the slightly pointed posterior end. In unworn shells, concentric ridges reach the edges at serrations on either side of the umbo.

Constricted macomas *(Austromacoma constricta)* have a moderately inflated shell with fine, irregular, concentric growth lines. Most are gray white and more chalky near the umbo.

Elongate macomas *(Macoploma tenta)* are thin shelled and elongate with abundant growth lines. Colors range from off-white to yellowish.

Short macomas *(Psammotreta brevifrons)* have shells with a blunt hind end and almost imperceptibly fine growth lines. Color may be translucent or opaque white, occasionally with an orange tint.

HABITAT: All live in sandy and muddy shallows. White-crested tellins are most common near inlets. Constricted and elongate macomas are abundant in bays.

DID YOU KNOW? The shell sturdiness of tellins and macomas reveals their home sediments. Thicker-shelled species live in sandy rubble, and delicate species live in soft mud.

Beach-worn white-crested tellin, max 1.5 in (3.8 cm)

Constricted macoma, max 1.5 in (3.8 cm)

Elongate macoma, max 0.7 in (1.8 cm)

Short macoma, max 1.5 in (38 cm)

135

Living, variable coquina clams

A spawning patch of variable coquina clams in spring

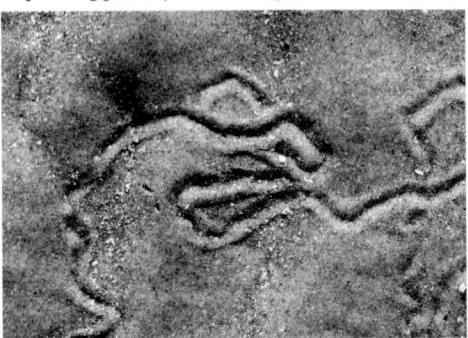

Coquina tracks at low tide

Variable (L) and Texas (R) coquinas

Coquina Clams

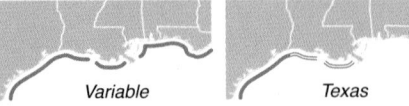

Variable   Texas

RELATIVES: Coquina clams (family Donacidae) are related to tellins and semeles.

IDENTIFYING FEATURES:

Variable coquina clams *(Donax variabilis)*, max 1 in (2.5 cm), have colorful, glossy, wedge-shape shells with faint riblets and groove-teeth lining their inner margins. Patterns vary between solids, radial rays, and concentric bands.

Texas coquina clams *(Donax texasianus)*, max 0.3 in (0.8 cm), are smaller than variable coquinas and have an arched ventral shell margin. Their colors include solid cream, pink, pale blue, and yellowish.

HABITAT: Variable and Texas coquinas live in the intertidal and swash zones.

DID YOU KNOW? Variable coquinas are one of the most abundant and ecologically important mollusks on Gulf beaches. Specialized for life in wave-washed sand, they filter-feed on algae and bacteria that the waves sweep ashore. Variable coquinas are a critical food for shorebirds, surf fishes, and crabs. These little clams undergo complex migrations—daily movements up and down the beach with the tide, developmental movement from deep to shallow as they grow, and aggregation in the swash during spawning. The occurrence of **spawning patches** peaks in the spring, when as many as 12,000 coquinas can share a square meter (11 square feet) of lower beach. Coquina clams develop from larva to adult in about four months and live an average of one year.

Semeles

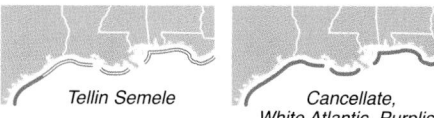

Tellin Semele

Cancellate,
White Atlantic, Purplish

RELATIVES: Semeles (family Semelidae) are related to tellins and coquinas.

IDENTIFYING FEATURES: Semeles have oval shells that are thin but strong. Like tellins, a semele's hind end bends slightly right. Their hinges have a triangular or diagonal depression angling back from the umbo.

Tellin semeles *(Cumingia tellinoides)* are dirty white and have a distinct point at the rear shell. At their hinge, a spoonlike hinge-ligament pit beneath the umbo protrudes into the inner shell.

Cancellate semeles *(Semele bellastriata)* are cream or gray, occasionally with purplish markings. Their sculpture has concentric ridges crossing radial riblets front and rear.

White Atlantic semeles *(Semele proficua)* have a central umbo and are cream with occasional nervous purple lines, which show both inside and out.

Purplish semeles *(Semele purpurascens)* have oval shells with an umbo toward the rear. They have smudge-streaks of blurry purple, brown, or orange, and are darkest at the umbo. Beach-worn shells are smooth, but fresh shells show concentric lines that are stronger at the anterior margin.

HABITAT: Tellin and white Atlantic semeles live in sediments of inlet areas and shallow bays open to the sea. Cancellate and purplish semeles occupy sand banks off beaches out to moderate depths.

DID YOU KNOW? A semele's slashlike or spoonlike hinge depression (chondrophore) bears a cushiony pad that springs the valves open when the animal's adductor muscles relax.

Tellin semele, max 1 in (2.5 cm)

Cancellate semele, max 0.8 in (2 cm)

White Atlantic semele, max 1.5 in (3.8 cm)

Purplish semele, max 1.5 in (3.8 cm)

137

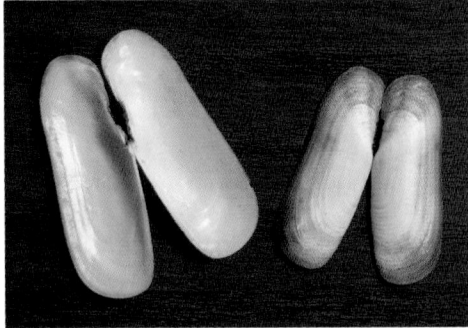

Stout tagelus, max 3.9 in (10 cm)

Purplish tagelus, max 1.6 in (4.0 cm)

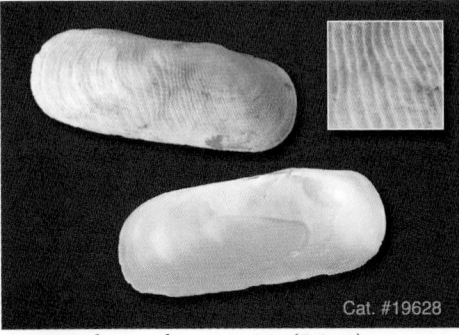

Cat. #19628

Corrugated razor clam, max 3 in (7.6 cm)

Green jackknife clam, max 1.5 in (3.8 cm)

Tagelus Clams, Razor Clam, and **Jackknife Clam** (Green)

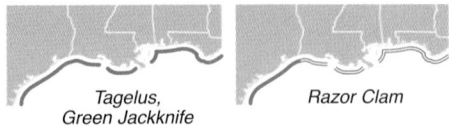

Tagelus, Green Jackknife

Razor Clam

RELATIVES: Tagelus clams (family Solecurtidae) are related to semeles and tellins, and more distantly to jackknife clams (Pharidae).

IDENTIFYING FEATURES: These clams have central umbones and elongate shells that gape at each end.

Stout tagelus clams *(Tagelus plebeius)* have thick, lumpy shells with smooth growth lines. They are ivory, pale purple, or light gray with a greenish brown periostracum on the margins of freshly beached shells. Rusty yellow staining is common.

Purplish tagelus clams *(Tagelus divisus)* have smooth, thin shells that are tinted purple inside and out. A darker purple ray from the umbo marks a slightly raised internal rib. Small shells may have a covering of brown periostracum.

Corrugated razor clams *(Solecurtus cumingianus)* are whitish with a sculpture of **wavy ridges** (image inset) that cross thin growth lines.

Green jackknife clams *(Solen viridis)* have a straight upper-shell edge and are about four and a half times as long as they are wide. They are whitish with a shiny, greenish periostracum.

HABITAT: Corrugated razor clams live in offshore sediments. These tagelus and jackknife clams live in the sand or mud of shallow embayments. Stout tagelus clams prefer closed lagoons and purplish tagelus clams prefer bays open to the sea.

DID YOU KNOW? Tagelus and jackknife clams live with only their siphons exposed and feed on suspended particles. Jackknife clams dig fast and can also swim.

Jackknife Clam *(Minor)* and **Duckclams**

RELATIVES: Jackknife clams (family Pharidae) are not directly related to duckclams (Anatinellidae), which are closer to mactras, rangias, and surfclams.

IDENTIFYING FEATURES:

Minor jackknife clams *(Ensis minor)* have fragile shells with a curved, straight-razor shape, almost six times as long as they are wide. They are purplish inside and yellowish outside with arches of growth lines between two long radial rays.

Channeled duckclams *(Raeta plicatella)* have white, thin, ear-shape shells with strong concentric growth ridges and a highly flared hind end bearing a furrow leading from the umbo.

Smooth duckclams *(Anatina anatina)* are similar to channeled duckclams, but are off-white with relatively smooth growth lines. The furrow from the umbo to posterior end is more distinct.

HABITAT: Minor jackknife clams live in shallow bays and sandy inlet tidal flats. Channeled duckclams live in sand just outside the surf zone. Smooth duckclams live in offshore sands to moderate depths of about 250 ft (75 m).

DID YOU KNOW? Duckclams gape at their flared hind end, from which inhalant and exhalant siphons protrude. The siphons are connected along their length and as long as the shell. Inhalant water is drawn in by gill action, extracting both oxygen and food. The nutrition comes in the form of suspended phytoplankton filtered by the gills. Indigestible grit is assembled into mucus-bound strings (pseudofeces) that are expelled without having passed through the digestive tract.

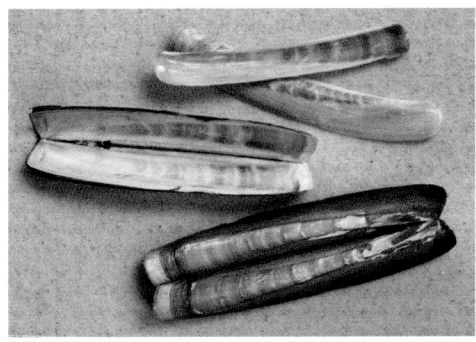

Minor jackknife clam, max 4 in (10.2 cm)

A pile of minor jackknife clams after storm erosion

Channeled duckclam, max 3.2 in (8.1 cm)

Smooth duckclam, max 3 in (7.6 cm)

139

Fragile Atlantic mactra clam, max 4 in (10.2 cm)

Brown rangia, max 3 in (7.5 cm)

Atlantic rangia clam, max 2.7 in (7 cm)

Atlantic rangia clams show varied staining histories

Mactra and Rangia Clams

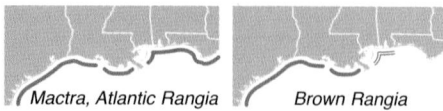

Mactra, Atlantic Rangia *Brown Rangia*

RELATIVES: Mactras and rangia clams are allied with surfclams in the family Mactridae.

IDENTIFYING FEATURES: These clams have a distinct hinge-ligament pit (chondrophore) behind the central hinge teeth under the umbo.

Fragile Atlantic mactras (*Mactrotoma fragilis*) are relatively thin-shelled with a forward umbo. Beached shells are cream to whitish and often have some remaining brownish periostracum behind a ridgeline on the hind end.

Brown rangia clams (*Rangia flexuosa*) have thick shells with inflated, foward-pointing umbones, and a distinct ridge from the umbo to the pointed hind end. Their rear lateral hinge tooth extends halfway to the ventral margin. Shells are white with a persistent brown periostracum.

Atlantic rangia clams (*Rangia cuneata*) are similar to brown rangias, but are not as pointed and have a rear lateral hinge tooth extending almost to the ventral margin. Beached shells are white, yellow, variously stained, and generally worn.

HABITAT: Mactras live in sand from just off the beach out to moderate depths. Rangia clams live in brackish water estuaries. Atlantic rangias tolerate lower salinities than brown rangias.

DID YOU KNOW? Atlantic rangias were an important food source for prehistoric peoples. This bivalve composes more than 80 percent of the mollusk shells found in middens and other archeological sites along the upper Texas and Louisiana coast. More recently, these abundant shells have been used in the construction of roadways, levees, and cement.

Surfclams, Rose Corbula,
and **Angelwing** *(Fallen-)*

RELATIVES: Surfclams are allied with mactras and rangias in the family Mactridae, not directly related to corbulas (Corbulidae), which are more closely allied with angelwings (Pholadidae).

IDENTIFYING FEATURES:

Dwarf surfclams *(Mulinia lateralis)* have an umbo forward of center and a tapered hind end. There is a triangular hinge-ligament pit (chondrophore) behind the central hinge teeth under the umbo. Colors may be white, cream, gray, or purple-gray, with highlighted growth bands.

Southern surfclams *(Spisula raveneli)* have strong shells with a central umbo and fine growth lines. Like the dwarf surfclam, there is a distinct triangular chondrophore. Colors are white or cream with a light brown periostracum, which may be absent in beach-worn shells.

Rose corbulas *(Corbula dietziana)* have shells with highly inflated right valves and smaller left valves, each with a narrowly squared off posterior end. Their sculpture shows thickened irregular concentric ridges. Colors are light gray to rose.

Fallen-angelwings (Atlantic mud-piddocks) *(Barnea truncata)* have fragile, whitish, winglike shells with radial ribs. These are stubby for an angelwing, with pronounced shell gapes both front and rear.

HABITAT: These surfclams live in sandy shallows. Rose corbulas are in offshore rubble. Fallen-angelwings bore into peat and rotten wood on bay bottoms.

DID YOU KNOW? The abundant dwarf surfclam feeds many estuarine animals including ducks and black drum.

Dwarf surfclam, max 0.8 in (2 cm)

Southern surfclam, max 5.1 in (13 cm)

Rose corbula, max 0.5 in (1.2 cm)

Fallen-angelwing, max 2.8 in (7.1 cm)

Angelwing, max 6.7 in (17 cm)

Angelwing with apophysis spoon

Angelwing inhalant (L) and exhalant (R) siphons

Campeche angelwing, max 5 in (12.7 cm)

Angelwings *(Angelwing, Campeche)*

RELATIVES: Angelwings, including fallen-angelwings, are in the family Pholadidae with shipworms.

IDENTIFYING FEATURES:

Angelwings *(Cyrtopleura costata)* have fragile, whitish, winglike shells with radial ribs and a flared shell margin near the hinge that curves out at the umbo.

Campeche angelwings *(Pholas campechiensis)* are similar to angelwings, but have weaker radial ribs and membranous shell over the umbo divided into several delicate compartments.

HABITAT: Angelwings burrow deep within silty sands in shallow waters and on tidal flats. Campeche angelwings burrow into compacted clay, wood, and even soft rocks.

DID YOU KNOW? These angelwings live with much of their soft parts outside their shells. They have no hinge teeth or ligament, with valves held together and opened only by their bulging body. The angelwing's muscular foot attaches to a spoonlike shell protrusion called an **apophysis**, which provides leverage to open and close the shell. This fragile shell part is often missing in beached shells. On exposed tidal flats, the fused **inhalant and exhalant siphons** of angelwings mark the locations where the bivalves live, 3 ft (0.9 m) or deeper below. The bivalves begin burrowing as juveniles. As they grow, they rotate in the burrow, rasp with their anterior shell ridges, and puff out sediment in an excavation process that takes years. As adults, angelwings are unable to rebury themselves if removed from their burrow. Their flared inhalant siphon draws in water for respiration and filter-feeding.

Shipworms and Piddocks

RELATIVES: Shipworms (family Teredinidae) are bivalves (not worms) related to angelwings and piddocks (Pholadidae).

IDENTIFYING FEATURES:

Shipworms (Teredinidae) leave traces that include snaking **tunnels** through beached driftwood. The expanding, pencil-width tunnels follow the wood's grain and are lined with white, fragile, shell material. The animal is a worm shape, with small, winglike shell valves (generally deep in wood and not easily seen). Opposite the valves are paddlelike "pallets." *Teredo* shipworms have these pallets hollowed like a vase, whereas *Bankia* shipworms have funnel-shape pallets stacked in a set of 16.

Wedge piddocks (*Martesia cuneiformis*), max 0.7 in (1.8 cm), are found in driftwood riddled with their boreholes. Their delicate shells are white with concentric ridges that, anterior to an oblique furrow, are serrated. The dorsal plate covering the hinge (**mesoplax**) is a teardrop shape with a center groove. The similar striate piddock (*M. striata*) is more elongate and has a circular mesoplax with no groove.

Oyster piddocks (*Diplothyra curta*) are found bored into rocks and thick shells. They are similar to the "wood" piddocks above, but have weaker ridges and a pear-shape mesoplax.

HABITAT: Shipworms and wedge piddocks bore into wood. Oyster piddocks bore into limestone and shells.

DID YOU KNOW? Piddocks burrow into hard substrates for protection. Shipworms have steered historical marine events including the failure of Christopher Columbus's final New World expedition and the defeat of the Spanish Armada.

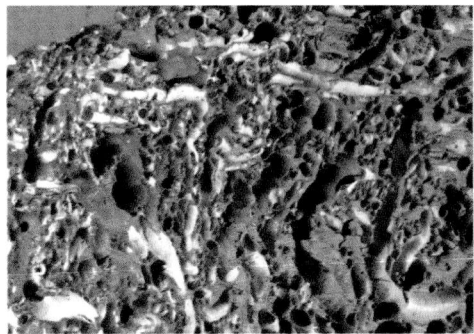

Shipworm tunnels, max 0.4 in (1 cm) diameter

Shipworm tunnel shell

Wedge piddock showing its teardrop-shape mesoplax

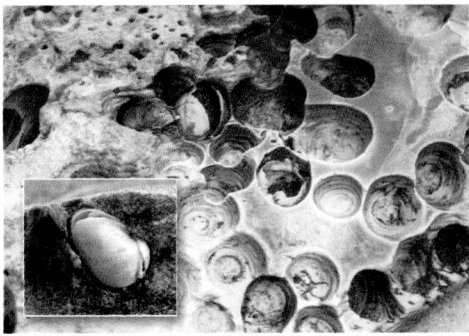

Oyster piddock, max 0.7 in (1.7 cm), in shell

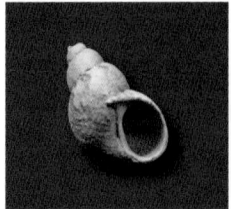
Turtlegrass pheasant

Spotted pheasant

Grass cerith

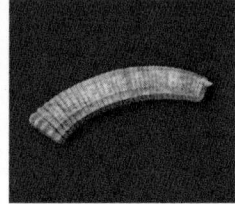
Imbricate caecum

Cancellate risso

Parasitic scalesnail

Circular Chinese-hat

Oyster dovesnail

Lunar dovesnail

Dwarf olive

Hemphill's mangelia

Waxy mangelia

Micro Shells

RELATIVES: Gastropods and bivalves

IDENTIFYING FEATURES: These shells are too small to be seen on a casual beach stroll and are all less than about 1/4 inch (6 mm) as adults. The path into the amazing world of itty-bitty shells is traveled by those on their hands and knees. Peering into drift piles at the recent strand line will reveal many of these petite species in addition to smaller (younger) versions of the larger species shown on previous pages.

Gastropods:

Turtlegrass pheasant *(Eulithidium thalassicola),* Florida and south Texas, family Phasianellidae

Spotted pheasant *(Eulithidium affine),* all beaches, family Phasianellidae

Grass cerith *(Bittiolum varium),* all beaches, family Cerithiidae

Imbricate caecum *(Caecum imbricatum),* all beaches, family Caecidae

Cancellate risso *(Phosinella cancellata),* Florida and southern Texas, family Rissoinidae

Parasitic scalesnail *(Cochliolepis parasitica),* all beaches, family Tornidae

Circular Chinese-hat *(Calyptraea centralis),* all beaches, family Calyptraeidae

Oyster dovesnail *(Parvanachis ostreicola),* all beaches, family Columbellidae

Lunar dovesnail *(Astyris lunata),* all beaches, family Columbellidae

Dwarf olive *(Olivella lactea),* Florida and south Texas, family Olividae

Hemphill's mangelia *(Pyrgocythara hemphilli),* Florida beaches, family Mangeliidae

Waxy mangelia *(Cryoturris cerinella),* all beaches, family Mangeliidae

Gastropods continued …

Brown-tip mangelia *(Kurtziella atrostyla)*, all beaches, family Mangeliidae

Impressed odostome *(Boonea impressa)*, all beaches, family Pyramidellidae

Candé's barrel-bubble *(Acteocina candei)*, all beaches, family Tornatinidae

Hemphill's turbonilla *(Turbonilla hemphilli)*, all beaches, family Pyramidellidae

Crenulated pyram *(Longchaeus suturalis)*, all beaches, family Pyramidellidae

Bivalves:

Lunate crassinella *(Crassinella lunulata)*, all beaches, family Crassatellidae

Miniature lucine *(Radiolucina amianta)*, all beaches, family Lucinidae

Many-lined lucine *(Parvilucina crenella)*, all beaches, family Lucinidae

Amethyst gemclam *(Gemma gemma)*, all beaches, family Veneridae

White strigilla *(Strigilla mirabilis)*, all beaches, family Tellinidae

Atlantic abra *(Abra aequalis)*, all beaches, family Semelidae

Caribbean corbula *(Caryocorbula swiftiana)*, all beaches, family Corbulidae

Brown-tip mangelia

Impressed odostome

Candé's barrel-bubble

Hemphill's turbonilla

Crenulated pyram

Lunate crassinella

Miniature lucine

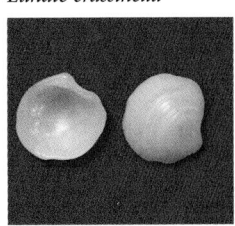

Many-lined lucine

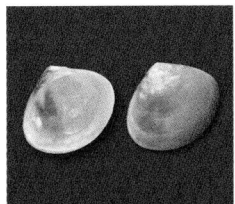

Amethyst gemclam

White strigilla

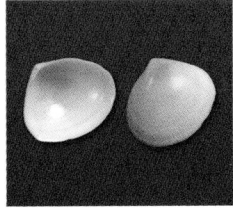

Atlantic abra

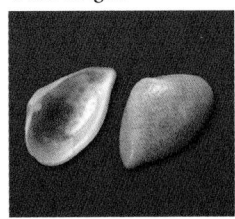

Caribbean corbula

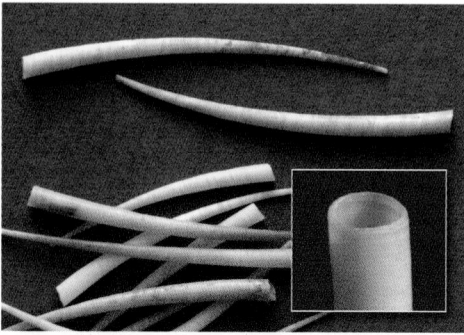

Ivory tuskshells, max 2 in (5 cm)

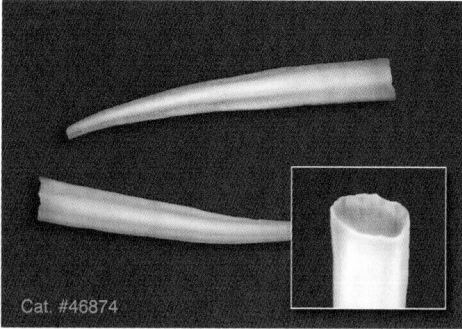

Cat. #46874

Reticulate tuskshell, max 2 in (5 cm)

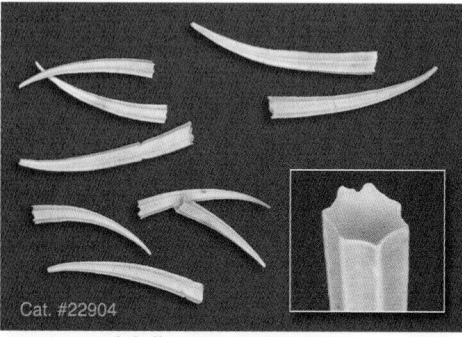

Cat. #22904

American tuskshells, max 1 in (2.5 cm)

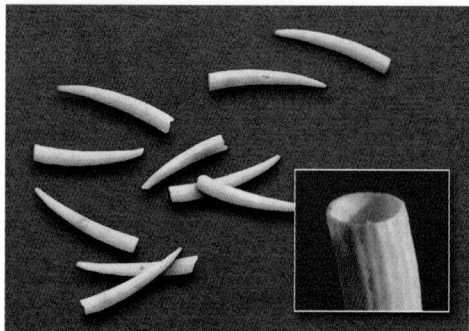

Pilsbry tuskshells, max 1.1 in (2.8 cm)

Tuskshells

Ivory, Reticulate,
American

Pilsbry

RELATIVES: Tuskshells are scaphopod mollusks (class Scaphopoda). These tuskshells are in the family Dentaliidae.

IDENTIFYING FEATURES: Most tuskshells are pale, delicate, curved, tapered tubes open at each end. The foot and mouth of the living tuskshell were formerly located at the wide end. The smaller rear opening passed respiratory currents.

Ivory tuskshells (*Graptacme eborea*) are off-white, gray, or pale peach with a relatively sharp tip. They are smooth and round in cross-section.

Reticulate tuskshells (*Dentalium laqueatum*) are straight except for curved apex, and have 9–12 faint axial ribs.

American (Texas) tuskshells (*Paradentalium americanum*) are whitish, with six (sometimes seven), evenly spaced longitudinal (axial) ridges that impart a hexagonal outline to the shell's cross-section (more pronounced in smaller shells).

Pilsbry tuskshells (*Antalis pilsbryi*) are round in cross-section and have nine evenly spaced axial ribs, with other ribs that do not run the length of the shell.

HABITAT: American tuskshells live in bay sediments. The other species live in sand out to moderate depths.

DID YOU KNOW? Scaphopod is Greek for "boat-footed." A tuskshell uses its boat-shape foot to burrow through sediment. The foot extends from their wide (anterior) end, along with oral tentacles that gather forams (p. 51) and other tiny invertebrates as food. The tuskshell's pointed rear is often their only exposed feature, and allows them to breathe.

Paper Nautilus, Octopus, and **Ram's Horn Squid**

Paper nautilus, max 12 in (30 cm)

RELATIVES: These are cephalopods, a separate mollusk class from the gastropods, bivalves, and scaphopods. Argonauts (family Argonautidae) are with common octopuses (Octopodidae) in the order Octopoda. The ram's horn squid is in the order Spirulida, family Spirulidae.

IDENTIFYING FEATURES:

Paper nautili are the eggcases of the **greater argonaut** *(Argonauta argo)*, a species of open-ocean octopus. The female argonaut creates this paper-thin eggcase, which coils around the octopus similar to a wrinkled nautilus shell.

Common octopuses *(Octopus vulgaris)* are shell-less, reddish-brown cephalopods with eight arms that extend about three fourths of the animal's length. Arm suckers typically have dark rings.

A living female argonaut stranded with her eggcase

Ram's horn squid *(Spirula spirula)* are beached as white, chambered coils. In life, the coil is within the squid's back end opposite two large eyes and 10 tentacles.

HABITAT: Argonauts and ram's horn squid live in the deep open ocean. Common octopuses are in coastal waters.

DID YOU KNOW? Paper nautili are not true shells because they are not produced by a mantle. The eggcase hardens from carbonates and proteins secreted by webbing between the female's arms, and it functions only in egg protection. The argonaut holds herself within the case, but is not attached to it. Ram's horn squid (inset) use their buoyant, chambered coil to suspend themselves head-up (buoyant coil down!). For protection, the squid puckers up by withdrawing its head and tentacles.

Common octopus, max 20 in (50 cm)

Ram's horn squid shells, max 1 in (2.5 cm)

147

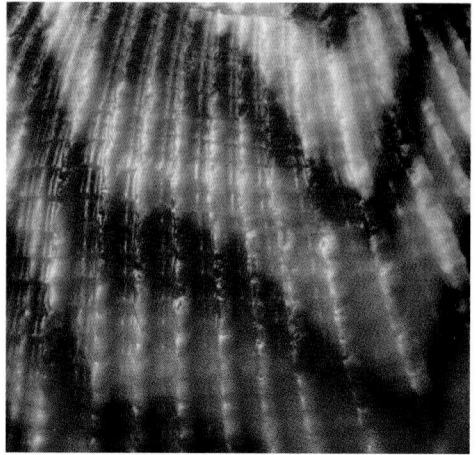

Color pattern laid by the mantle of a turkey wing ark

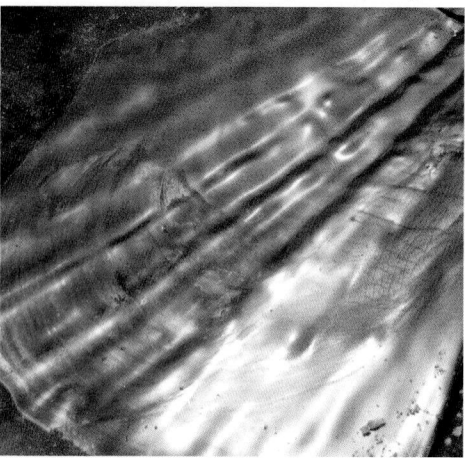

The interior nacre of a penshell

Eastern oyster shells with different afterlife experiences

Shell Color Variation

A seashell's colors are determined by their mollusk genetics, from the animals' diet, from the environment they occupied in life, and from their surroundings in the afterlife. **Shell color patterns** like rays, blotches, and zigzags, came from pigment-secreting zones in the animal's mantle. The pigments were laid down within layers of calcium carbonate and glyco-proteins as the mantle secreted the shell. Pigments include melanins (yellow to dark brown), carotenoids (yellow, orange, red, or purple), and tetrapyrroles (red, brown, or purple).

How mollusks benefit from their shell colors is a bit of a mystery. Potentially, patterns may act to camouflage or send a conspicuous message. But colors also may have nothing to do with how they appear to us. Shell pigments may assist with the animal's immune response, protect from UV damage, act as antioxidants, strengthen the shell, or serve as reference markers in organized shell-making.

Some of the prettiest shell colors are completely hidden while the shell is alive, like the iridescence of **nacre**. These nacreous colors are not from pigments; they are structural. Nacre is a composite material made of layered aragonite (a crystal of calcium carbonate) and organic materials. The colors we see come from interference of light wavelengths passing through and reflected by the thin layers, which have a thickness close to the wavelength of visible light.

Shell colors acquired in the afterlife continue to tell a mollusk's story. **Black shells** were likely darkened by iron sulfide after burial in muck. A beach with numerous black shells indicates that the surf zone was once a lagoon behind the barrier island. Pink, rust, and brown are the colors most shells turn after decades of exposure to oxidized minerals in water. Although glossy white shells are probably recent, bone-white shells may be ancient fossils.

Mollusk Bits and Pieces

For every whole shell found on a beach there are thousands of bits and pieces. Some surf-worn shards have the clear distinguishing features of the original shell. Do you recognize these?

A. **Giant tun shell** (p. 84)

B. **Giant eastern murex** (p. 89)

C. **Scotch bonnet** (p. 83)

D. **Florida rocksnail** (p. 90)

E. **Lettered olive** (p. 101)

F. **Shark eye moonsnail,** inside spire whorls (pp. 81–82)

G. **Lightning whelk,** (clockwise from top) **spire, body whorl, columella** (p. 96)

H. **Apple murex** (p. 88)

I. **Lace murex** (p. 89)

J. **True tulip, columella** (p. 93)

K. **Eastern banded tulip** (p. 93)

L. **Lion's paw scallop** (p. 119)

M. **Calico scallop** (p. 118)

N. **Horse conch, operculum** (p. 94)

O. **Atlantic rangia clam** (p. 140)

P. **Anglewing** (p. 142)

Q. **Turkey wing** (p. 111)

R. **Atlantic giant cockle** (p. 127)

S. **Eastern oyster, muscle scar** (p. 116)

T. **Blood ark** (p. 110)

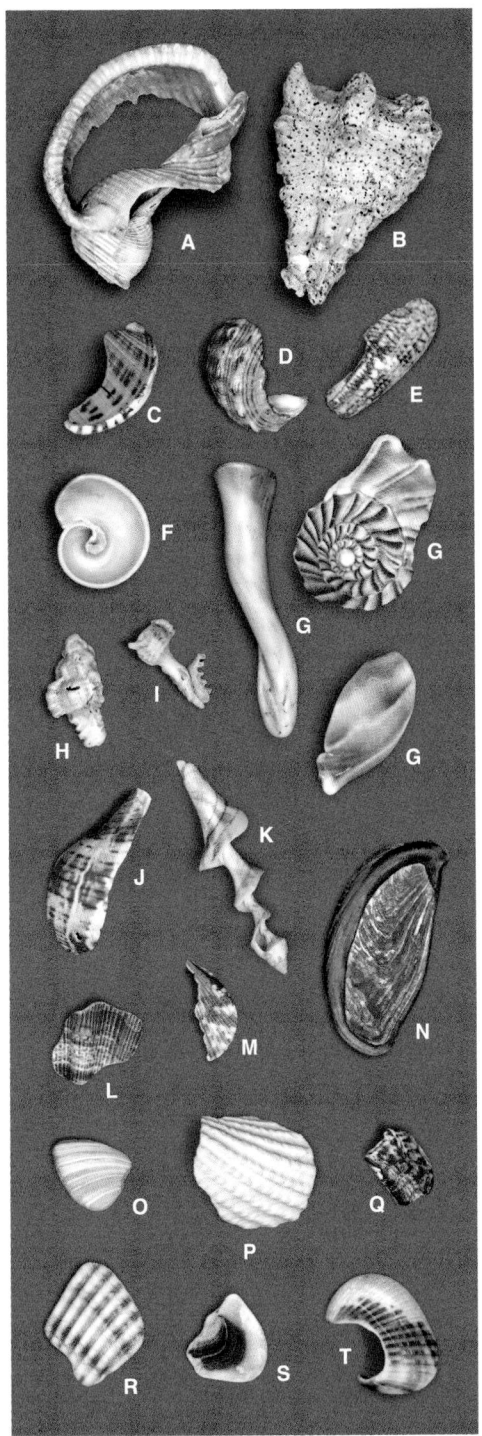

Some commonly found shell fragments

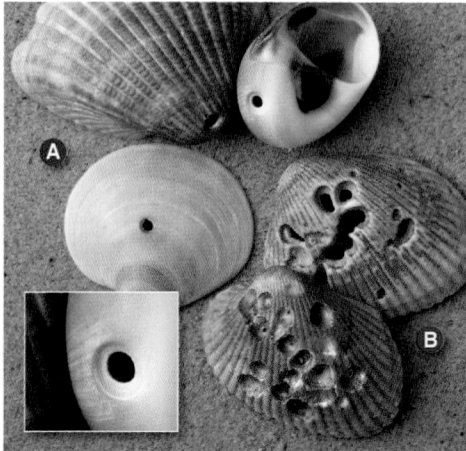

Boreholes from gastropods (A) and bivalves (B)

Boring sponge perforations in a quahog

Polychaete worm grooves in ark shells

Shell Wars (Shell Bioerosion)

Mollusk shells often bear clues to how they met their demise and who made use of them after their death. This evidence includes boreholes, perforations, and grooves.

Shells with single, circular **boreholes** (top image, **A**, and inset) were likely eaten by a predatory gastropod. Atlantic oyster drills (p. 89) tend to leave a cylindrical hole, whereas shark eye moonsnails (pp. 81–82) leave a countersunk, circular borehole with an outer diameter about twice the inner diameter. These snails penetrate shells using alternating periods of shell-desolving chemical application from their "accessory boring organ," and rasping by their file-like tongue (radula). Two tactics for hole-boring into bivalves include edge and umbo boring. Boring at a valve edge is fastest (because the shell is thinner) but is risky because closing valves could pinch the snail's proboscis. Umbo boring is safer, but in the time it takes to bore through this thickest part of the shell, a snail may have its prey stolen by a larger gastropod or become a meal itself. Note the evidence of homicide in the bore-holed shark eye, top image.

Scattered **perforations** in a shell were likely made by boring sponges (p. 57). These sponges partially acid-digest living and dead shells to invade them as occupied living space.

Other animals that use shells as living space include polychaete "blister" worms like *Polydora* (family Spionidae), which leave snaking **groove marks.** The router-tool style indentations are made as the worm rasps with its bristled body, aided by acids it secretes. Bivalve shells may also be penetrated by other bivalves. Chimney clams (p. 126) leave **oblong boreholes** (top image, **B**) in either shell or rock. These bean-shape clams live out their lives within the pit they form, cemented over with chalky calcium carbonate.

How Shells Come to Be

The seashells that strand on a beach represent former lives. But how does this life come to be? As it turns out, making little mollusks is different for our two major mollusk groups, and development takes slightly different routes.

Marine Gastropods

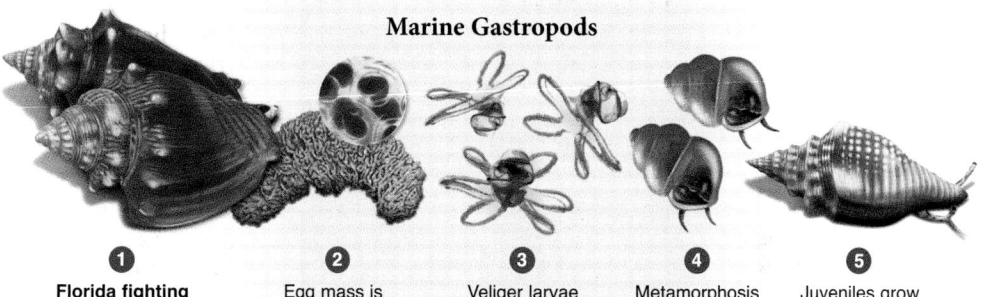

❶ Florida fighting conchs have separate sexes and reproduce by internal fertilization

❷ Egg mass is 2–3 inches long and contains about 90,000 eggs

❸ Veliger larvae hatch in about 3 days and are less than 1 mm

❹ Metamorphosis into tiny juveniles in 18–24 days

❺ Juveniles grow to adults in about 4 years

Most marine gastropods with shells have separate sexes (they are dioecious). But in an odd twist, some can switch from male to female depending on the need (for example, p. 79). In sea slugs, individuals have both male and female capabilities, but do not fertilize themselves. Fertilization is either internal or external depending on the species.

A fertilized egg develops into a veliger larva that is either free swimming (as in conchs) or within an egg capsule (as in whelks). Free swimming veligers feed on plankton until they become juveniles, which look like miniature versions of the adult snail. Small gastropods become adults within a year, but larger gastropods may take many years to mature.

Marine Bivalves

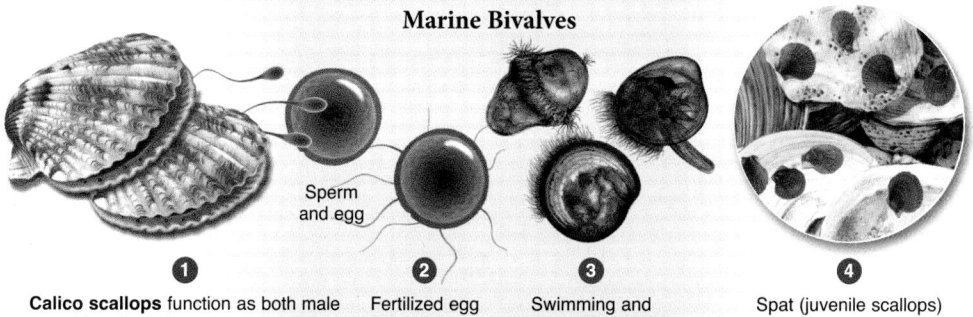

Sperm and egg

❶ Calico scallops function as both male and female in the same individual. Sperm are released into the water first, followed by eggs.

❷ Fertilized egg (microscopic)

❸ Swimming and settling larval stages 2 weeks

❹ Spat (juvenile scallops) settle on substrate

Some bivalves maintain separate sexes, some change their sex, and some are hermaphrodites (express both sexes simultaneously). Most fertilization is external, with sperm released, then eggs. Sperm fertilize eggs from many individuals. Eggs hatch into trochophore larvae, which develop into veliger larvae that eventually settle onto the sea bottom. There, they undergo metamorphosis into juvenile bivalves known as spat. Bivalves vary tremendously in their growth, maturation period, and lifespan. Calico scallops mature in months and seldom survive more than a year, but some cold-water clams mature late and can live for hundreds of years.

Fresh and stranded (inset) sargassum sea lace

Sea lace encrusting a cancellate cantharus shell

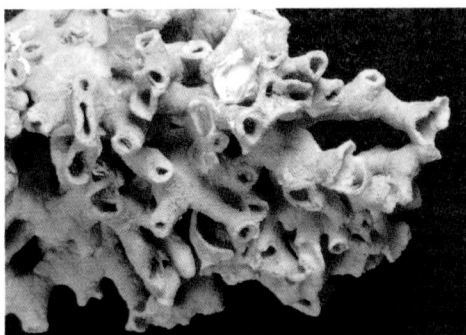

Gulf staghorn bryozoan

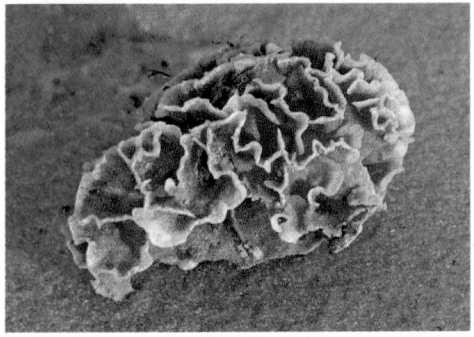

Lettuce bryozoan, max 7 in (18 cm)

Bryozoans *(Sea Laces, Staghorn, Lettuce)*

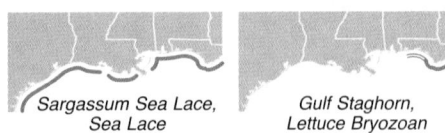

Sargassum Sea Lace,
Sea Lace

Gulf Staghorn,
Lettuce Bryozoan

RELATIVES: These "moss animals" are in the phylum Bryozoa and are remotely related to annelid worms and mollusks. Sea laces, and staghorn and lettuce bryozoans share the order Cheilostomatida.

IDENTIFYING FEATURES: All are colonies of individual animals called zooids.

Sargassum sea lace *(Jellyella tuberculata)* is a lacy crust covering sargassum algae (p. 310) and other drifters. Colonies reach about an inch wide (2.5 cm) and are composed of tiny rectangular compartments, each with two, tiny blunt projections.

Sea lace *(Biflustra* spp.) is similar to sargassum sea lace, but grows on hard objects in coastal waters and has no blunt projections.

Gulf staghorn bryozoan *(Schizoporella pungens)* colonies encrust sea whips (pp. 66–67), but also stand free as 6-in (15-cm) branched tubes. They are brownish-white when beached. Zooids live in compartments surrounding the tubes.

Lettuce bryozoan *(Thalamoporella floridana)* colonies look like an oblong bouquet of potato chips. They are brittle, but will keep their shape if protected.

HABITAT: Most bryozoans live attached to substrates in shallow waters. Sargassum sea lace lives at the ocean surface encrusting surface-drifting sargassum and flotsam.

DID YOU KNOW? Zooids in bryozoan colonies filter-feed using a semicircle of mouth tentacles. Although the calcified crusts of sargassum sea lace look brittle, their compartments have flexible joints, which allow the colony to bend with their wave-tossed, sargasso-weed substrate.

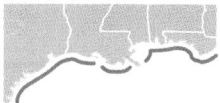

Bryozoans
(Bugula, Alternating, Bushy, Spaghetti)

Common bugula, max 8 in (20 cm)

RELATIVES: Common bugula shares the order Cheilostomatida with sea lace, staghorn, and lettuce bryozoans, and is only distantly related to *Amanthia* bryozoans in the order Ctenostomatida.

IDENTIFYING FEATURES:

Common bugula *(Bugula neritina)* colonies are composed of tough, burgundy branches that turn brownish in the sun. The zooids live in alternating positions along the branches.

Alternating bryozoan *(Amathia alternata)* colonies strand as bushy, brownish to yellowish clumps with bifurcate branches (two branches per node) that appear to zigzag due to alternating zooid rows.

Alternating bryozoan, max 12 in (30 cm)

Bushy bryozoan *(Amathia distans)* colonies are pale brown, soft, and fragile, with numerous bifurcate branches. The branch (stolon) tips have spiral rows of zooids.

Spaghetti bryozoan *(Amathia verticillata)* colonies have limp, translucent, triple or quadruple branching stolons with fuzzy zooids clumped in opposite rows along segments of stolon. Old colonies look like a clump of thin rice noodles.

HABITAT: These bryozoans live attached to hard substrates in shallow bay and Gulf waters.

Bushy bryozoan, max 8 in (20 cm)

DID YOU KNOW? Although each of these species is likely native to the Gulf, they have been introduced to warm-water regions around the globe. The principal way these animals invaded was accidental transport by ocean-crossing ships, which had hulls fouled with bryozoans or carried larvae in their ballast water. These colonies occasionally go through population booms when they are super abundant.

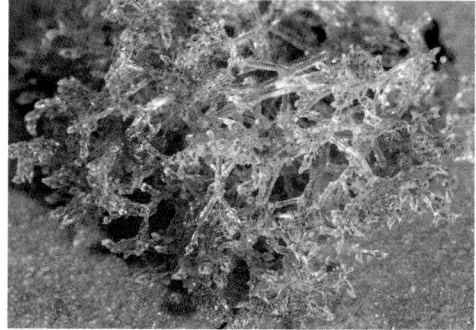

Spaghetti bryozoan, max 3 ft (0.9 m)

Parchment tube worm tubes, arrow shows worm

Calcareous tube worm tubes on a cockle shell

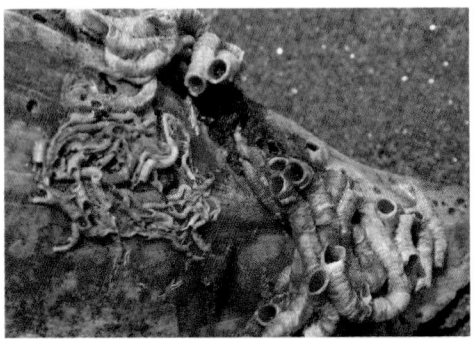
Fan worm tubes on a stained, lightning whelk shell

Unearthed, plumed worm tubes

Polychaete Tube Worms
(Parchment, Calcareous, Plumed)

RELATIVES: Polychaete tube worms are in the phylum Annelida, class Polychaeta, which includes segmented bristle worms.

IDENTIFYING FEATURES:

Parchment tube worms (*Chaetopterus variopedatus,* family Chaetopteridae) leave whitish, curved, limp, paperlike tubes up to 12 in (30 cm) long. Each tube was formerly a U-shape beneath the sand and was home to a **worm** with specialized segments like paddle-shape flaps. Storms often fill the beach with their tubes.

Calcareous tube worms (family Serpulidae) secrete whitish, calcareous, tubes adhering to shells, rocks, or debris. Some groups form masses of parallel or intertwined tubes. In this worm family, **fan worms** (*Hydroides* spp.) form tubes to about 0.3 in (8 mm) diameter.

Plumed (shaggy parchment-tube) worms (*Diopatra cuprea,* family Onuphidae) live within pinky-finger-thick tubes that project from the sand near the low-tide line. Layers of glued shell bits and other debris armor the end of each tube. The worm inside is iridescent.

HABITAT: Parchment tube worms live near the surf zone. Calcareous tube worms encrust a variety of objects out to moderate depths and attach to floating objects adrift on the open ocean. Plumed worms live in shallow, silty sands.

DID YOU KNOW? Although blind, parchment tube worms glow in the dark, emitting a luminous blue cloud of mucous when disturbed. Shells covering plumed worm tubes angle upward to anchor the tube in sand.

Polychaete Tube Worms
(Spiral, Soda Straw)

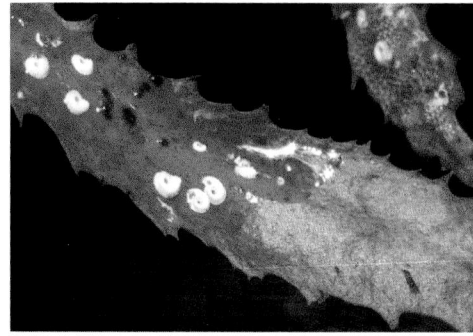

Spiral tube worms on sargassum algae

RELATIVES: These are polychaete worms (class Polychaeta). Spiral tube worms share the family Serpulidae with other calcareous tube worms. Soda straw worms (Onuphidae) are related to plumed worms, and are distantly related to encrusting sand tube worms (Sabellariidae).

IDENTIFYING FEATURES:

Spiral tube worms *(Spirorbis* spp.) live within tiny, pinhead-size, shell-coils stuck to rocks, shells, seagrass, or sargassum algae. A handful of stranded seagrass or sargassum weed may have hundreds of worm coils covering surfaces with tiny white dots.

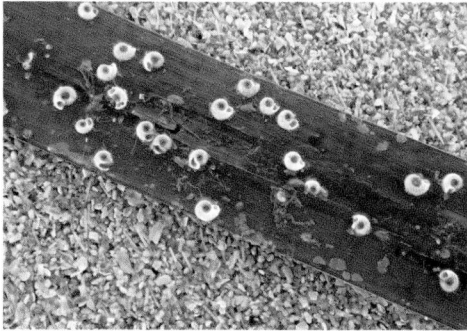

Spiral tube worms on seagrass

Soda straw worms *(Onuphis eremita)* live within 1/4-in- (6-mm) wide, sandy gray tubes that are tough and stretchy above the sand and limp below. Eroding surf can fill the tide line with piles of **limp tubes**.

Encrusting sand tube worms, *(Sabellaria floridensis)* cover shells and shell bits with course-grained tubes. The worm inside is pink and fragile. Colonies comprising dozens of tubes may reach about three inches (7.6 cm) in height.

Soda straw worm tubes beached after a storm

HABITAT: Spiral tube worms live on offshore floating algae and in shallow seagrass. Encrusting tube worms and soda straw worms live in sandy shallows.

DID YOU KNOW? Spiral tube worms are good parents. A mother broods her larvae within her shell until they are mature enough to be released. Soda straw and other tube worms that penetrate the seabottom are important transporters of dissolved chemicals between deep sediments and surface waters, and provide sites for elevated microbial activity.

Encrusting sand tube worm colony

Palp worm, max 1.4 in (3.5 cm)

Dumeril's clam worm on encrusted sargassum

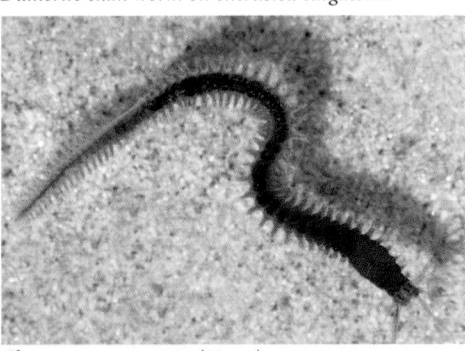

Clam worm, max 6 in (15 cm)

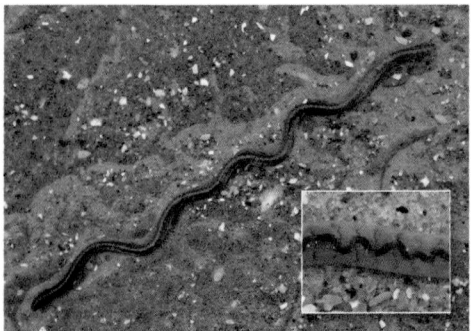

Capitella polychaete worm, max 4 in (10 cm)

Polychaete Worms
(Palp, Clam, Capitella)

RELATIVES: Palp worms (family Spionidae) are more closely related to sand-builder and serpulid worms than to clam worms (Nereididae) or Capitella polychaetes (Capitellidae).

IDENTIFYING FEATURES:

Palp worms *(Scolelepis squamata)* live within inconspicuous subsurface tubes and have two long feeding palps.

Dumeril's clam worms *(Platynereis dumerilii)*, 1.5 in (4 cm), have multiple pairs of palps and antennae, and numerous, two-lobed "feet" (parapodia). Their four eyes are in a trapezoid arrangement.

Clam (pile) worms *(Alitta succinea)* have up to 160 segments, four inconspicuous eyes, two antennae, and a pair of stout conical palps. Segments have paired parapodia that are longest at the thinnest (rear) portion of the worm.

Capitella polychaetes *(Capitella capitata)* are blood red with a flexible, fragile body.

HABITAT: Palp worms and capitella polychaetes live in the swash zone. Dumeril's clam worms live in tubes made of mucus and fine threads in habitats ranging from pelagic sargassum (p. 310) to shallow seagrass. Clam worms live in shallow seagrass, algae, and rubble.

DID YOU KNOW? Palp worms are the most common worm in the surf zone and may occur at densities greater than 200 per square meter. They are an important food for surf fishes and shorebirds. Dumeril's clam worm has the simplest eyes in the animal kingdom, is considered to be a "living fossil," and is used for laboratory studies of genetics. An abundance of capitella worms indicates polluted waters.

Polychaete Worms
(Lugworm, Bamboo, Ice Cream Cone)

RELATIVES: Lugworms (family Areni-colidae) and bamboo worms (Maldani-dae) are distantly related to Capitella worms, and are less closely related to ice cream cone worms (Pectinariidae).

IDENTIFYING FEATURES:

Southern lugworms *(Arenicola cristata)* are recognized by their **fecal castings,** which are in coils as if squeezed from a tube no wider than 0.4 in (1 cm). Com-pare these with the thicker feces of acorn worms (p. 192). In spring, lugworm **egg masses** are conspicuous as gelatinous blobs up to a yard long. The **living lug-worms** are up to 12 in (30 cm), and stout, with bushy red gills at mid body.

Bamboo worms *(Clymenella mucosa)* are orange polychaetes with elongate segments. They live in vertical, sandy mucus tubes, and are betrayed by their jellylike **egg sac**, which is attached to their tube at the sur-face and is seen in spring.

Ice cream cone worms *(Pectinaria gouldi)* build themselves a narrow, cone-shape tube out of a single layer of sand grains cemented with a protein glue. Typically, only vacant tubes are found. The head of the worm is at the larger end of the tube and has long ciliated feeding tentacles.

HABITAT: These worms live on silty intertidal flats near protected beaches.

DID YOU KNOW? Worms like these take part in a critical process on tidal flats—bioturbation, which is the turnover of surface sediments. Regardless of sand composition, an ice cream cone worm chooses consistent grain sizes to make its tube—fine grains for the narrow end and coarse grains at the wide end.

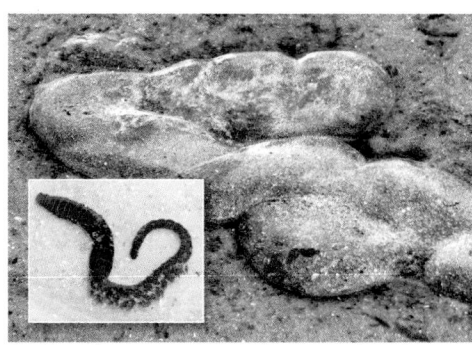

Southern lugworm egg mass, live worm (inset)

Fecal castings from lugworms seen at low tide

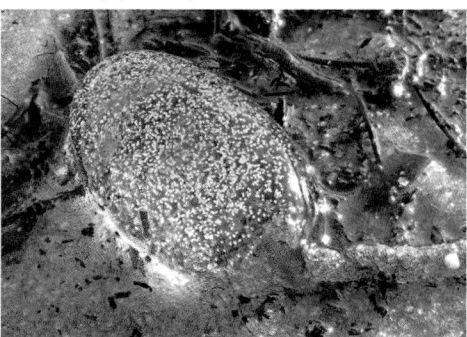

Bamboo worm egg sac, max 1.5 in (3.8 cm)

Ice cream cone worm tubes, 2 in (5 cm) long

157

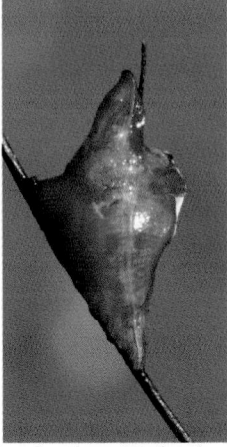

Sea whip barnacles, with (L) and without (R) sheath

Fragile barnacles on a Texas granite jetty

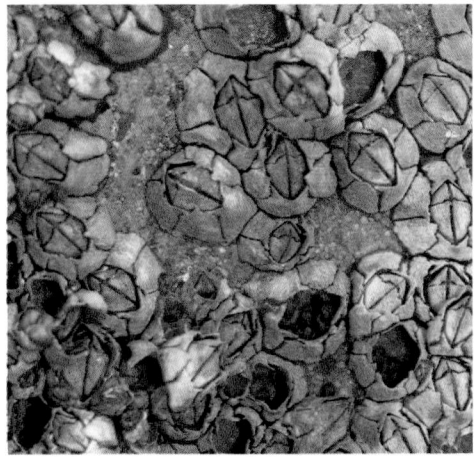

Fragile barnacles on a Florida limestone jetty

Barnacles Without Stalks
(Sea Whip, Fragile)

RELATIVES: Barnacles are crustacean arthropods (phylum Arthropoda, subphylum Crustacea) in the subclass Cirripedia. These barnacles share the suborder Balanomorpha—stalkless "acorn" barnacles. Sea whip barnacles are in the family Archaeobalanidae, and fragile barnacles are in the family Chthamalidae (tham-AL-a-dee).

IDENTIFYING FEATURES:

Sea whip barnacles *(Conopea galeata)*, 0.6 in (1.5 cm), are golden brown with a diamond-shape profile and are found on sea whip (p. 66–67) branches. Living barnacles are covered by a sheath of living seawhip tissue.

Fragile barnacles *(Chthamalus fragilis)*, 0.4 in (1 cm), are dull gray-white and are often crowded together on intertidal rocks. They have six plates surrounding a kite-shape orifice with rounded corners. Plates are smooth, but become ribbed and buttressed in older and uncrowded barnacles. Orifice plates join at a wavy suture line. Another species in the Gulf, *Chthamalus proteus*, is distinguishable from *C. fragilis* only through molecular analyses, and is considered a "cryptic species."

HABITAT: Sea whip barnacles live with sea whips in coastal waters. Fragile barnacles live on rocks and other hard substrates in the high intertidal zone.

DID YOU KNOW? Larval sea whip barnacles are thought to settle on a sea whip where a one-tooth simnia (p. 80) has eaten away a spot, exposing the gorgonian's axial skeleton. As the barnacle grows, it gets covered by the host's tissue and colonial polyps, forming a gall with a wide opening that allows the barnacle to feed.

Barnacles Without Stalks
(Triangle, Striped, Venustus, Titan Acorn)

RELATIVES: These are balanomorph "acorn" barnacles in the family Balanidae.

IDENTIFYING FEATURES: All have six shell plates with distinct, triangular radii between plates.

Triangle barnacles *(Balanus trigonus)* have ribbed plates with whitish stripes on a purple background surrounding a distinct triangular orifice.

Striped acorn barnacles *(Amphibalanus amphitrite)* have smooth, whitish plates with purplish stripes of irregular width and spacing. The orifice is diamond shape, but is flat or rounded at one end.

Venustus (charming) acorn barnacles *(Amphibalanus venustus)* look like striped acorn barnacles, but have regularly spaced reddish purple stripes of equal width. The plates are mostly smooth, and the orifice is oval with one pointed end.

Titan acorn barnacles *(Megabalanus coccopoma)*, 2 in (5 cm), are large, domelike, and pinkish purple, with a circular orifice.

HABITAT: Triangle and titan acorn barnacles live in coastal waters out to moderate depths. Striped and venustus acorn barnacles are in bay and shallow Gulf waters. All attach to hard substrates.

DID YOU KNOW? Barnacles live attached for life, head down within their shells. They feed by gathering plankton with their feathery legs, which also act as gills. At low tide, their orifice plates close to conserve water until the sea returns. Titan acorn barnacles are native to the northern Pacific, and were likely introduced to the Atlantic through inter-ocean shipping.

Shells from triangle barnacles, max 0.8 in (2 cm)

Striped acorn barnacle, max 0.7 in (1.8 cm)

Venustus acorn barnacle, max 0.5 in (1.3 cm)

Titan acorn barnacles on a granite jetty

159

Ivory acorn barnacle, max 1 in (2.5 cm)

Whitish acorn barnacle shells, max 0.5 in (1.3 cm)

Thin-shelled turtle barnacles on a speckled crab

Thick-shelled turtle barnacle on a juvenile green turtle

Barnacles Without Stalks
(Ivory Acorn, Whitish Acorn, Turtle)

RELATIVES: These are balanomorph barnacles. Acorn barnacles (Balanidae) and turtle barnacles (Chelonibiidae) are in different families.

IDENTIFYING FEATURES:

Ivory acorn barnacles *(Amphibalanus eburneus)* are white and steep-sided with pointed plates and no radial grooves.

Whitish (bay) acorn barnacles *(Amphibalanus subalbidus)* are white and dome-like with blunt-topped plates and no radial grooves. The similar barnacle, *A. improvisus,* is smaller than 0.4 in (1 cm) and has radial grooves.

Turtle barnacles *(Chelonibia testudinaria)* reach 3 inches (7.6 cm) on sea turtles, and 1 inch (2.5 cm) on crabs, terrapins, and other small, shelled, mobile, marine animals. The barnacle has a smoothly domed shell and an oval orifice.

HABITAT: Ivory and whitish acorn barnacles reach beaches attached to driftwood or flotsam. Turtle barnacles strand along with the animals they lived on.

DID YOU KNOW? Unlike most barnacles, turtle barnacles can move to a better spot—not fast, the width of a pinhead (1.4 mm) in a day. Compared to turtle barnacles on crabs, turtle barnacles on turtles have thicker, more streamlined, and more pitted shells. Toughness and streamlining may help, because turtles bump into things and swim faster than crabs. Given that a sea turtle is a moving target, pitted barnacle shells may aid settling of male larvae onto larger barnacles, which act as females when a male is present, thus facilitating reproduction. Males are tiny, less than 0.4 in (1 cm).

Barnacles With Stalks

RELATIVES: Stalked barnacles share the subclass Cirripedia with acorn barnacles, but are in a separate group (order Lepadiformes, family Lepadidae).

IDENTIFYING FEATURES: All have leaf-shape shells on fleshy stalks.

Pelagic goose barnacle, max 1.5 in (3.8 cm)

Pelagic goose barnacles *(Lepas anatifera)* have a purple-brown stalk and no radial grooves, but do show faint concentric growth lines. Some have radiating burgundy spots.

Gooseneck barnacles *(Lepas hillii)* have a dark stalk that is orange near the shell. The shell plates have growth lines but no radiating grooves. The lines between plates often have an orange tint.

Goose barnacles *(Lepas anserifera)* have an orange stalk and radial grooves on their shell.

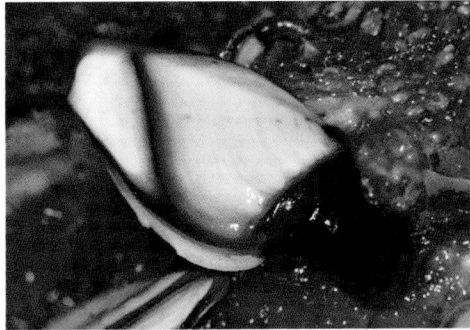

Gooseneck barnacle, max 1.5 in (3.8 cm)

Duck barnacles *(Lepas pectinata)* are small, max 1 in (2.5 cm), pale purple or gray, and have rough radial ribs.

HABITAT: These barnacles live on wood, seabeans, and other flotsam adrift on the open ocean. They reach the beach when their ride does. For some reason, the duck barnacle is most common on hard objects like pumice, light bulbs, and the floating spiral of the **ram's horn squid** (p. 147).

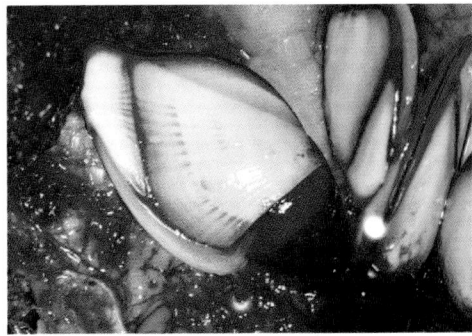

Goose barnacle, max 1.5 in (3.8 cm)

DID YOU KNOW? These stalked barnacles feed on plankton and other tiny drifting food bits. They grow from a swimming larva to adult size in a matter of weeks. Loggerhead sea turtles eat goose barnacles, and also provide a home for them on their carapace. Some little loggerheads adrift in the open Atlantic are known to host masses of barnacles heavier than themselves.

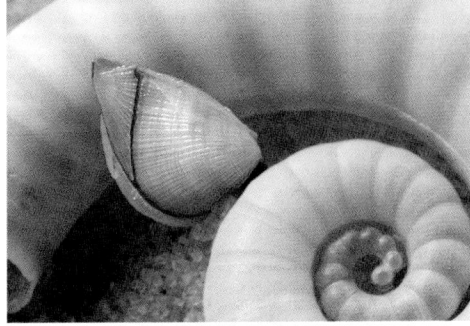

Duck barnacle on a ram's horn squid spiral

Beachhoppers, most species max at 0.3 in (0.8 cm)

Big-eyed sandhopper, max 0.7 in (1.8 cm)

Digger amphipod curled up, max 0.4 in (1 cm)

Dock roach isopod, max 1.5 in (3.8 cm)

Seapill, Sphaeroma walkeri, *max 0.2 in (0.5 cm)*

Amphipods and Isopods

All but Sandhopper — *Big-eyed Sandhopper*

RELATIVES: All are crustaceans. Beachhoppers and digger amphipods are in the order Amphipoda. Dock roaches and seapills are in the order Isopoda.

IDENTIFYING FEATURES: Amphipods are flattened side-to-side, and isopods are flattened top-to-bottom.

Beachhoppers (family Talitridae) are sand-color critters that bounce by the dozens as moist wrack is disturbed. Some family members are large and white, like **big-eyed sandhoppers** (*Americorchestia* spp.).

Digger amphipods (family Haustoriidae) are abundant but seldom seen. Although they actively dig in wet sand, when sieved out, they curl into a tiny ball. The Gulf has many species.

Dock roaches (wharf lice) (*Ligia exotica*) are charcoal-color, swift-running, insect-like isopods.

Seapills (marine roly-polies) (family Sphaeromatidae) reach 0.4 in (1 cm), and are brownish isopods with flattened tails (uropods). They curl into a ball when handled. A small species, ***Sphaeroma walkeri,*** is common on marine wood.

HABITAT: Beachhoppers live under seaweed piles, or bury under matchhead-size sand clumps on the lower beach. Dock roaches scurry about on jetties and wrack above the tide. Digger amphipods are within intertidal sands. Seapills live in intertidal rubble and driftwood.

DID YOU KNOW? Beachhoppers eat seaweed remnants, and although they look like fleas, they don't bite. Dock roaches live in the shadows by day and feed at night. Digger amphipods are one of the most diverse and abundant animals living within swash-zone sands.

Prawns and Shrimps

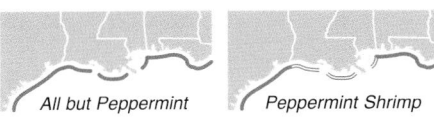

All but Peppermint   Peppermint Shrimp

RELATIVES: All are decapod crustaceans, but grass shrimp and peppermint shrimp are caridean (true) shrimps, which are more closely related to crabs and lobsters than to white and brown shrimps (suborder Dendrobranchiata, prawns, a sister group to all other decapods).

IDENTIFYING FEATURES: Caridean shrimps have a double overlapping second abdominal segment that gives their "tail" an angular bend. Prawn tails flex into a gentle curve.

Atlantic white shrimp (*Penaeus setiferus*) are white prawns with pinkish sides, a green-edged tail, and reddish antennae almost three times their body length.

Brown shrimp (*Penaeus aztecus*) are prawns with brown or reddish mottling, most pronounced at the end of their tail fans, and have antennae that are nearly as long as their bodies.

Grass shrimp (*Palaemon* spp.) are translucent except for their yellowish eyestalks. Several similar species live in Gulf waters.

Peppermint shrimp (*Lysmata wurdemanni*) have candycane red stripes on a translucent body.

Sargassum shrimp (*Latreutes fucorum*) are a fresh sargassum color with a broad (high) rostrum that is dorsally smooth.

Brown grass shrimp (*Leander tenuicornis*) range in color from sargassum gold to sargassum brown, with a pointed, spine-topped rostrum.

HABITAT: Juvenile brown and white shrimps live in bays, and move to offshore waters as adults. Grass and peppermint shrimps live in coastal waters. Brown grass and sargassum shrimps live in offshore sargassum seaweed.

White shrimp, max 10 in (25 cm)

Brown shrimp, max 7 in (18 cm)

Grass shrimp, max 2 in (5 cm)

Peppermint shrimp, max 2.8 in (7 cm)

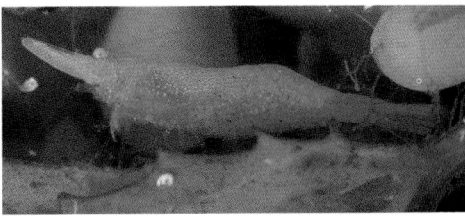

Sargassum shrimp, max 0.6 in (1.6 cm)

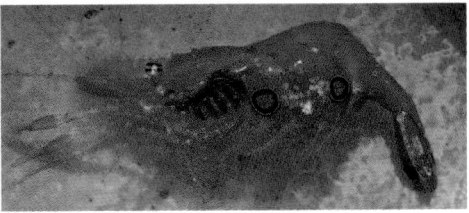

Brown grass shrimp, max 2 in (4.7 cm)

163

Carolinian ghost shrimp, max 3.2 in (8 cm), male

Carolinian ghost shrimp, female

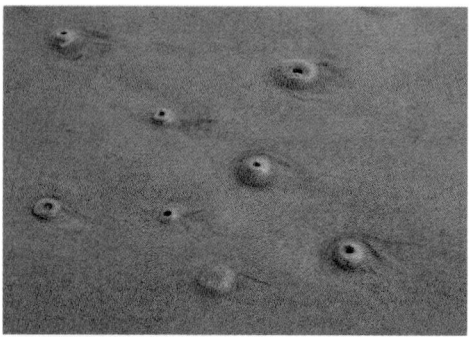

Ghost shrimp burrow openings in the swash zone

Ghost shrimp burrow and fecal pellets

Ghost Shrimps

RELATIVES: Ghost shrimps are decapod crustaceans in a separate group (Axiidea) from true shrimps and other decapods

IDENTIFYING FEATURES:

Carolinian ghost shrimp *(Callichirus major)* are identified by their pencil-size **burrow openings**, which are often surrounded by **fecal pellets** that look like chocolate ice cream sprinkles. The animal itself is elongate, pale, and soft-bodied, with one claw (first pereopod, right or left) larger than the other (especially in males) and bearing an unforked, curved finger. Exposed burrows on an eroded sand flat often show their dark, muddy, reinforced lining. Another species, the beach ghost shrimp *(Callichirus islagrande),* is most common in Texas and Louisiana. This species differs in having pointed eyestalks and a fork-tipped hook on the largest claw.

HABITAT: Ghost shrimps burrow in fine, intertidal sands on protected beaches and inlet margins.

DID YOU KNOW? These crustaceans have burrows as deep as six feet (1.8 m) that house commensal animals such as bivalves, polychaete worms, pea crabs, and copepods. Ghost shrimps are deposit feeders that produce about 500 fecal pellets a day, which feed hermit crabs and other intertidal critters. The digging from ghost shrimp circulates sand in an important ecological process called bioturbation. This ecological engineering makes the sand and inhabiting fauna of a ghost shrimp colony different from a sand flat without these crustaceans. Because extracting ghost shrimp for fishing bait is ecologically harmful, this practice is regulated in Texas and illegal at National Seashore beaches.

Mole Crabs and Sand Crab

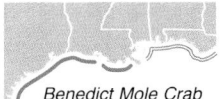

Benedict Mole Crab

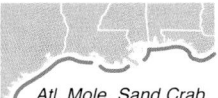

Atl. Mole, Sand Crab

RELATIVES: Mole (family Hippidae) and sand (Albuneidae) crabs are decapod crustaceans in the group Hippoidea, and are more closely related to hermit crabs than to true (Brachyuran) crabs.

IDENTIFYING FEATURES:

Benedict mole crabs *(Emerita benedicti)* are similar to Atlantic mole crabs, but their first pereopod dactylus has a blunt point, with its length more than twice its width.

Atlantic mole crabs *(Emerita talpoida),* also known as sand fleas, have a teardrop-shape and short, fuzzy antennae. They are mottled gray or pale above, and white below. The end segment of the front digging leg (dactylus of first pereopod) is oval with a rounded tip.

Sand crabs *(Lepidopa benedicti)* have a squarish carapace and two whip antennas (antennal flagella) much longer than their body. Their colors may be iridescent or chalky white, gray, or pink.

HABITAT: Mole crabs can be locally abundant in the swash zone. Sand crabs burrow within intertidal fine sands at inlet areas and low-energy beaches.

DID YOU KNOW? Mole and sand crabs feed on plankton and detritus driven into the surf. Atlantic mole crabs **swim-dig** backwards through flooded sands but are helpless if placed above the receding swash. They position themselves tail down and head seaward so their netlike antennae can inflate with the backwash from each wave. Surf fishers scoop mole crabs as preferred bait for whiting and pompano. Mole crab **molts** (old shells) in wrack are abundant when the local population goes through synchronous molting, driven by previously abundant food.

Benedict mole crab, max 1 in (2.5 cm)

Atlantic mole crab, max 1.5 in (3 cm)

As the wave recedes (A), the mole crab digs (B)

Sand crab, max 0.7 in (1.7 cm)

Molts from Atlantic mole crabs

165

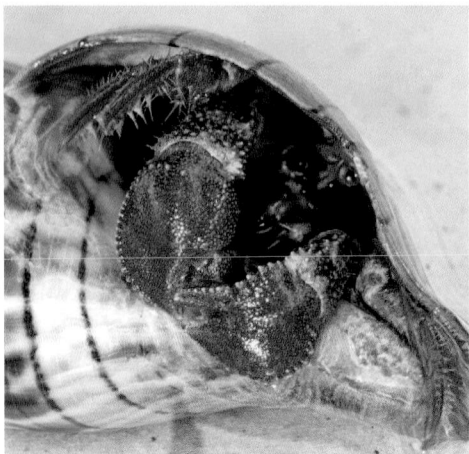
Dimpled hermit crab in a banded tulip shell

Flatclaw hermit crab in a sharkeye moonsnail shell

Long-wristed hermit in a banded tulip shell

Hermit Crabs, Righthanded

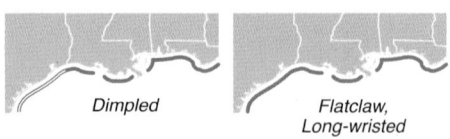
Dimpled   Flatclaw, Long-wristed

RELATIVES: Hermit crabs (not true crabs) are decapod crustaceans in the infraorder Anomura, shared with mole crabs. These hermits are in the family Paguridae—righthanded hermits.

IDENTIFYING FEATURES: Hermits have soft, curled abdomens and live in snail shells. Body lengths (sizes given here) are about as long as the shell they inhabit. These species have a right (major) claw larger than the left.

Dimpled (palmate) hermit crabs *(Pagurus impressus),* 3 in (7.6 cm), are rusty brown and have wide, granular claws edged with small blunt spines.

Flatclaw hermit crabs *(Pagurus pollicaris),* 1.5 in (3.8 cm), have pale, granular claws with the largest bearing wide, flat fingers.

Long-wristed hermit crabs *(Pagurus longicarpus),* 1 in (2.5 cm), have pale, striped legs and an elongate right claw.

HABITAT: These hermits stay seaward of the high-tide line but can become exposed at low tide.

DID YOU KNOW? Hermit crabs eat detritus and carrion. The way that hermit crabs switch shells has been used in case studies of "vacancy chain theory." It's a concept developed by sociologists to explain how reusable but limited resources like apartments or jobs get transferred within human populations. As it turns out, leaving an old shell (home, job) for a new one sends ripples through the socio-economic order with many benefits to those "downstream" waiting for their chance at a better opportunity.

Hermit Crabs, Lefthanded

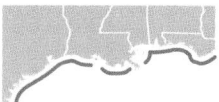

RELATIVES: Lefthanded hermits are allied within the family Diogenidae.

IDENTIFYING FEATURES: These species have a left (major) claw larger than the right.

Surf hermit crabs *(Isocheles wurdemanni),* 1 in (2.5 cm), show dirty-white claws with scattered hairs (setae) and claw fingers that are short and curved at their tips. The left claw is only slightly larger than right.

Giant red hermit crabs *(Petrochirus diogenes),* 7 in (18 cm), are red with heavy, knobby, unequal claws.

Thinstripe hermit crabs *(Clibanarius vittatus),* 2.5 in (6.4 cm), have brown-green legs with light stripes. Their claws are hairy, bumpy, and almost equal in size.

HABITAT: Giant red hermits live on reefs and rubble and can become beached after storms. Surf hermits inhabit the surf zone, and thinstripe hermits occupy a variety of bay and shallow coastal waters.

DID YOU KNOW? Hermit crabs are not as reclusive as their name suggests. The shells they occupy are often shared with a variety of commensal critters, including slipper snails, spotted porcelain crabs, and the hermit crab anemone. Ironically, hermits occasionally congregate. This may be at a food source, or to trade shells. Hermit crabs respond to the scent of injured fellow hermits differently based on their size. Growing hermits in need of a larger shell crawl toward the smell of tragedy, whereas hermits happy with their shell move away from possible danger. Striped hermits walk about exposed on a low-tide flat but require seawater replenishment to breathe.

Surf hermit crab in a cancellate cantharus shell

Giant red hermit in a true tulip shell

Thinstripe hermit crab in a crown conch shell

An ironic gathering of thinstripe hermits at low tide

167

Green porcelain crab, max 0.4 in (1 cm) body width

Olive pit porcelain crab, 0.5 in (1.3 cm) body width

Spotted porcelain crab, 0.5 in (1.3 cm) body width

Female squatter pea crab, in a penshell

Porcelain Crabs and Pea Crab

RELATIVES: Porcelain crabs (family Porcellanidae) are not true crabs, but are anomuran decapods related to mole and hermit crabs. Pea crabs are true (brachyuran) crabs in the family Pinnotheridae.

IDENTIFYING FEATURES:

Green porcelain crabs *(Petrolisthes armatus)*, like all porcelain crabs, have three pairs of walking legs (true crabs have four pairs). This species has long antennae and is greenish brown with blue mouthparts.

Olive pit porcelain crabs *(Euceramus praelongus)* have a three-pointed rostrum and an indented rear carapace.

Spotted porcelain crabs *(Porcellana sayana)* are pink with blue-white spots.

Squatter (mussel) pea crabs *(Tumidotheres maculatus)* are pale brown with weak legs and an oval carapace to about 0.5 in (1.3 cm) wide.

HABITAT: Green, olive pit, and spotted porcelain crabs live in shallow-water rubble. Spotted porcelain crabs also share shells with hermit crabs. Squatter pea crabs squat inside living mussels, scallops, or pen shells. Other local pea crab species live similarly close to lugworms, ghost shrimp, and tube worms.

DID YOU KNOW? Green porcelain crabs are alien invaders from Brazil. Their large claws are used in territorial fights, not to feed. Their food is gathered by their feathery mouthparts, which filter particles from the water. Porcelain crabs get their name from an ability to lose appendages, which facilitates escape from predators. Pea crabs eat particles of food filtered out by the various invertebrates they live with.

Spider Crabs and Calico Crab

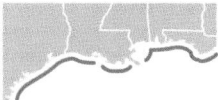

Longnose spider crab, max 4 in (10 cm) body width

RELATIVES: These are true (brachyuran) crabs in the families, Epialtidae (spider crabs) and Hepatidae (box crabs).

IDENTIFYING FEATURES:

Spider crabs (*Libinia* spp.) are sluggish and often covered with sea-bottom growth. **Longnose spider crabs** (*L. dubia*) have seven or fewer bumps on the carapace midline, whereas **portly spider crabs** (*L. emarginata*) have more than seven, although these may be worn.

Calico box (Dolly Varden) crabs (*Hepatus epheliticus*) are stout with irregular spots, short legs, and wide claws that partially hide their faces.

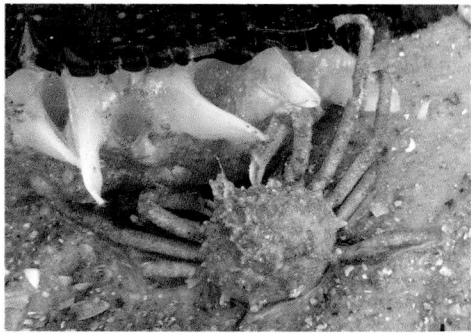
Juvenile longnose spider crab on canonball jelly

HABITAT: Spider crabs are common in shallow waters and are swept onto beaches after rough weather. Calico box crabs bury themselves in surf-zone sands.

DID YOU KNOW? Some juvenile longnose spider crabs less than 1 in (2.5 cm) body width ride and nibble on **cannonball jellies** (p. 60). This commensal relationship ends when the crab matures and begins life on the bottom. Demersal (bottom-dwelling) spider crabs of all sizes camouflage themselves with sponges and algae. The concealing debris is attached to hooked, velcro-like hairs on the crab's body. Box crabs hide their faces with their broad claws to keep sand out of their gill chambers. Like all crustaceans, these crabs molt their exoskeleton (shell) each time they incrementally grow. Early in its molting cycle, a crab forms a new, pliable shell beneath the old one. The soft new crab exits its old shell through a split in the back of the carapace. These old shells (molts) often wash ashore, retaining the former owner's pattern and colors.

Portly spider crab, max 4 in (10 cm) body width

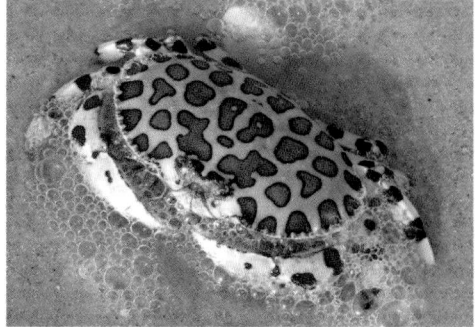
Calico box crab, max 4 in (10 cm) body width

Mottled purse crab, max 2.5 in (6 cm) body width

Female blue crab, max 8 in (20 cm) body width

Juvenile lesser blue crab, max 4.7 in (12 cm) body width

Speckled crab, max 5.5 in (14 cm) body width

Purse Crab and Swimming Crabs
(Blue, Lesser Blue, Speckled)

RELATIVES: These are true (brachyuran) crabs in the distantly related families, Leucosiidae (purse crabs) and Portunidae (swimming crabs).

IDENTIFYING FEATURES:

Mottled purse crabs (*Persephona mediterranea*) have bumpy, rounded bodies, long legs, and narrow claws.

Blue crabs (*Callinectes sapidus*), like all swimming crabs, have pointed projections on the sides of their carapace and swimming paddles on their hind legs. Blue crabs are greenish and blue. Females have orange highlights and clawtips, and are more commonly beached than male crabs.

Lesser blue crabs (*Callinectes similis*) are similar to blue crabs, but have a smoother, evenly granulated carapace. The inner surfaces (palms) of the claws are blue.

Speckled swimming crabs (*Arenaeus cribrarius*) are shaped like blue crabs but are brownish with crowded light spots.

HABITAT: Mottled purse crabs bury themselves in surf-zone sands. Blue crabs and lesser blue crabs live in estuaries. The females occasionally beach as they migrate through inlets to spawn at sea. Speckled crabs bury themselves in swash-zone sands.

DID YOU KNOW? Purse crabs may get their name from Persephone (winter-bringing Greek goddess) or from their egg-protecting, purselike, under-curled abdomen. These swimming crabs are swift predators that eat worms, mollusks, small fish, and other crabs. Although their claws are fast and sharp, blue crabs can be safely handled by firmly holding the base of the swimming leg. It's one of the only places they can't reach.

Swimming Crabs
(Iridescent, Blotched, Lady, Sargassum)

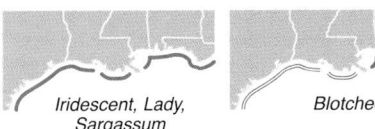

Iridescent, Lady, Sargassum — Blotched

RELATIVES: Lady and sargassum crabs share the family Portunidae with other swimming crabs.

IDENTIFYING FEATURES:

Iridescent swimming crabs *(Achelous gibbesii)* have slender claws, purple legs, and iridescent patches above their toothed carapace margin.

Blotched swimming crabs *(Achelous spinimanus)* have a comparatively narrow carapace with small lateral projections, which are tipped in red-brown, as are the white-blotched claw fingers.

Florida lady crabs *(Ovalipes floridanus)* are golden and purplish with a circular carapace that has similar length spines at its front margin.

Sargassum swimming crabs *(Portunus sayi)* have a golden brown carapace with light blotches.

HABITAT: Iridescent and blotched swimming crabs live on a variety of sea-bottom habitats in inlets and offshore. Lady crabs bury in shallow sand. Sargassum crabs live only on floating sargassum in the open sea.

DID YOU KNOW? Crab sexes differ in the shape of their tail (pleon), which is pressed underneath them. A male's tail is wide only at the base, and the female's is like a half circle with a short projection. Sargassum swimming crabs blend in with the colors and patterns of the pelagic algae that compose their habitat. They change colors by fluctuating color-bearing cells (chromatophores) under their transparent carapace. Thus, fresh golden algae harbors golden mottled crabs, and older brown algae holds brownish crabs.

Iridescent swimming crab, 3 in (7.6 cm) body width

Blotched swimming crab, 4.5 in (11 cm) body width

Florida lady crab, max 3.5 in (8.9) body width

Sargassum swimming crab, 2 in (5 cm) body width

Harris mud crab, max 0.8 in (2.0 cm) body width

Atlantic mud crab, max 2.5 in (6.4 cm) body width

Florida stone crab, max 5 in (13 cm) body width

Western stone crab, max 4 in (10 cm) body width

Mud Crabs and Stone Crabs

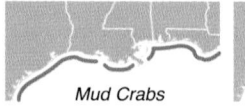

Mud Crabs

Florida Stone Crab

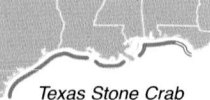

Texas Stone Crab

RELATIVES: All are true (brachyuran) crabs. Mud crabs (family Panopeidae) are only distantly related to stone crabs (family Menippidae).

IDENTIFYING FEATURES:

Harris mud crabs *(Rhithropanopeus harrisii)* have grayish claws with pale fingers.

Atlantic mud crabs *(Panopeus herbstii)* are dark and without spots. Their black-fingered claws are of unequal size and have light tips.

Florida stone crabs *(Menippe mercenaria)* are stout-bodied with robust, unequal claws like Popeye's forearms. The larger claw is a crusher with blunt fingers, and the smaller is a pincer claw with numerous small cutting teeth. Juveniles are deep purple with light spots, and adults are olive or gray-purple with black spots. Most have banded legs.

Western stone crabs *(Menippe adina)* are similar to their Florida cousins, but are olive- or maroon-brown with yellowish spots or mottling. The legs are not banded.

HABITAT: Mud crabs prefer muddy bottom and oyster reefs. Harris mud crabs are mostly estuarine and occasionally enter fresh water. Stone crabs live in shallow rocky areas and offshore. All are most common near inlets.

DID YOU KNOW? Mud crabs bury within polychaete worm colonies to avoid being eaten by blue crabs. Stone crabs eat barnacles, oysters, and clams by crushing them with their claws.

Hairy Crab and Grapsoid Crabs
(Tidal Spray, Squareback Marsh, Mottled)

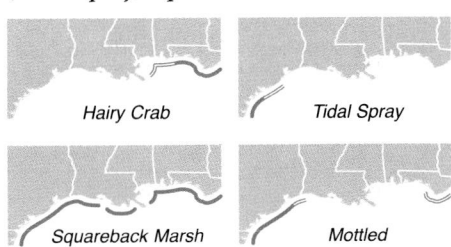

Hairy Crab

Tidal Spray

Squareback Marsh

Mottled

Spineback hairy crab, max 1.2 in (3 cm) body width

RELATIVES: Hairy crabs (family Pilumnidae) are pilumnoid crabs, remotely related to mud crabs, and less related to the grapsoid crab group containing tidal spray crabs (Plagusiidae), squareback marsh crabs (family Sesarmidae), and mottled shore crabs (Grapsidae).

IDENTIFYING FEATURES:

Spineback hairy crabs *(Pilumnus sayi)* are covered with messy hairs, which conceal prominent spines on the carapace sides and claws.

Tidal spray crab, max 1.7 in (4.4 cm) body width

Tidal spray crabs *(Plagusia depressa)* have broad legs and a hexagonal carapace dotted with reddish bumps. The claws have purple stripes.

Squareback marsh crabs *(Armases cinereum)* are mottled dark brown to olive. They have widely set eyes with no spine behind the eye socket.

Mottled shore crabs *(Pachygrapsus transversus)* are similar to squareback crabs, but have larger claws, each with a spiny "biceps" joint (merus) and pinkish fingers. The carapace is mottled green with dark lines in front and spines behind the eyes.

Squareback marsh crab, 1.1 in (2.8 cm) body width

HABITAT: Hairy crabs live in shallow rubble. Tidal spray and mottled shore crabs cling to shoreline debris and rock jetties at the water line. Squarback marsh crabs favor intertidal rubble.

DID YOU KNOW? Tidal spray crabs have a pantropical distribution thanks to their habit of rafting on floating debris.

Mottled shore crab, max 1.7 in (4.4 cm) body width

173

Gulfweed crab, max 0.8 in (2.0 cm) body width

Male blue land crab, max 4.5 in (11 cm) body width

Female blue land crab with eggs

Black-backed land crab, max 3.5 in (9 cm) body width

Grapsoid Crabs
(Gulfweed, Blue Land, Black-backed Land)

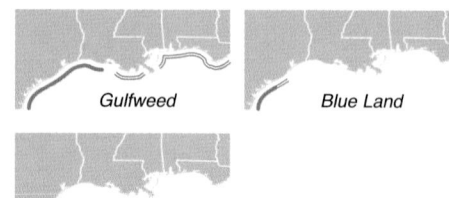

Gulfweed   Blue Land

Black-backed Land

RELATIVES: These are grapsoid crabs. Gulfweed crabs share the family Grapsidae with mottled shore crabs, and land crabs are allied within the family Gecarcinidae.

IDENTIFYING FEATURES:

Gulfweed crabs *(Planes minutus)* have a circular, untoothed carapace and are variably colored orange, brown, or pale, with patches of white ranging small to large.

Blue land crabs *(Cardisoma guanhumi)* move on land and have one claw (left or right) larger than the other. **Adult males** have an extra large claw and are blueish. **Females** are blue and orange, or pale gray.

Black-backed land crabs *(Gecarcinus lateralis)* are smaller than blue land crabs and have a black patch covering most of their upper carapace. Their remaining upper surface and legs are orange.

HABITAT: Gulfweed crabs cling to pelagic sargassum and floating debris. Land crabs live in burrows near coastal lagoons and migrate to beaches to breed.

DID YOU KNOW? Gulfweed crabs alter their colors to blend into the background. Most look like golden sargassum, and many mimic the algae's spattering of tiny white tube worms. A crab on floating white plastic will turn mostly white. Land crabs eat leaves and berries. They are mostly nocturnal, and retreat to their burrows during the day. They breathe with internal gills that must stay wet from water gathered anywhere they can find it.

174

Fiddler Crabs

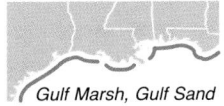

Gulf Marsh, Gulf Sand   *Atlantic Sand*

RELATIVES: Fiddlers are ocypodoid crabs that share the family Ocypodidae with ghost crabs.

IDENTIFYING FEATURES: These crabs live in colonies and seldom stray far from their burrow. Male fiddlers have an over-size claw, either right or left, for combat and mating rituals. Female claws are roughly equal in size.

Gulf marsh fiddlers *(Minuca longisignalis)* have a mottled olive carapace and are commonly fronted in bright turquoise. Their walking legs are densely bristled. The male's large claw is orange and gray, and is used to "beckon" females in repeated, long (5-sec), jerky, robotic waves.

Gulf sand fiddlers *(Leptuca panacea)* have a slate gray to pale gray carapace with varied symmetrical patches. The male's large claw is purplish in front, and its palm side is smooth. Males wave repeatedly at about one wave per second.

Atlantic sand fiddlers *(Leptuca pugilator)*, 0.8 in (2.0 cm), are similar to Gulf sand fiddlers, but have a yellowish or white carapace with dark or purple symmetrical patches and brown or orange edge-highlights. The male's large claw has yellowish fingers and is pale purple and orange at the base. His wave is typically one every six seconds.

HABITAT: Fiddler crabs burrow in intertidal muddy sand.

DID YOU KNOW? Fiddlers glean bacteria, fungi, and algae from sediments. In addition to waving at females, males also "rap" with their claw, sending sounds unique to each species. Attracted females mate in the male's burrow.

Male Gulf marsh fiddler, 1.1 in (2.7 cm) body width

Male Gulf sand fiddler, max 0.9 in (2.3 cm) body width

Atlantic sand fiddlers, females have equal claw sizes

175

Female ghost crab with eggs, 2 in (5 cm) body width

Juvenile ghost crab, on Florida Panhandle sands

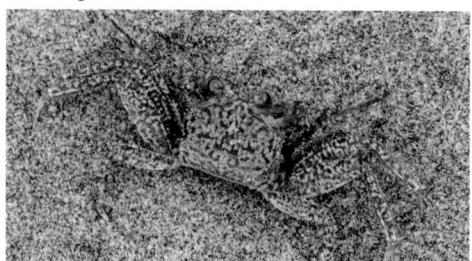

Juvenile ghost crab, on a southern Texas beach

Ghost crab tracks and burrow

A well-planned burrow ensures several herring meals

Ghost Crab

RELATIVES: Ghost crabs share the family Ocypodidae with fiddler crabs.

IDENTIFYING FEATURES:

Atlantic ghost crabs (*Ocypode quadrata*) are lightweight, swift-footed crabs with pale, square bodies and yellowish legs. **Juveniles** are a mottled, sand color. **Gravid females** carry their dark egg masses to the swash where the eggs are released. **Ghost crab tracks** appear as six rows of widely spaced commas. **Burrows** from juvenile ghost crabs are finger-size, and those from the largest adult burrows might accommodate a tennis ball. Upper beach burrows have spoil-mound aprons, often with radiating tracks from the rolling of excavated sand balls. Tides that have washed over burrows leave only a hole.

HABITAT: Adult ghost crabs use the entire beach and dune, and young ghost crabs prefer the lower beach. Favorite burrow locations include moist sand away from waves; next to driftwood, dead fish, and other groceries; and sand tilled by nesting sea turtles. Over-wintering crabs have burrows behind the dune. Burrows from males are closer to the swash zone, which puts them on the route that potential mates take as they moisten their gills.

DID YOU KNOW? Ghost crabs eat a wide variety of stranded items, including plants, insects, and fish. They also prey on mole crabs, clams, and sea turtle eggs. These mostly nocturnal crabs keep their internal gills moist by making regular trips to the water. The burrows they hide in during the day are up to 4 ft (1.2 m) deep. At mid-day and over winter, crabs seal themselves in behind a plug of sand.

Horseshoe Crab

RELATIVES: These arthropods are not crustaceans, but are chelicerates, more closely related to spiders than to crabs. Horseshoe crabs are in the class Merostomata, and are the closest living relatives to the ancient trilobites.

Female horseshoe crab

IDENTIFYING FEATURES:

Horseshoe crab (*Limulus polyphemus*) males reach 8.3 in (21.3 cm) width, and females reach 13.8 in (35 cm). Both sexes have a domed, U-shape head (cephalothorax), a spine-edged abdomen, and a stiff, pointed, **tail spine (telson)**. Live animals are chestnut brown or olive. A **molted exoskeleton** is a lightweight, tan, complete facsimile of the animal that shed it.

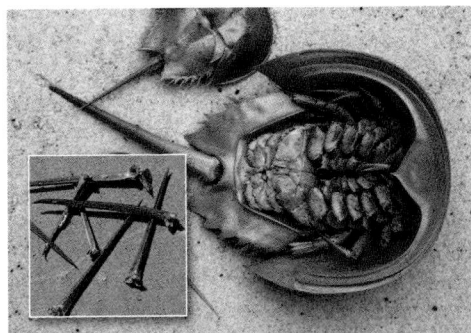

Molts. Separated tail spines (inset)

HABITAT: Horseshoe crabs live in coastal lagoons and bays. Their habitats vary between sand, mud, rubble, and seagrass. Adults mate and nestle eggs into the upper swash zone of low-energy beaches. In winter, horseshoe crabs move into deeper waters to avoid rough seas and the cold.

A male clings to a larger nesting female at low tide

DID YOU KNOW? Breeding occurs during the highest tides of summer. Females emerge with males attached. Both attached males and interloping "satellite" males fertilize eggs the female lays just under the sand. Nesting sites are typically protected beaches. Horseshoe crabs nearly identical to those seen today preceded the dinosaurs by over a hundred million years. Although they look a little creepy, horseshoe crabs are gentle and harmless. Females take a decade or more to mature and live 20–40 years. Intense harvest for eel and whelk bait caused severe declines in the Gulf horseshoe crab population, but harvest is now regulated and many **nesting areas** are protected.

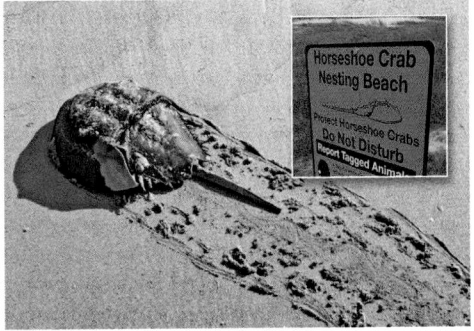

A horseshoe crab makes tracks. Nesting sign (inset)

Seaside dragonlet, 1.4 in (3.5 cm) body length, male

Seaside dragonlet, female

Carolina saddlebags, max 2.1 in (5.3 cm) body length

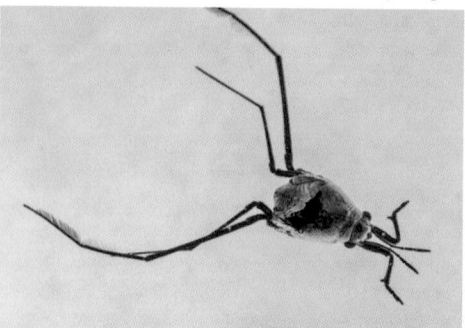

Common sea skater, max 0.2 in (0.5 cm) body length

Dragonflies and Sea Skater

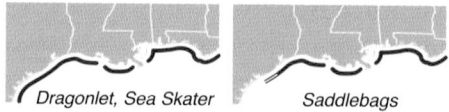

Dragonlet, Sea Skater   *Saddlebags*

RELATIVES: Insects share the arthropod subphylum Hexapoda. Dragonflies are in the order Odonata, and sea skaters are true bugs in the order Hemiptera.

IDENTIFYING FEATURES:

Seaside dragonlets (*Erythrodiplax berenice*) are clear-winged dragonflies with deep red eyes. **Males** are indigo and **females** are orange with black. Immature sexes have an indigo thorax and orange tail (abdomen).

Carolina saddlebags (*Tramea carolina*) are dragonflies with clear wings except for a conspicuous red patch at the base of each hind wing. The body is dark reddish in males and paler in females.

Common sea skaters (*Halobates micans*) are wingless, long-legged, silvery blue bugs that strand in the wrack.

HABITAT: Seaside dragonlets live as nymphs in marshy estuaries and as adults are found in many coastal habitats. Carolina saddlebags nymphs live in fresh water, but adults migrate along coasts. Sea skaters live on the surface of the open sea.

DID YOU KNOW? Dragonflies hunt flying insects. Sea skaters hunt floating zooplankton on foot and can stride the water's surface tension at more than a yard per second. Of the million or so insect species known, sea skaters are the only ones living exclusively in the open sea. Their pale cerulean color comes from a dense layer of hairs that create a light-interference effect. But their function is to prevent the insect from being wetted when accidentally submerged. If momentarily dunked, a sea skater carries an air layer held by the hydrophobic coat of hair, giving the bug buoyancy to quickly surface.

Cricket, Grasshopper, and **Rove Beetles**

Southern mole cricket, max 1.3 in (3.2 cm)

RELATIVES: Crickets (family Gryllotalpidae) and grasshoppers (Acrididae) are insects in the order Orthoptera. Rove beetles (family Staphylinidae) are related to other beetles in the order Coleoptera.

IDENTIFYING FEATURES:

Southern mole crickets (*Neoscapteriscus borellii*) have broad front legs with large digging "claws," and hind legs that are not enlarged for jumping. Their **burrows** are in the dune and upper beach.

Southern mole cricket burrow, 0.7 in (1.8 cm) wide

Seaside grasshoppers (*Trimerotropis maritima*) are speckled to blend in with the dune sand where they live. When moving, they show reddish hind shins (tibiae) and yellow wings crossed by a broad black band. If disturbed, this grasshopper will take flight and crepitate with a continuous harsh buzz.

Rove beetles in beach wrack tend to be tiny and brown, and do not resemble beetles. They have shortened elytra (wing coverings) and slink quickly beneath the sand grains. **Littoral rove beetles** (*Bledius punctatissimus*) have short pearly elytra that cover a third of their dark abdomen. Their locally abundant **surface trails** are conspicuous on wet sand.

Seaside grasshopper, max 1.7 in (4.3 cm)

HABITAT: These crickets and grasshoppers live in the dune. Rove beetles live in the upper intertidal zone.

DID YOU KNOW? Littoral rove beetles feed at night on diatoms (single-celled algae) washed in from the sea. The beetles often live in assemblages of over 2,000 adults per square yard. They remain buried in fine sands during high tide and breathe air trapped between sand grains.

The tiny littoral rove beetle, max 0.1 in (3 mm)

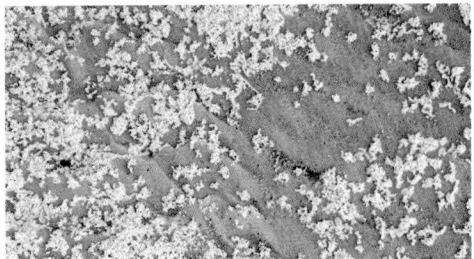

Surface trails from dozens of littoral rove beetles

179

Eastern beach tiger beetle, max 0.5 in (1.3 cm)

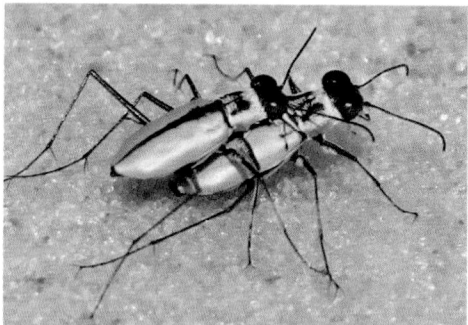

Mating pair of eastern beach tiger beetles

S-banded tiger beetle, max 0.5 in (1.3 cm)

Coastal tiger beetle, max 0.5 in (1.3 cm)

Tiger Beetles

RELATIVES: Beetles are insects in the order Coleoptera. Tiger beetles share the family Carabidae with other ground beetles.

IDENTIFYING FEATURES: When not flying, beetles have their wings hidden by hard wing-covers (elytra). Tiger beetles have long legs and large eyes. They make short, awkward flights and are among few beach insects active in daylight.

Eastern beach tiger beetles (*Habrosceli-morpha dorsalis*) have pearly elytra with faint or distinct, dark scrawled lines, which are more pronounced west of the Mississippi Delta.

S-banded tiger beetles (*Cicindela trifasciata*) are dark gray-brown to light brown, sometimes with a greenish tint, and have a light S-shape mark on each elytron.

Coastal tiger beetles (*Ellipsoptera hamata*) are dark and tan with a complex symmetrical pattern of connected hooks that look like a Rorschach ink blot.

HABITAT: Eastern beach tiger beetles hunt the wrack line. Their larvae ambush prey from tiny burrows between the highest tide and the dune. S-banded and coastal tiger beetles are most common on sandy marsh bars and mudflats.

DID YOU KNOW? There are two subspecies of eastern beach tiger beetles in the Gulf, east (*H.d. saulcyi*) and west (*H.d. venusta*) of the Mississippi. Glaring lights, which attract and kill adults, and vehicle traffic, which destroys larval burrows, have eliminated tiger beetles from many beaches. Adults eat beachhoppers, flies, and tiny carrion bits. Some tiger beetles sprint at 125 body lengths per second. If you could do that, you would be running at more than 500 miles per hour.

Biting Flies *(Mosquitoes, Midges, Stable Fly, Deer Flies, and Horse Flies)*

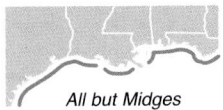
All but Midges

Biting Midges

RELATIVES: These insects are flies in the order Diptera. Mosquitoes (family Culicidae) are more closely allied with biting midges (Ceratopogonidae) than to stable flies (Muscidae) or to deer and horse flies (Tabanidae).

IDENTIFYING FEATURES: Flies have one pair of wings (other insects have two pairs). Each of these species has a taste for blood and is willing to hurt you to get it.

Saltmarsh mosquitoes (*Ochlerotatus* spp.), 1/4 in (6 mm), have long, striped legs and are either golden-brown *(O. sollicitans)* or black *(O. taeniorhynchus).*

Biting midges (*Culicoides* spp.), also known as sandflies or no-see'ums, are tiny and gray with black eyes.

Stable (dog) flies (*Stomoxys calcitrans*) look like a pale housefly (they are in the same family). These biting flies have a light spot behind their eyes and several light patches on the abdomen.

Deer flies (*Chrysops* spp.) have separated eyes and are about a half inch (1.3 cm) long with dark and yellow-green markings and dark-banded wings.

Saltmarsh horse flies (*Tabanus lineola*), 0.8 in (2 cm), have close-set eyes, clear wings, and a striped abdomen. Other horse fly species (*Tabanus* spp.) may have bright green eyes.

HABITAT: All these flies have aquatic or semi-aquatic larvae that develop along estuarine shorelines. Stable flies also lay eggs in livestock manure.

DID YOU KNOW? Good news—frequent mosquito bites can offer resistance to the injected saliva that causes itching.

Saltmarsh mosquito, Ochlerotatus sollicitans

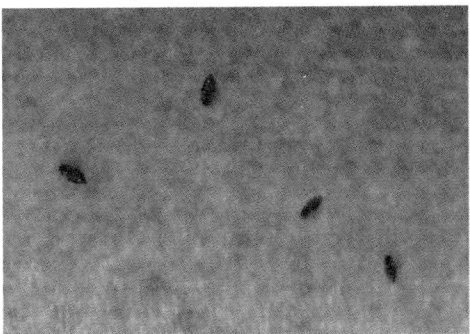
Biting midges or no-see'ums, max 0.1 in (3 mm)

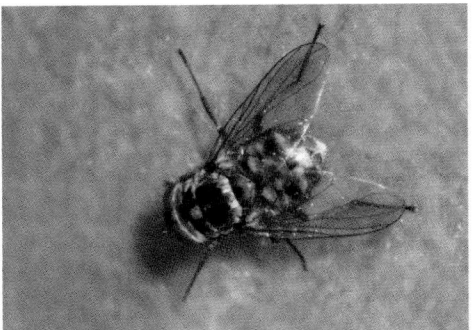
Stable fly (dog fly), max 0.25 in (6 mm)

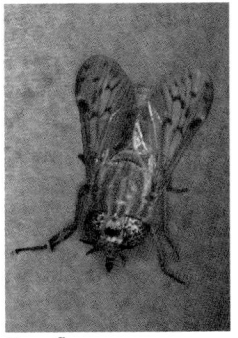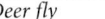
Deer fly

Saltmarsh horse fly

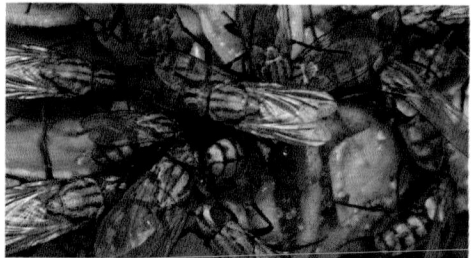

Flesh flies on a dead ghost crab

Beach robber fly, max 0.5 in (1.3 cm)

Bearded robber fly, max 1 in (2.5 cm)

Seaweed fly, max 0.25 in, (0.6 cm) on a crab carapace

Brine flies on a sandy Texas mudflat

Flies *(Flesh, Robber, Seaweed, Brine)*

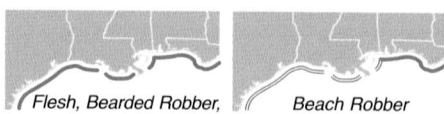

Flesh, Bearded Robber,
Seaweed, Brine

Beach Robber

RELATIVES: Flesh flies (family Sarcophagidae), robber flies (Asilidae), seaweed flies (Anthomyiidae), and brine flies (Ephydridae) are true flies (order Diptera).

IDENTIFYING FEATURES:

Flesh flies (*Sarcophaga* spp.), 1/4 in (6 mm), have red eyes and are black and gray with longitudinal stripes on the thorax.

Beach robber flies *(Laphystia litoralis)* have a plump, banded abdomen and dark green eyes.

Bearded robber flies (*Efferia* spp.) are large with brown eyes, a thin abdomen, and legs of orange and black. A close look reveals the dense head bristles for which the insect was named.

Seaweed flies (*Fucellia* spp.) look like thin house flies with broadly separated, sepia eyes, and smoky tinted wings.

Brine (shore) flies, 0.3 in (0.8 cm), have a smooth, gray body and small eyes astride a bulging forehead. The bottom right image shows a southern Texas species, **Lipochaeta slossonae.**

HABITAT: Flesh fly eggs hatch within the female fly, which bears maggots directly on the carrion food source. Robber flies patrol the upper beach and dune for prey. Seaweed flies are in the wrack, and brine flies feed on mudflats.

DID YOU KNOW? Robber flies capture arthropods many times their own size, including wasps and spiders. These flies seize their prey in the air, clutch it with powerful legs, and jab it with pointed mouthparts, which inject a neurotoxin with proteolytic (flesh-desolving) enzymes. Swarms of seaweed and brine flies provide important food for coastal birds.

Sand Wasp and Butterflies

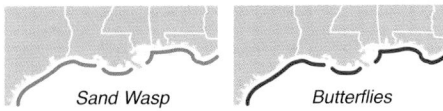

Sand Wasp   Butterflies

RELATIVES: Sand wasps (family Crabro-
nidae) are in the insect order Hymenop-
tera with ants and bees. Butterflies are in
the order Lepidoptera. Gulf fritillaries and
white peacocks are brushfoot butterflies in
the family Nymphalidae. Southern whites
are related to sulphur butterflies in the
family Pieridae.

IDENTIFYING FEATURES:

Sand wasps (*Microbembex monodonta*)
are yellow-green with black markings.
They have green eyes and a triangular
upper lip (labrum) that looks like a beak.

Gulf fritillaries (*Agraulis vanillae*) are
orange with black and silver highlights.

White peacocks (*Anartia jatrophae*) have
pale wings with a double row of light-
brown, edge crescents, and six black spots
(two on hind wings are hidden in image).

Great southern whites (*Ascia monuste*)
are either white (males) or gray (females)
with black forewing tips.

HABITAT: Solitary sand wasps can be
seen in the dune, hovering, then disappearing into their burrows. Butterflies
flutter by in the dune where flowers provide nectar sources. These butterflies have
larval host plants in coastal areas—passion
vines (Gulf fritillary); frog fruit (white
peacock); and pepper grass, saltwort, and
searocket (great southern white).

DID YOU KNOW? Sand wasps scavenge
dead and dying insects to provision burrows for their developing young, which
emerge as adults in summer. The highest
densities of Gulf fritillaries occur in fall,
when the butterflies migrate southeast to
overwinter in southern Florida. Great
southern whites move in mass, one-way
migrations during spring and fall.

Sand wasp, max 0.5 in (1.3 cm) length

Gulf fritillary, wingspan max 3 in (7.5 cm)

White peacock, wingspan max 2.7 in (7 cm)

Great southern white, wingspan max 2.7 in (7 cm)

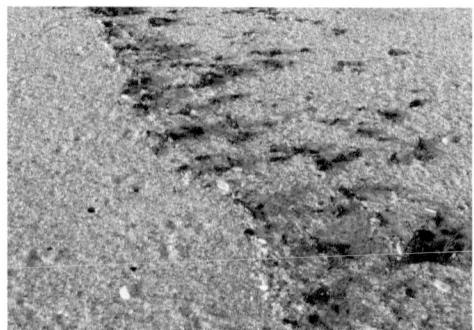

Strong offshore winds account for mass drownings

A winged fire ant among other drowned insects

A drowned dragonfly (green darner, Anax junius)

Many Gulf fritillary butterflies don't survive migration

Drowned Insects

IDENTIFYING FEATURES: Occasionally, lots of dead insects wash in with the tide. These **mass-drowning** events commonly involve **red imported fire ants** (*Solenopsis invicta,* which have wings during reproductive flights), honeybees (*Apis mellifera),* and love bugs (*Plecia nearctica,* flies in the family Bibionidae). Some individual insects found include waterbeetles (order Coleoptera, family Dytiscidae), **dragonflies** (order Odonata), and butterflies, especially the monarch (*Danaus plexippus),* and **Gulf fritillary** (*Agraulis vanillae).*

Drowned insects are found after strong winds have blown offshore, or after heavy rains. Fire ants are found when these winds follow rains that prompt winged sexuals (alates) to fly. Dragonflies and butterflies pelted by raindrops are often forced onto the sea.

HABITAT: Nearly all these insects live in terrestrial or freshwater habitats away from the beach. They reach the sea as lost members of the aerial plankton.

DID YOU KNOW? Insects were the first animals to take flight and are still the only invertebrates to truly fly. It may not be a coincidence that the insects most involved in mass drownings are not native to North America (for example, fire ants, honeybees, and lovebugs). Insects familiar with (that evolved in) the local neighborhood are believed to hug the coast rather than disperse seaward when offshore winds blow. This coast-hugging explains the clustering of migrating insects along beaches, including the unpleasant massing of biting stable flies occasionally seen on Florida Panhandle beaches.

Sea Stars (Starfish)
(Lined, Banded, Two-spined)

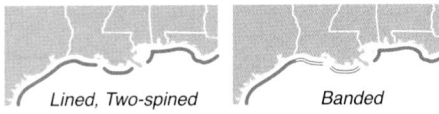

Lined, Two-spined   Banded

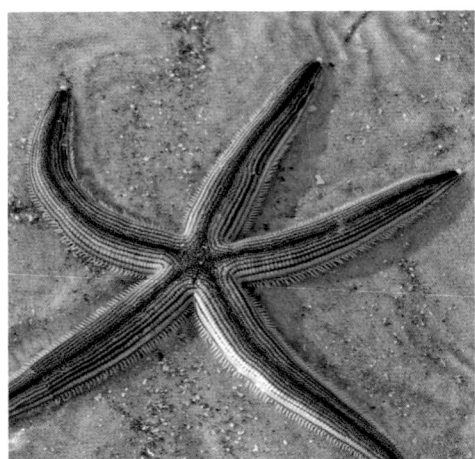

Lined sea star, max 6 in (15 cm)

RELATIVES: Sea stars are echinoderms (phylum Echinodermata) in the class Asteroidea. Lined and banded sea stars (family Luidiidae) share the order Paxillosida with two-spined sea stars (Astropectinidae).

IDENTIFYING FEATURES:

Lined sea stars *(Luidia clathrata)* are grayish, brownish, or salmon, with a dark stripe down each of their five arms.

Banded sea stars *(Luidia alternata)* are cream with bands of dark green, purple, or brown. Their underside is lined with orange tube feet, and their arms are fringed with slender spines.

Two-spined sea stars *(Astropecten duplicatus)* range pale gray to reddish-brown, are relatively flat, and have tapered arms lined with marginal plates. Plates where the arms join the central disc have short, blunt spines. The undersides of marginal plates have abundant, flat **spines** that project away from the arm.

Banded sea star, max 8 in (21 cm)

HABITAT: All live in sandy shallows or seagrass but are common near the swash zone after storms and at low tide.

DID YOU KNOW? Like all echinoderms, sea stars have an internal skeleton of calcium carbonate. Their movement is aided by a water-vascular system that provides hydraulic pressure for their many tube feet. These sea stars feed mostly on small bivalves. Unlike many other sea stars, these species swallow their prey whole, including the shell, which is ejected from the mouth after digestion. Should one of these sea stars lose an arm, the missing appendage will regenerate in about three months.

Two-spined sea star, max 8 in (21 cm). Spines (inset)

Royal sea star, max 6 in (15 cm). Purple form (inset)

Common sea star, max 6 in (15 cm). Juvenile (inset)

Small-spine sea star, max 6 in (15 cm)

Sea Stars *(Royal, Common, Small-spine)*

Royal   Common, Small-spine

RELATIVES: Royal sea stars (order Paxillosida, family Astropectinidae), common sea stars (order Forcipulatida, Asteriidae), and small-spine sea stars (order Spinulosida, Echinasteridae) are remotely related.

IDENTIFYING FEATURES:

Royal (flat) sea stars *(Astropecten articulatus),* are similar in shape to two-spined sea stars (previous page), but are purplish blue and lined with marginal plates that are often bright orange (but sometimes purple). The plates where each arm joins the central disc have no upper spines.

Common sea stars *(Asterias forbesi)* have stout, inflated arms and range in color from yellow-orange to purple. They are textured with pale spines and have a bright orange madreporite (filter for water input to the vascular system) above one armpit.

Small-spine (shiny) sea stars *(Echinaster spinulosus)* are red-orange to purple with stubby arms and widely spaced spines on the upper surface.

HABITAT: All live in sandy shallows near the beach, but common and small-spine sea stars also occupy jetties, pilings and other rubble.

DID YOU KNOW? Common sea stars eat large bivalves using their hydraulic tube feet to pull apart the closed valves. The sea star can do this without fatigue, outlasting the endurance of the clam's closing (adductor) muscles. When the bivalve surrenders, the sea star pokes its stomach into the open shell and digests its prey. Small-spine sea stars are both predator and scavenger, feeding on sessile invertebrates such as sponges and tunicates.

Brittle Stars

RELATIVES: Brittle stars (class Ophiuroidea) are only remotely related to sea stars. Elegant brittle stars (family Ophiolepididae), long-armed burrowing brittle stars (Amphiuridae), and blood brittle stars (Ophiactidae) share the order Amphilepidida.

IDENTIFYING FEATURES:

Elegant brittle stars (*Ophiolepis elegans*) have five serpentine arms attached to a pentagon-shape disc with non-overlapping scales in a rosette pattern. They are green or gray with darker arm bands.

Long-armed burrowing brittle stars have a finely scaled disc to about 0.2 in (0.5 cm) wide with prominent radial shields (paired scales on disc at each arm). Their arms are extremely long (to 14 times disc width) and flexible with projecting spines. Species include **Ophiophragmus spp.** and **Amphiodia spp.**

Blood brittle stars (*Hemipholis elongata*) have five arms (to 10 times disc width) with small spines. Their tube feet are red. The circular disc is up to 0.4 in (1.0 cm) wide and covered with small overlapping scales radiating from a central scale.

HABITAT: These brittle stars live in sandy shallows and rubble near the beach. Blood brittle stars often burrow into subtidal sediments where plumed tube worms (p. 154) live. All prefer Gulf waters that vary little in salinity.

DID YOU KNOW? Blood brittle stars wave their arms to capture zooplankton with sticky **tube feet.** The feet are red because of hemoglobin-containing cells within their water vascular system. The hemoglobin helps them respire beneath oxygen-poor sediments.

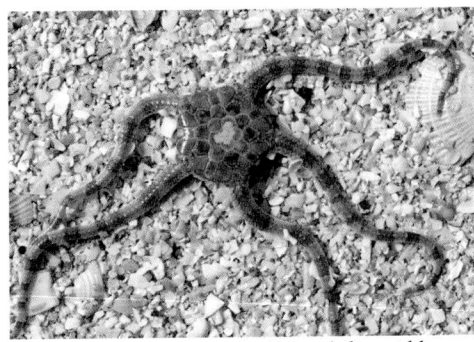

Elegant brittle star, max 1 in (2.5 cm) disc width

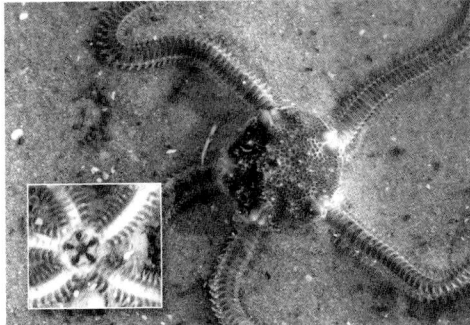

Ophiophragmus, *disc underside (inset)*

Amphiodia, *long-armed burrowing brittle star*

Blood brittle star, red tube feet (inset)

187

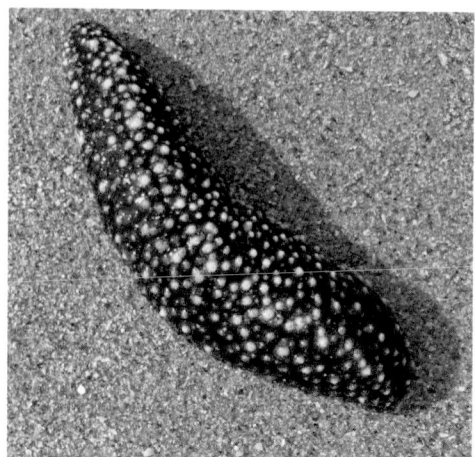

Noble sea cucumber, max 12 in (30 cm)

Noble sea cucumber showing unbranched tentacles

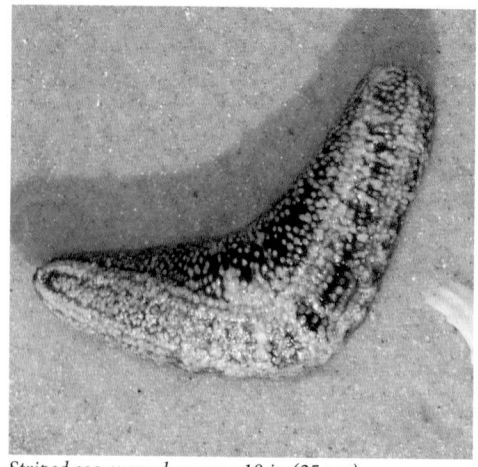

Striped sea cucumber, max 10 in (25 cm)

Sea Cucumbers

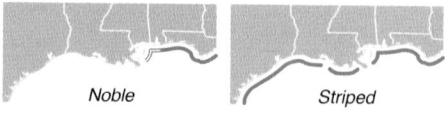

Noble   Striped

RELATIVES: Sea cucumbers are echinoderms in the class Holothuroidea, and are more closely related to sea urchins than to sea stars.

IDENTIFYING FEATURES: Shapes range from straight or bent cylinders, to distorted footballs in varied states of inflation. Colors and texture are pronounced in fresh specimens but pale and absent in beachworn animals.

Noble sea cucumbers (*Holothuria princeps*) are brownish with conspicuous light warts. They have a slightly curved shape, a dorsal covering of papillae, and ventral rows of tube feet. Their mouth is surrounded by 20 short, leaf-shape **tentacles**.

Striped (green) sea cucumbers (*Thyonella gemmata*) are gray or green with their tube feet in five relatively organized rows (called radii). They often retain the boomerang shape they had while in their burrows. Their mouth tentacles are treelike. A smaller sea cucumber in the same family, (*Allothyone mexicana*), has tube feet all over its body and has indistinct radii.

HABITAT: The sea cucumbers here live within burrows in shallow sandy areas and are swept onto beaches following rough weather or harmful algal blooms.

DID YOU KNOW? Sea cucumbers gather plankton using the retractable tentacles circling their mouth. These animals are notorious for self-eviscerating. When sufficiently stressed, they voluntarily expel their digestive, respiratory, circulatory, and reproductive organs. This drastic move may function to fully occupy a potential predator, or maybe just disgust them. The organs of the sea cucumber grow back in several weeks.

Sea Urchins and Sand Dollar

RELATIVES: Short-spined urchins (order Camarodonta), purple sea urchins (order Arbacioida), and keyhole urchins (order Clypeasteroida) are echinoderms in the class Echinoidea.

IDENTIFYING FEATURES: Sea urchins have a circular body supported by skeletal plates that bear many movable spines.

Short-spined (variegated) urchins (*Lytechinus variegatus*), 4 in (10 cm), are whitish, greenish, brown, or mauve, with relatively blunt, short, tubular spines that are lighter at their base.

Purple sea urchins (*Arbacia punctulata*) have medium-long, purplish spines that are sharp cones above and flattened paddles underneath.

Five-holed keyhole urchins (sand dollars) (*Mellita quinquiesperforata*), 4 in (10 cm), have five slotlike holes and a coating of fine, brown, movable spines.

HABITAT: These sea urchins live in shallow coastal waters. Sand dollars are locally abundant near beaches and can be seen slowly escaping low tide.

DID YOU KNOW? Short-spined urchins often hold shells and other debris using their suckered tube feet. They are important grazers on seagrasses. Purple sea urchins graze on sponges and algae using strong beaklike teeth. The slots in a sand dollar provide shortcuts for food bits traveling to the mouth from the animal's topsides. The holes also allow these disclike animals to sink into the sand. The conspicuous petals on the sand dollar's upper surface (ambulacra) are traced by dual lines of pores for their tube feet, which the animal uses only for breathing.

A short-spined urchin adorned with shell bits

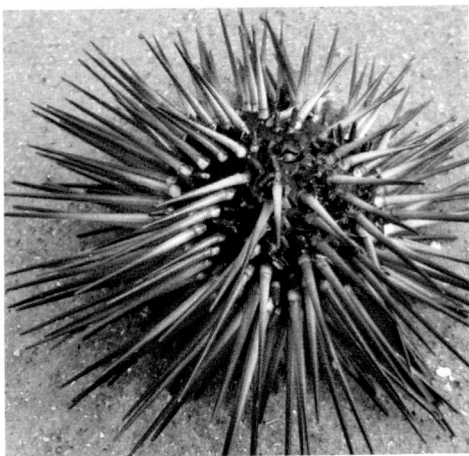

Purple sea urchin, max 4 in (10 cm)

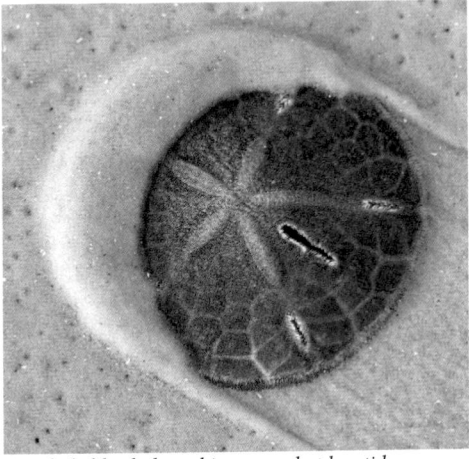

Five-holed keyhole urchin exposed at low tide

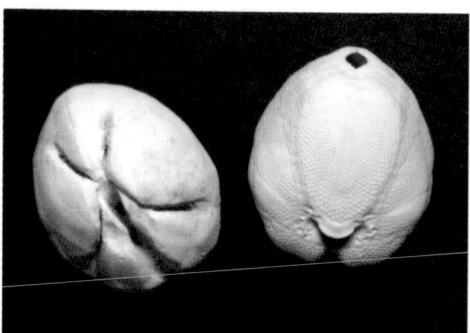
Heart urchin tests, aboral (L) and oral (R) sides

Rock-boring urchin tests, aboral (L) and oral (R) sides

Purple sea urchin tests, aboral view, and spines

Short-spined urchin tests and spines, aboral (L), oral (R)

Sea Urchin Tests

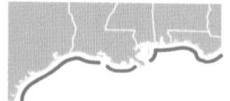

IDENTIFYING FEATURES:

A sea urchin's round, inner shell is called a **test**. In life, the test was covered with pincers (pedicellariae), tube feet, and **spines** that moved on ball-and-socket joints. The test's top (aboral) side is domed, and the bottom (oral) side is flat.

Heart (mud) urchin tests (*Moira atropos*), 2.5 in (6 cm), are an egg shape with five radiating grooves. Most beached tests are bone white. In life, mud urchins are covered with short, delicate, tan spines.

Rock-boring urchin tests (*Echinometra lucunter*), 2.5 in (6.3 cm), are oval and white. Fresh tests may have brownish highlights. Their thick, purple-brown spines are about an inch (2.5 cm) long.

Purple sea urchin tests (*Arbacia punctulata*), 1.5 in (3.8 cm), are whitish with purple highlights.

Short-spined urchin tests (*Lytechinus variegatus*), 3 in (7.5 cm), are greenish when fresh and bone-white when sun-bleached.

HABITAT: Heart urchins live in offshore muddy sediment. Rock-boring urchins live in shallow, rocky rubble and use their teeth to bore into reefs. See previous page for purple and short-spined urchins.

DID YOU KNOW? Sea urchin tests are made of 10 fused plates scattered with tiny holes for the urchin's tube feet. Each bump (tubercle) on the test is a former spine attachment point. There seems to be an immutable law that urchin tests can survive the pounding surf but are crushed into dust in a beachcomber's pocket. If you can get an urchin test home safely, it will remain as pictured for decades. Sea urchins have been on Earth since the upper Ordovician period, 450 million years ago.

Sand Dollar Tests

RELATIVES: Sand dollars (keyhole urchins) are echinoderms in the class Echinoidea.

IDENTIFYING FEATURES:

Notched (arrowhead) sand dollar tests (*Encope michelini*), 5.5 in (14 cm), have five marginal notches, one at the tip of each petal (ambulacrum) and one hole between petals in an interambulacrum. The edge with the corresponding hole is the straightest.

Five-holed keyhole urchin tests (*Mellita quinquiesperforata*), 4 in (10 cm), are mostly round with five slotlike holes—four at the tips of petals (ambulacra) and one between petals.

Sand dollar bits are broken pieces of sand dollar tests (skeletons), and look like worn, membranous, bone shards. Beachcombers are likely to find hundreds of these bits for each whole sand dollar.

HABITAT: These sand dollar species live just off Gulf beaches. The notched sand dollar occurs mostly in the eastern Gulf. Other sand dollar species live farther offshore, and are unlikely to reach the beach in recognizable pieces.

DID YOU KNOW? A sand dollar transfers food bits to the mouth on its underside. Within the mouth, the bits are crunched by a chewing apparatus made of 40 skeletal pieces, including five bird-shape teeth. Collectively, the apparatus is called **Aristotle's lantern**. The light-but-strong, double-curved shell construction of sand dollars, braced by complex spacing of internal pillars, has inspired architectural designs for large concrete structures.

Notched sand dollars with Aristotle's lantern pieces

Five-holed keyhole urchin tests

Bits of surfworn sand dollar tests

191

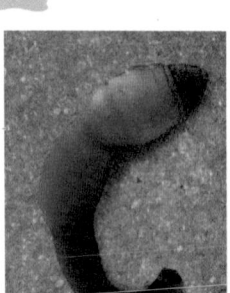 

Golden acorn worm (L), fecal cast (R)

A freshly stranded sea pork colony

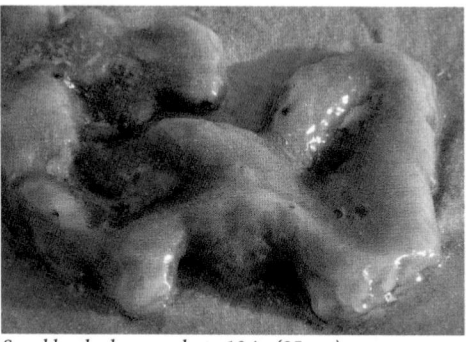

Sun-bleached sea pork, to 10 in (25 cm)

Sea liver tunicates, to 10 in (25 cm)

Acorn Worm and Tunicates
(Sea Pork, Sea Liver)

Acorn Worm   Sea Pork, Sea Liver

RELATIVES: Acorn worms (not really worms) are in the phylum Hemichordata, remotely related to these colonial tunicates (class Ascidiacea, order Aplousobranchia), which are chordates (phylum Chordata), a group including fishes, birds, and us.

IDENTIFYING FEATURES:

Golden acorn worms (*Balanoglossus aurantiacus*) are evident from their finger-width, sandy poops (fecal casts) that lie in coiled piles. The burrowing animals range in size from several inches to 3.5 ft (1.1 m), and are amber, wormlike creatures with an acorn-shape proboscis.

Sea pork (*Aplidium stellatum*) is a colony of tunicate animals (zooids) within a rubbery tunic. Displaced colonies are orange or pink when fresh, and become yellowish or gray after stranding. A fresh colony is dotted with embedded circles of 6–20, orange zooids.

Sea liver (*Eudistoma hepaticum*) colonies are similar to sea pork, but the rubbery tunic is softer and more wrinkled, with a purple color that fades to off-white.

HABITAT: Acorn worms live within intertidal, U-shape burrows on low-energy beaches. Sea pork grows on the bottom in seagrass, and sea liver commonly encrusts rocks near eastern Gulf inlets.

DID YOU KNOW? Acorn worms swallow sand just below the surface, creating a depression that accumulates bits of organic stuff. The animal takes in this food along with a lot of sand. The area surrounding an acorn worm burrow smells like medicinal iodine due to bromine compounds that may protect the animal's soft, naked body from infections.

Tunicates *(Sandy-skinned, Leathery)*

Sandy-skinned

Leathery

RELATIVES: These tunicates are chordates in the class Ascidiacea and the order Stolidobranchia.

IDENTIFYING FEATURES:

Sandy-skinned tunicates *(Molgula occidentalis)* look like soft potatoes rolled in sand. Their thin test is embedded with mud, sand, and shell bits, and they are often still attached to shells and rocks.

Sandy-skinned tunicates, max 2 in (5 cm)

Leathery (rough, pleated) sea squirts *(Styela plicata)*, each 4 in (10 cm), are firm, wrinkled lumps with a basal (formerly attached) end opposite a puckered end with 4 lobes around a siphon, which **squirts** when squeezed. They wash ashore in singles and in attached groups.

HABITAT: Sandy-skinned tunicates and leathery sea squirts live in shallow waters attached to rocks, docks, shells, and debris.

DID YOU KNOW? Although sea squirts may be barely recognizable as a living animal, they have a lot in common with humans, including gill slits, a rigid notochord, and a hollow nerve cord (traits we had during early development). Larval tunicates swim like tadpoles before settling into an attached existence. As adults, they are wrapped in tunics made of tough, fibrous cellulose.

A cluster of leathery sea squirts

All tunicates make a living by filtering particles from seawater. A plum-size sea squirt can filter bathtub-volumes of water each day, removing (and consuming) about 95 percent of the suspended bacteria. Tunicates are an acquired taste appreciated by tulip snails, stingrays, and Kemp's ridley sea turtles.

Sea squirts squirt when squeezed

193

Sharks enter the surf to feed on schooling fish

Mermaid's purse bundle, clearnose skate (inset)

Skate egg showing yolk and embryo (left of yolk)

Sharks and Skate Egg Case

RELATIVES: Fishes are vertebrates—chordates with a backbone. Sharks (order Carcharhiniformes), skates (Rajiformes), and rays are in the class Elasmobranchii with other cartilaginous fishes.

IDENTIFYING FEATURES:

Sharks in the surf zone include: Bonnethead sharks *(Sphyrna tiburo)*, 3.2 ft (1 m), with a shovel-shape head; Atlantic sharpnose sharks *(Rhizoprionodon terraenovae)*, 3.5 ft (1.1 m), with a relatively long snout; lemon sharks *(Negaprion brevirostris)*, 9 ft (2.7 m), with dorsal fins of similar size; and blacktip sharks *(Carcharhinus limbatus)*, to 8 ft (2.4 m), with black fin tips.

Skate egg cases (mermaid's purses), 3.5 in (9 cm), are commonly from the **clearnose skate** *(Raja eglanteria)*, which breeds December–May. The plasticlike egg cases are black with four tendrils. Fresh ones without exit-slits may have a spherical **yolk** or wiggling embryo.

HABITAT: Sharks are most common in deeper waters but will enter the surf when small fish are plentiful. Skates attach their egg cases, sometimes in clusters, to soft corals and other anchored objects.

DID YOU KNOW? Accidental deaths tabulated by the National Safety Council show a 1-in-4-million lifetime risk of succumbing to shark bite. Fireworks are 10 times riskier, lightning is 50 times more deadly, and sun/heat exposure kills about 300 times as many people. But just the same, exit the water when small fish abound. Although skating kills 30 times more people than sharks do, skates (the fish) are harmless. Their embryo pumps water through its egg case by beating its tail down one of the hollow tendrils.

Stingrays

RELATIVES: Stingrays are Elasmobranch fishes in the order Myliobatiformes.

IDENTIFYING FEATURES:

Bluntnose stingrays *(Hypanus say)*, 3 ft (0.9 m) width, have a disc with rounded corners and a blunt snout, which is shorter than the distance between the eyes. Their whiplike tail has dorsal and ventral finfolds and a hardened spine (barb) near the base. The harpoon-serrated barb is covered in venomous mucus.

Atlantic stingrays *(Hypanus sabinus)* resemble small bluntnose stingrays but have a pointed snout that extends longer than the distance between their eyes.

Southern stingrays *(Hypanus americanus)* are similar to bluntnose rays but have a diamond-shape disc and a tail with no dorsal finfold.

Cownose rays *(Rhinoptera bonasus)* have a batlike shape with sharp pectoral corners and a deep groove around their head below the eyes. Colors range from golden olive to brown, with no spots. They have one or two spines at the base of their cordlike tail.

HABITAT: Bluntnose, southern, and cownose rays live in the surf and offshore. Atlantic stingrays live mostly in estuaries.

DID YOU KNOW? Stingrays don't attack, but will thrust their spine in defense if stepped on. Injuries from stingray spines are painful. Avoid stings when wading by shuffling your feet to flush buried rays. Stingrays caught on hook and line can be handled like a bowling ball, with a thumb-and-finger grip by the spiracles (holes behind their eyes). Keep the ray's topside away, de-hook, and release.

Bluntnose stingray (image arrows indicate tail spine)

Atlantic stingray, max 15 in (38 cm) width

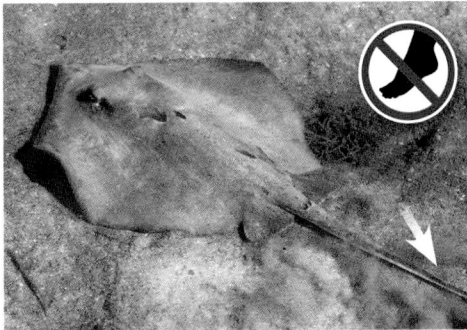

Southern stingray, 5 ft (150 cm) width. Males smaller

Cownose ray, max 45 in (110 cm) width

Bay anchovy, max 2.4 in (6 cm)

Male striped killifish, max 6 in (15 cm)

Pinfish, typically 4.5 in (11 cm)

Barred grunt, max 8 in (20 cm)

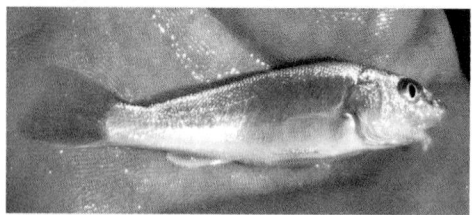

Juvenile whiting (Gulf kingfish), typically 4 in (10 cm)

Juvenile Florida pompano, typically 6 in (15 cm)

Swash Zone Fishes

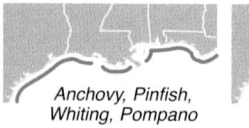

Anchovy, Pinfish,
Whiting, Pompano

Striped Killifish

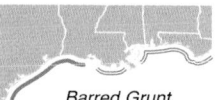

Barred Grunt

RELATIVES: These are bony fishes in the superclass Osteichthyes. Anchovies (order Clupeiformes, family Engraulidae) and killifishes (order Cyprinodontiformes, family Fundulidae) are only distantly related to the perciform fishes (order Perciformes), pinfish (family Sparidae), grunts (Haemulidae), Whiting (Sciaenidae), and pompano (Carangidae).

IDENTIFYING FEATURES:

Bay anchovies *(Anchoa mitchilli)* have a large gape and a silver stripe the width of their pupil. The similar, striped anchovy *(Anchoa hepsetus)* has a wider stripe almost the width of the entire eye.

Striped killifish *(Fundulus majalis)* are silvery gray with 10–15 black side bars (male) or dark longitudinal lines with bars near the tail (female).

Pinfish *(Lagodon rhomboides)* are silvery with yellow stripes and 5–6 vertical side bars. Its dorsal spines are stiff and sharp.

Barred grunts *(Conodon nobilis)* have yellow lines and eight dark bars on their upper sides.

Juvenile Whiting *(Menticirrhus littoralis)* are silvery with no bars, a low-slung mouth, and a single, stout chin barbel.

Juvenile Florida pompano *(Trachinotus carolinus)* are smooth and silvery. Their forked tail and anal fin are yellowish.

HABITAT: Shallow coastal waters

DID YOU KNOW? Small fish often risk stranding themselves to avoid predators that can't follow into shallow waters.

Surf Catches

RELATIVES: These surf catches are related to other bony fishes in the superclass Osteichthys. Catfishes (order Siluriformes, family Ariidae) are only distantly related to the perciform fishes (order Perciformes), bluefish (family Pomatomidae), silver perch (Sciaenidae), and the jacks, pompano and jack crevalle (Carangidae).

IDENTIFYING FEATURES:

Hardhead (sea) catfish *(Ariopsis felis)* have harpoonlike dorsal and pectoral spines, and medium-length whiskers (barbels). The related gafftopsail catfish *(Bagre marinus)* has long barbels and extended rays from its forward fins.

Bluefish *(Pomatomus saltatrix)* are blue-green above and light below with a large toothy mouth.

Florida pompano *(Trachinotus carolinus)* are smooth and silvery with elongated dorsal and anal fin rays.

Jack crevalle *(Caranx hippos)* are similar to pompano but have a larger eye and mouth, and a keeled caudal peduncle (tail-fin base).

Silver perch *(Bairdiella chrysoura)* are silver with yellow fins and no chin barbels. They have spines on their preopercle (forward cheek) and a sharp anal fin spine.

HABITAT: These are generally coastal fishes.

DID YOU KNOW? Surf-zone fishes often occupy the nearshore trough (p. 3) inside the sandbar. They prey on coquina clams (p. 136), mole crabs (p. 165), and small fish. Don't want that fish you caught? Release it alive! Feeding wading birds is bad for them, and littering the beach with carcasses is totally uncool.

Hardhead catfish, max 18 in (46 cm)

Bluefish, to 18 in (46 cm) near shore

Florida pompano, max 17 in (43 cm)

Jack Crevalle, to 18 in (46 cm) near shore

Silver perch, max 8 in (20 cm)

197

Spot croaker, max 10 in (25 cm)

Atlantic croaker, max 12 in (30 cm)

Whiting, max 12 in (30 cm)

Southern kingfish, max 16 in (41 cm)

Red drum, 39 in (100 cm)

Surf Catches

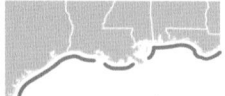

RELATIVES: These perch-like fishes (superclass Osteichthys, order Perciformes) are in the family of drums and croakers (Sciaenidae).

IDENTIFYING FEATURES:

Spot croakers *(Leiostomus xanthurus)* are silvery gray with oblique streaks on their upper sides and a dark spot behind the gills. Their tail fin is concave.

Atlantic croakers *(Micropogonias undulatus)* are golden to grayish with irregular dark streaks. They have 3–5 chin barbels and a convex tail fin.

Whiting (Gulf kingfish) *(Menticirrhus littoralis)* are silvery above and white below with a single chin barbel.

Southern (king croaker) kingfish *(Menticirrhus americanus)* have a single chin barbel, dusky side blotches, and dark-tipped pectoral fins.

Red drum (redfish) *(Sciaenops ocellatus)* are red-bronze above with one or more black eye-spots on the upper tail base.

HABITAT: Surf and coastal waters

DID YOU KNOW? These Sciaenids communicate with sound. Their voices vary by species, and are described as drumming, croaking, purring, knocking, or clucking, all at the low-frequency (bass) range of human hearing. The sounds are produced by muscles twitching the fish's resonating, air-filled, swim bladder. For most species, only males have these bladder muscles. The sounds are used for communication during courtship, which takes place at night in fish choruses. Sciaenids also make "disturbance calls" indicating alarm, pain, or annoyance. These sounds of distress are typically a long series of rapid pulses.

Migrating "Baitfish"

RELATIVES: These small fishes include mullets (family Mugilidae) in the order Mugiliformes, and two fish families in the order Clupeiformes—anchovies (family Engraulidae) and herrings (Clupeidae).

IDENTIFYING FEATURES: The term "baitfish" refers to small fish in large schools that attract predatory fishes and birds. The **"fall mullet run"** and other fish migration events are characterized by directed movement of massive schools as the weather changes. Migrations frequently draw other fishes and diving birds into a grand biological spectacle. Small fish herded by predator fish often form "bait balls" in which the prey fish swarm within a tightly packed spherical formation. The most conspicuous baitfish migrations take place mid-August through mid-December.

Striped (black) mullet *(Mugil cephalus)* have blurry horizontal side stripes, and **white mullet** *(Mugil curema)* show plain sides with a gold spot on their gill opercle. Shadowy schools may each contain thousands of "finger mullet" that periodically burst into the air like exploding fireworks. Mullet aerials commonly precede the rushes and boils of predatory fishes like jack crevalle or **tarpon.**

Anchovy species form coast-hugging schools that look like dark clouds. These finger-length, translucent fish spend the summer in coastal lagoons and the northern Gulf off the Mississippi River. In late summer, they migrate west to arrive in southern Texas.

Herrings, sardines, and menhadens (clupeids) are silvery with no lateral line, soft fins (no fin spines), and a short, deep, lower jaw hidden by their wide, rounded upper jaw. Common Gulf clupeids include

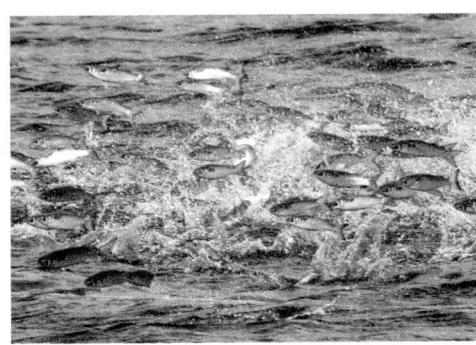
Juvenile striped mullet migrate south in the fall

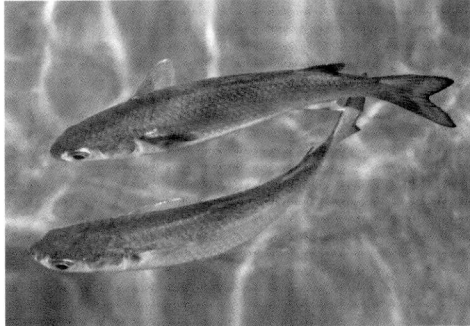
White mullet in the surf

A tarpon slashes through desperate mullet

Pelicans and gannets feast on fingerlings

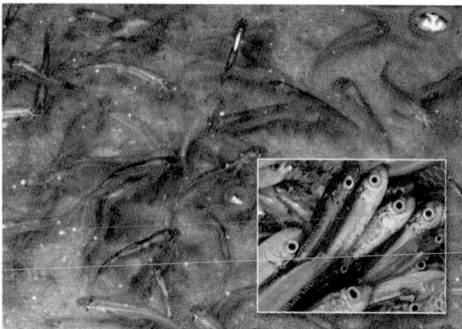

Dwarf herring schooling near a Texas jetty

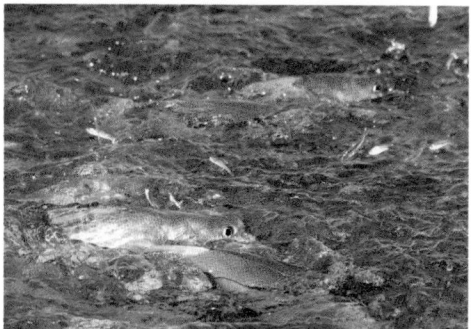

Open-mouthed hardhead catfish hoping for a herring

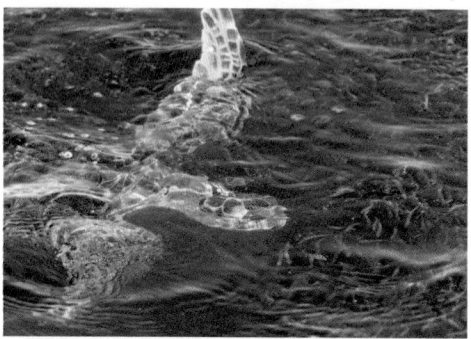

A green turtle tries to catch small schooling herring

Gulls target herring leaping away from fish predators

Atlantic thread herring (threadfins, greenies, greenbacks) *(Opisthonema oglinum)*, 8 in (20 cm), which are deep-bodied with a keeled belly. Their colors are green-blue above with silver sides and a faint spot behind the gills, and the last ray of their dorsal fin is conspicuously elongate. This fin thread is not present in the similar, scaled sardine (pilchard, greenback, whitebait) *(Harengula jaguana)*, 6 in (15 cm). Both of these herrings form large schools just outside the surf. Look for them as they flick the surface.

Dwarf round herring *(Jenkinsia lamprotaenia)* are small (3 in, 8 cm), silvery, shallow-bodied herring that form crowded schools near rock jetties. Their schools are so dense that slow-moving **catfish** and **sea turtles** attracted to the swarm stand a chance at catching a meal simply by swimming with their mouth open.

HABITAT: Mullet live most of the year in coastal lagoons, marshes, and freshwater streams. The surf serves as a migration corridor for fingerling (juvenile) mullet seeking more southern waters to overwinter, and for adult mullet (16 in, 41 cm) moving offshore to spawn. Spawning occurs far offshore between November and mid-January. These anchovy and herring species hug the shoreline during seasonal migrations. Adults spawn offshore.

DID YOU KNOW? Herrings and anchovies filter feed by straining water through their comblike gill rakers. Filtered food includes planktonic crustaceans and larval fish. Mullet eat algae and tiny invertebrates slurped from bottom sediments. Fingerling mullet are nearing their first year. Mullet jump even when not leaping for their lives. A jump in waters with little disolved oxygen may allow them to gulp air. Or, sticking the landing might be the goal. A big landing smack could free up a clogged gizzard. Yes, mullet have a gizzard. It's a specialized, muscular stomach used to grind food along with ingested grit.

Fish out of Water

RELATIVES: Fishes found stranded on the beach belong to a diverse array of groups, including the orders Batrachoidiformes (midshipmen and toadfishes), Clupeiformes (anchovies and herrings), Tetraodontiformes (burr- and pufferfishes), Siluriformes (catfishes), Lophiiformes (frogfishes), Perciformes (drums and perchlike fishes), and Carcharhiniformes (sharks).

Atlantic midshipman, max 8 in (20 cm)

WHY IS THIS FISH ON THE BEACH?

A wide variety of predicaments can result in fish stranding on the beach. Nearshore spawning or migration often produces fish too weak to avoid being caught by waves. Small fish may enter the risky surf to avoid predators, only to be swept ashore. Other fish are so reliant on their offshore floating-algal habitat that when it reaches shore, the fish do too. Fish also occasionally succumb to toxins produced by harmful algal blooms (p. 43). But the most common reason why there are dead fish on a beach is fishing waste. Unfortunately, some shore fishers toss unwanted catches in the sand rather than in the surf. Commercial offshore trawl, purse-seine, and gillnet fisheries result in abundant bycatch (discarded non-target species).

Bay anchovies succumb quickly after stranding

Atlantic midshipmen (*Porichthys plectrodon*) have a large head and a limp body lined with pearly dots (photophores). The related toadfish (*Opsanus* spp.) is darker with no dots.

Anchovies (*Anchoa* spp.) are frail, finger-size fishes that commonly strand on the beach due to their habit of entering the risky swash to flee predators.

Striped burrfish (*Chilomycterus schoepfii*) are pufferfish covered with sharp spines and have a distinct beak.

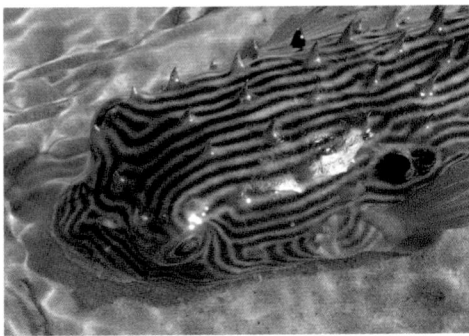

Striped burrfish, max 10 in (25 cm)

Hardhead catfish (*Ariopsis felis*) have dorsal and pectoral spines that could cause a painful foot injury. They are common unwanted surf catches.

A stranded hardhead catfish, max 18 in (46 cm)

201

Sargassum frogfish, max 4.5 in (11 cm)

Black drum, max 50 in (1.3 m)

Red drum, max 39 in (100 cm)

A discarded bonnethead shark

Sargassum frogfish *(Histrio histrio)* are sargassum-camouflaged (mottled brown, orange, yellow, and white) with smooth, scaleless skin decorated by fleshy tabs.

Black drum *(Pogonias cromis)* have a body depth greatest just behind the head. They are gray with 4–5 dark side bars and numerous chin barbels.

Red drum *(Sciaenops ocellatus)* are deepest at mid-body, have no barbels, and show one or more black tail spots.

Bonnethead sharks *(Sphyrna tiburo)*, typically to 3.2 ft (1 m), have shovel-shape heads. You may not want this fish, but the marine ecosystem definitely does.

HABITAT: Atlantic midshipmen strand after spawning in nearshore waters. Anchovies live in coastal waters. Sargassum frogfish live in open-sea sargassum. The largest black and red drum are found offshore to moderate depths. Burrfish, catfishes, and sharks often enter the surf.

DID YOU KNOW? The male Atlantic midshipman sings using air-bladder contractions and flashes with luminescent photophores to court potential mates. Females become spent following egg-laying, and males exhaust themselves guarding eggs and fry. Both sexes have venomous cheek spines; live fish should be handled carefully. Although among the many "trash fish" caught by surf fishers, taste tests have shown sea catfish to be as good or better than freshwater catfish as tablefare. Sharks may not be as good to eat, but they are ecologically valuable as living fish. Black and red drum are valuable fishes, but are discarded as non-target bycatch in offshore net fisheries. Many stranded fish are recognizable only as bones. To identify these "verte-bits," see p. 268.

Alligator and Gopher Tortoise

Alligator   Gopher Tortoise

RELATIVES: These reptiles (class Reptilia) are in the distantly related orders, Crocodilia (alligators and crocodiles) and Testudines (turtles and tortoises).

IDENTIFYING FEATURES:

American alligators *(Alligator mississippiensis)* reach 14.5 ft (4.4 m) in males, and 9.8 ft (3 m) in females. **Gator tracks** on beaches are alternating, clawed footprints, with or without foot-drag swirls, aside a wavy tail-drag mark.

Gopher tortoises *(Gopherus polyphemus)* are dark brown or grayish with shovel-like forelimbs and elephantine hind limbs. **Tortoise tracks** are a parallel set of stubby footprints up to 8 in (20 cm) wide. Unlike most other turtles in the area, gopher tortoises don't often drag their plastron (lower shell).

HABITAT: Alligators live in wetlands including coastal swamps, marshes, and bays. They often inhabit brackish estuaries and occasionally enter the sea. Gopher tortoises dig burrows in sandy scrub habitat including coastal dunes.

DID YOU KNOW? Small alligators near the coast eat crabs, shrimps, insects, and small fishes. Adults eat almost exclusively fish. Due to commercial hunting, American alligators were close to extinction in the 1950s, but federal protection allowed them to come back. The species was declared "recovered" in 1987 and is a prominent example of conservation success. Conversely, our gopher tortoises are in decline. They are protected by state laws throughout their range, and are federally protected as Threatened in coastal areas of Louisiana, Mississippi, and Alabama.

A basking American alligator smiles in profile

Gopher tortoise, max 14 in (36 cm) shell-length

Gopher tortoise young of the year, 2 in (5.1 cm)

Alligator tracks

Gopher tortoise tracks

203

Female Texas diamondback terrapin

Juvenile Mississippi diamondback terrapin

Ornate diamondback terrapin with barnacles (p. 160)

Nesting female terrapin tracks on a saltmarsh beach

Diamondback Terrapins

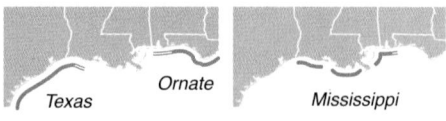

Texas   Ornate   Mississippi

RELATIVES: Terrapins are turtles (order Testudines) in the family Emydidae, shared with pond turtles.

IDENTIFYING FEATURES: Terrapins have a carapace sculptured by thick, concentric growth layers in the scutes (scales) covering the shell. Young terrapins also have a row of dorsal knobs. Adult females range 6–9 in (15–23 cm) in carapace length, and males are 4–5.5 in (10–14 cm).

Texas diamondback terrapins (*Malaclemys terrapin* ssp. *littoralis*) have carapace scutes without light centers. Their upper beak is light, and their neck and legs are densely spotted with black.

Mississippi diamondback terrapins (*Malaclemys terrapin* ssp. *pileata*) have an oval carapace that lacks light scute centers and has orange, upturned edges to the carapace margins. The top of the head, upper beak, neck, and limbs are dark.

Ornate diamondback terrapins (*Malaclemys terrapin* ssp. *macrospilota*) have carapace scutes with orange and yellow centers. The head and limbs are pale gray with black spots.

HABITAT: Terrapins depend on coastal estuaries—lagoons, tidal creeks, and saltmarsh. They lay their eggs in nests dug into low-energy sandy beaches above the tide, mid-May through mid-July.

DID YOU KNOW? For their small size, these turtles have strong jaws, which they use to crunch marsh periwinkles, plicate horn snails, ribbed mussels, fiddler crabs, and marsh crabs. Populations are declining, with a key threat being drowning in crab traps. Simple "terrapin-excluding devices" are available to retrofit traps so that turtles can escape while crabs are retained.

Loggerhead Sea Turtle

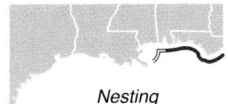

Nesting

RELATIVES: Loggerheads are reptiles in the order Testudines and share the family Cheloniidae with other "hard-shelled" sea turtles.

IDENTIFYING FEATURES:

Loggerhead sea turtles *(Caretta caretta)* as adults, 3 ft (1 m) shell length, are orange, yellow, and brown, with a large head and a stout, tapered shell, which typically has scattered barnacles. Their **nest** is a circular or slightly elongate mound with an adjacent shallow pit. The **hatchling** has a lumpy, walnut-size shell that is gray, tan, or charcoal above.

HABITAT: Loggerheads lay their eggs in nests between the wrack line and the dune toe. Most loggerheads nesting on northern Gulf beaches have migrated from foraging waters off southwestern Florida and the Yucatan Peninsula, Mexico.

DID YOU KNOW? Although loggerheads are Threatened, they are the most common nesting sea turtle in the US Gulf of Mexico. Nests have 70–150 (average 115) ping-pong-ball-size eggs buried about 18 in (46 cm) beneath the sand. Each female loggerhead makes an average of five nests separated by two-week intervals during the May–August nesting season. Migrating hundreds of miles and laying hundreds of eggs is strenuous, which is why loggerheads typically take off 1–3 years between nesting trips. Hatchlings emerge in groups from nests, July–October, 45–60 days after eggs were laid. Both nesting and hatchling emergence occurs mostly at night. Adult loggerheads eat large, hard-shelled invertebrates like marine snails and crabs.

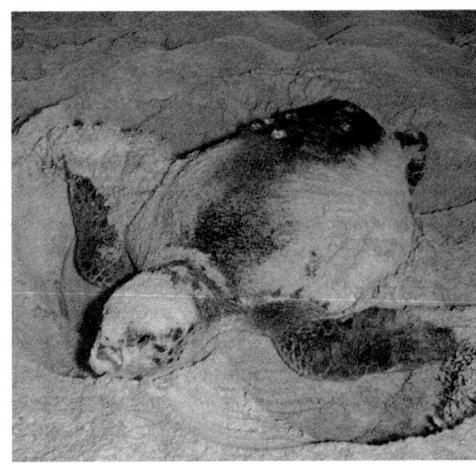

A loggerhead nesting at dawn scatters sand

A typical loggerhead nest the morning after

Hatchling loggerhead, 1.8 in (4.5 cm) shell length

A green turtle camouflages her nest

Green turtle nest and return track

Hatchling green turtle, 2.2 in (5.5 cm) shell length

Green Turtle

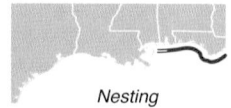

Nesting

RELATIVES: Green turtles share the family Cheloniidae with loggerhead and Kemp's ridley sea turtles.

IDENTIFYING FEATURES:

Green turtles *(Chelonia mydas)* as adults are 3.5 ft (1.1 m) in shell length, and are olive or brownish with a relatively small head and a smooth, oval, domelike shell, which may have scattered dark spots. Their **nest** is a large, elongate mound with an adjacent deep pit. The 3-inch (7.6-cm) **hatchlings** have a smooth, dark shell, outlined in white, as are the flippers.

HABITAT: Most green turtle nests are near the toe of the dune. Although there is regular nesting in Florida's Panhandle, green turtles nest in greater abundance in the southern Gulf and on Florida's Atlantic beaches. Juvenile green turtles are common near jetties (p. 392) and occasionally strand on the beach (p. 210).

DID YOU KNOW? Green turtles are federally listed as Threatened, a status that recently changed from Endangered due to increases seen in North Atlantic populations. This species is Florida's second most common nesting sea turtle. Nests have 100–180 (average 135) eggs buried about 21 in (53 cm) beneath the sand. The females make an average of four nests separated by two-week intervals during the June–September nesting season. Green turtles take a year off between nesting migrations. Hatchlings emerge from nests late July through November, 50–65 days after eggs were laid. Nesting is nocturnal, but the two-hour process occasionally leaves females on the beach past dawn. Adult and coastal juvenile green turtles eat seagrasses and algae.

Kemp's Ridley

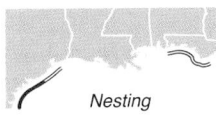

Nesting

RELATIVES: Kemp's ridleys are allied with loggerheads and green turtles in the family Cheloniidae.

IDENTIFYING FEATURES:

Kemp's ridley turtles (*Lepidochelys kempii*) have an adult shell length of 26 in (65 cm), which makes them the smallest sea turtle. These turtles are pale- to medium-gray or olive with a discus-shape shell and a cusped, parrotlike beak. Unlike other sea turtles, Kemp's ridleys nest during the day. Their **nest** is a circular or slightly elongate mound with an adjacent shallow pit. It is smaller than a loggerhead nest, and is easily erased by blowing wind. **Hatchlings** are dark gray all over and have three interrupted ridges.

HABITAT: Ridleys lay their eggs in nests on the upper beach. The majority of US nesting is in southern Texas, but the largest worldwide nesting assemblage is in northeastern Mexico. Most Kemp's ridleys nesting in Texas have migrated from foraging waters off Louisiana, although their range spans the Gulf Coastline from northern Mexico to the Florida Keys.

DID YOU KNOW? Kemp's ridley is Endangered, and has become the world's rarest sea turtle. Most ridley nesting on Mexico beaches takes place in *arribadas* (arrivals) when many females emerge simultaneously. Texas arrivals are smaller, but like Mexico, take place April–June when gusty winds blow. Nests average 103 eggs, which in Texas are most commonly relocated into hatcheries due to beach-driving threats. Each female makes two or three nests separated by a 2–3 week interval. Hatchlings emerge from nests or hatcheries June–August. Kemp's ridleys grow up on a diet of crabs and tunicates.

A Kemp's ridley covers her eggs in a nest

A fresh ridley nest on the upper beach

Hatchling Kemp's ridley, 1.8 in (4.5 cm) shell length

207

Tracks from Loggerhead (left) and ridley (right)

A green turtle track down the beach

A loggerhead false crawl

Sea Turtle Tracks

IDENTIFYING FEATURES:

Loggerhead tracks are 39 in (100 cm) wide with a smooth, wavy center straddled by alternating commalike swooshes from the turtle's rear flippers. The rear flippers erase most of the front-flipper marks at the track's margins as the turtle crawls forward.

Kemp's ridley tracks are 30 in (76 cm) wide with alternating scratches from the turtle's rear flippers, which leave claw marks in wet sand. The tail leaves a thin, wavy drag mark in soft sand.

Green turtle tracks are 47 in (120 cm) wide and have a relatively straight set of central ridges (either low or pronounced) that straddle a thin, straight, tail-drag line, which is punctuated by regularly spaced tip-pokes. Both the rear flipper prints (either side of the center ridges) and front flipper slashes (at the margins) are in parallel sets. Front-flipper slash marks are conspicuous all along the track edges.

TURTLES AND THEIR TRACKS:

The direction a sea turtle crawled (**arrows in images**) is revealed by the way sand is pushed by their flippers. These species have tracks that correspond to the turtles' characteristic crawling gait and flipper form. Loggerheads and Kemp's ridley females crawl baby-style, leave alternating flippersteps, and have relatively short front flippers. Although loggerheads have tails too short to drag the sand, ridleys show an occasional, slight tail mark. Green turtles crawl with a butterfly stroke, leave parallel flipper marks, and have longer front flippers and tails. If undisturbed, loggerheads and green turtles will nest during about half of their beach visits. An abandoned nesting attempt is called a **"false crawl."** When a Kemp's ridley arrives, she will follow through with nesting about 90 percent of the time. Take care not to disturb sea turtle tracks and nests; they are counted by biologists each morning.

Other Sea Turtle Signs in the Sand

IDENTIFYING FEATURES:

Depredated nests are a hole in the beach with numerous scattered eggshells and animal footprints. Raccoons dig into nests from many angles and scatter broken eggs in all directions. Foxes, coyotes, and dogs dig from one side only. Ghost crabs leave a burrow with only a few eggshells near the entrance. Fish crows commonly feast on nests opened by other predators.

Washed-out eggs are white, pinkish, yellowish, or grayish spheres, or are gray-yellow and deflated (and stinky). Although some white eggs may be viable, most are not, due to jostling and exposure.

Hatchling emergence sign is typically a bowl-like depression with hatchling tracks fanning out toward the sea. In the Gulf region, most hatchlings emerge from nests that have been screened for protection or corralled in a hatchery (p. 397). Hatchlings leave their eggshells buried during a 1–5 day escape from the nest. The tracks of loggerhead and ridley hatchlings are miniature versions of the adult tracks, about 2–3 fingers wide. Unfortunately, it is common for hatchling tracks to circle, meander, and spread out toward the dune when artificial lighting visible from the beach at night confuses the orienting hatchlings.

DID YOU KNOW? In Florida, raccoons eat more sea turtle eggs than any other predator. Ghost crabs are a distant second, with coyotes growing in egg-predation prominence. Storm erosion washes out many eggs, but because nests are spread out, spring through fall and over hundreds of miles of Gulf beaches, only a minor fraction of eggs is destroyed by stormy surf during most years. Hatchlings emerge at night and orient toward the brightest horizon. Beachfront lights lure hatchlings landward and harm large numbers each year.

A raccoon-depredated loggerhead nest

A nest eroded by a severe storm

Tracks from dozens of emerged loggerhead hatchlings

A weakened green turtle with skin and eye tumors

A stranded juvenile Kemp's ridley

Loggerhead carcass. Paint means it was counted

Stranded Sea Turtles

IDENTIFYING FEATURES:

Sea turtle strandings occur when turtles are weakened, severely injured, stunned in cold water, or killed after being entangled, diseased, struck by boats, hooked by fishing gear, or drowned in nets. Live stranded sea turtles can be treated and released back into the wild. Dead stranded turtles provide valuable information to conservation biologists. Occasionally, sea turtles receive a necropsy (animal autopsy) in which additional information is gathered, but decomposed animals cannot be examined this way.

DID YOU KNOW? Juvenile green turtles (p. 200) are the most common sea turtles in the surf zone and near jetties. Turtles in coastal waters are struck and killed by boats, which is the most common recognizable cause of death in Gulf strandings. Turtles wrapped in fishing line and trap ropes also show likely causes of death, but for most strandings, there is insufficient evidence for a determination. This includes turtles that may have drowned in coastal fisheries. Sea turtles that strand are often weakened by diseases, such as **skin tumors** called fibropapillomas.

Stranded turtles should be reported to Sea Turtle Stranding and Salvage Networks in:

Texas (361-949-8173 x 226)
Louisiana (844-732-8785)
Mississippi (888-767-3657)
Alabama (866-732-8878)
Florida (888-404-3922)

Permitted conservation volunteers transport live turtles to rehabilitation facilities. Responders who have recorded data from a sea turtle carcass may mark it with paint. These turtles do not require additional calls.

Lizards

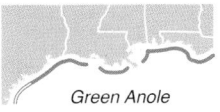
Green Anole

Brown Anole

Keeled Earless

A male green anole

RELATIVES: Lizards are reptiles sharing the order Squamata with snakes. Anoles are in the family Dactyloidae, and earless lizards are in the family Phrynosomatidae.

IDENTIFYING FEATURES:

Green anoles *(Anolis carolinensis)*, 7 in (18 cm), can change from flat brown to bright green and have a long tapered head. Males have a pink throat fan (dewlap).

Brown anoles *(Anolis sagrei)*, 6 in (15 cm), are gray to dark brown with a wedge-shape head. Females often have a rusty head and contrasting light-and-dark markings down their back. Males have an orange-red dewlap with a yellow border.

A male brown anole displaying his dewlap

Keeled earless lizards *(Holbrookia propinqua)*, 5.5 in (14 cm), have short snouts and no external ear openings. Patterns vary, but males consistently show black bands on their sides. A female's bands are less conspicuous. Egg-bearing females gain yellow and orange lower highlights.

HABITAT: Green anoles are slinky climbers and prefer leafy canopies and tall grass. Brown anoles are leaping runners that prefer lower levels. Earless lizards are found on the open dune face.

A plain-colored male keeled earless lizard

DID YOU KNOW? Green anoles are natives that have been pushed out of many areas by the brown anole, an invader from Cuba and The Bahamas. This invader is now the most common lizard seen on human structures near Gulf beaches. All of these lizards feed on small insects. Males impress potential mates and rivals by bobbing their heads, and doing "push-ups".

A darkly patterned male keeled earless lizard

211

Western diamond-backed rattlesnake in a Texas dune

Western coachwhip, closeup

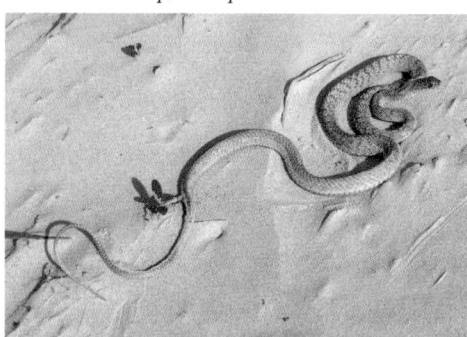

The slender, sand-colored western coachwhip

Eastern coachwhip eating a sanderling. Note scales

Snakes *(Rattlesnakes, Coachwhips)*

Western Diamonback

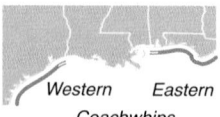
Western   Eastern
Coachwhips

RELATIVES: Snakes are reptiles that share the order Squamata with lizards. Rattlesnakes are in the family Viperidae, and coachwhips are in the family Colubridae.

IDENTIFYING FEATURES:

Western diamond-backed rattlesnakes *(Crotalus atrox)*, to 4.9 ft (1.5 m), have a broad triangular head and a short tail with alternating white and black rings and ending with one or more rattles. Diamond shapes down the snake's back have pale borders. The similar eastern diamond-backed rattlesnake *(Crotalus adamanteus)*, to 6 ft (1.8 m), has no tail rings and is only east of the Mississippi.

Western coachwhips *(Coluber flagellum* ssp. *testaceus)*, 8.2 ft (2.5 m), are slender snakes with large eyes and **scale outlines** that look like a braided whip. Coachwhips in Texas dunes are mostly sand colored, but some may be dark or pink.

Eastern coachwhips *(Coluber flagellum* ssp. *flagellum)*, are similar to western coachwhips but are distinctively tan with a darker head and neck.

HABITAT: These snakes live in the dune and occasionally hunt on the beach. Uncommon for a snake, coachwhips are active during the day.

DID YOU KNOW? Rattlesnakes have venomous fangs and will bite if harassed, but give plenty of warning. Do not approach a snake that rattles. Coachwhips have no venom and may bite if disturbed, but they are more likely to speedily streak away. Rattlesnakes patiently wait for small mammals to get within striking distance, and coachwhips actively hunt insects, lizards, birds, and other snakes.

Snakes *(Ribbonsnake, Watersnake)* and **Toad**

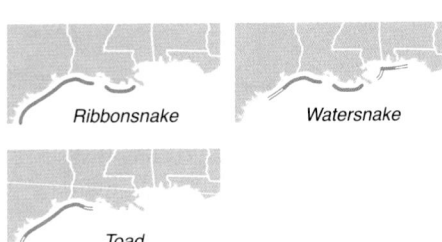

Ribbonsnake   Watersnake

Toad

Gulf Coast ribbon snake, max 36 in (91 cm)

RELATIVES: These snakes share the family Colubridae with coachwhips. Toads are amphibians (class Amphibia), with frogs in the order Anura, and with other toads in the family Bufonidae.

IDENTIFYING FEATURES:

Gulf Coast ribbon snakes *(Thamnophis proximus* ssp. *orarius)* are dark brown or olive above with three light stripes. Their belly is unmarked yellowish.

Mississippi green watersnakes *(Nerodia cyclopion)* are heavy-bodied with a thick neck. They are mottled dark green above, and below have a grade of patterns, from yellow in front, to dark-on-yellow mid-body, to yellow-on-dark toward the tail.

Mississippi green watersnake, max 55 in (140 cm)

Snake tracks are either sinusoidal, wavy slithers, or relatively straight caterpillar crawls. Snakes make caterpillar crawls as they compress their body in waves like an accordion, using belly scales for traction.

Coastal plain (Gulf Coast) toads *(Incilius nebulifer)* are mottled brown with occasional yellow flecks, a light back stripe, and a wider side stripes behind each eye. Cranial crests divide ahead of the bulging parotoid glands on either side of the neck.

Track where a snake scaled a low dune

HABITAT: Ribbon snakes live in brush near water. Green watersnakes often enter brackish water. Gulf Coast toads forage in lowlands and breed in brackish marshes.

DID YOU KNOW? These snakes are non-venomous. A toad's parotiod gland secretions are distasteful to predators.

Coastal plain toad, max 4 in (10 cm)

213

Adult common loon, winter

Red-breasted merganser

Adult red-breasted merganser, winter

Common Loon
and **Red-breasted Merganser**

RELATIVES: Loons (Gaviiformes) belong to a different order of birds from mergansers (Anseriformes), which are related to ducks and other waterfowl.

IDENTIFYING FEATURES:

Common loons *(Gavia immer)*, 24 in (61 cm), are diving birds the size of a large duck, sit low in the water, and have a straight, pointed bill. Most Gulf-coast loons are seen in winter, and in their winter plumage—a drab charcoal pattern with a whitish throat and breast. Summer adults in northern breeding areas are starkly contrasting black and white.

Red-breasted mergansers *(Mergus serrator)*, 16 in (41 cm), are diving ducks with a thin reddish bill and shaggy head-crest. All but breeding males have a cinnamon head and gray back. Common mergansers *(M. merganser)* are not common in saltwater habitats, and have lighter bodies with a white chin.

HABITAT: Loons and mergansers are occasionally found diving for fish in the surf but are more common in large coastal lagoons.

DID YOU KNOW? Loons from the northern US and Canada winter in the Gulf. Stressed by the long migration, they often strand on beaches. On land, even healthy loons can't walk. Their feet are behind them, like an outboard motor, and can only make swimming movements. In water, loons are impressive divers that can plunge more than 200 feet to catch bottom fish. Mergansers work in groups to herd fish and often paddle with their heads under water.

Northern Gannet and Frigatebird

Gannet

Frigatebird

Juvenile gannet in flight

RELATIVES: Gannets (family Sulidae) and frigatebirds (Fregatidae) share the order Suliformes with cormorants.

IDENTIFYING FEATURES:

Northern gannets *(Morus bassanus),* 31 in (79 cm), are big, sleek seabirds with thick bills. Their tapered wings span six feet (1.8 m). Adults are white with black wing-tips and have a wash of pastel yellow covering the head. Juveniles are brown with white flecks. They are recognizable at a distance due to their steep, forceful, folded-wing dives

Magnificent frigatebirds *(Fregata magnificens),* 35 in (89 cm), are dark seabirds with a hooked bill and thin, tapered wings spanning up to 8 ft (2.5 m). Their long forked tail is typically folded to a point. **Males** have an inflatable, red throat sac, and females have a white breast patch. **Juveniles** have a white head and breast.

A juvenile's flecks expand into solid, adult white

HABITAT: Gannets are oceanic birds but are near the coast when fishing is good. Sick and injured gannets often strand on the beach in winter. Summer breeding for northern gannets occurs on steep cliffs and rocky islands around Newfoundland. Frigatebirds soar offshore and only occasionally fly the coast. They nest in summer within mangrove trees in the Florida Keys and coastal Mexico.

Adult northern gannet

DID YOU KNOW? Gannets are supreme plunge divers: their eyes aim forward for binocular fish-spotting, they have no nostril holes, and their bills are watertight upon impact. The smack the birds make when entering the water is cushioned by a system of air cells beneath their skin. Frigatebirds can remain aloft for hours without a wingbeat.

Juvenile frigatebird (L), adult male (R)

215

Double-crested cormorant in flight

Double-crested cormorant drying its wings

A raft of cormorants

White pelicans, adults and juvenile (center)

Cormorant and White Pelican

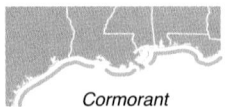

Cormorant

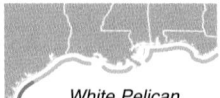

White Pelican

RELATIVES: Cormorants (family Phalacrocoracidae) share the order Suliformes with frigatebirds. Pelicans (family Pelecanidae) are allied with ibises and herons in the order Pelecaniformes.

IDENTIFYING FEATURES:

Double-crested cormorants *(Phalacrocorax auritus),* 27 in (69 cm), are snake-neck waterbirds with a hooked bill and eyes of vivid aquamarine. They fly with rapid wingbeats and often gather in large **"rafts."** Adults are dark, and juveniles are lighter, especially on the throat and breast.

American white pelicans *(Pelecanus erythrorhynchos)* are large white birds with a long, pouched bill. Their wingspan reaches 9.5 ft (2.9 m). The bird's black flight feathers are visible only when they spread their wings. Breeding adults develop a keel atop their bill.

HABITAT: Cormorants surface-dive for fish outside the surf but are most common in coastal lagoons. They visit in winter and breed in northern lakes. White pelicans fly near the beach, but their winter foraging takes place in coastal lagoons. In spring, most migrate to central Canada to forage and breed, but some white pelicans stay to nest on coastal lagoon islands in central Texas.

DID YOU KNOW? With poor oil glands, cormorants hang their wings out to dry before flying. White pelicans dip their bills to scoop small fish, then raise their head to drain the bill pouch and swallow captured prey. They often feed in groups, with several birds dipping in an elegant, synchronized-swimming exercise as they corral fish to make feeding more efficient.

Brown Pelican

RELATIVES: Brown pelicans share the family Pelecanidae with white pelicans.

IDENTIFYING FEATURES:

Brown pelicans *(Pelecanus occidentalis)*, 41 in (104 cm), with a wingspan to 7 ft (2.1 m), occupy the Gulf Coast year-round as the regions most common pelican. They are much darker and slightly smaller than white pelicans. Winter adults have grayish bodies, a white neck, and a pale yellow cap. During summer breeding, the back of the bird's long neck turns chestnut brown in both sexes. Juveniles are brown above and light below.

Juvenile in flight

HABITAT: Brown pelicans dive for fish off the beach, float just outside the surf, and use the beach for resting. All Gulf breeding occurs within a handful of coastal-island nesting colonies.

Juvenile diving

DID YOU KNOW? In 1966, Louisiana designated the brown pelican as the state bird. At that time, the pelican population of the northern Gulf had collapsed, leaving none nesting in "the Pelican State." The bird recovered following regulations banning DDT pesticides, and reintroduction of pelicans from Florida. Louisiana pelicans are now descended from those birds. Unlike white pelicans, brown pelicans plunge-dive to feed. The water-impact from the bird stuns herring, anchovies, and mullet, allowing the pelican to fill its pouch with water and fish. The bill is drained by tilting it downward. A following head raise to swallow means the pelican was successful. Yes, their bill can hold more than their belly can—about three gallons of fish and water, twice the capacity of their stomach. Fish are carried in the gullet, never in the pelican's bill pouch.

Adults in winter. Left bird is molting

Breeding male and female pair in the surf

217

Adult cattle egret in flight

Cattle egrets in breeding plumage on a Gulf jetty

Adult snowy egret prepares to land, foot (inset)

Juvenile snowy egret showing yellow leg-stripes.

Egrets *(Cattle, Snowy)*

RELATIVES: Egrets and herons share the family Ardeidae and are allied with pelicans in the order Pelecaniformes.

IDENTIFYING FEATURES: Egrets and herons are wading birds with a long neck, a spearlike bill, and long legs. These birds fly with their neck folded into an S shape over their back.

Cattle egrets *(Bubulcus ibis),* 20 in (51 cm), are stout for an egret (heron). Adults are are white with a yellow bill and legs. During breeding, both sexes have golden plumes on their head, breast, and back. Juveniles have gray legs and a dark bill.

Snowy egrets *(Egretta thula),* 20 in (51 cm), are delicate, medium-size wading birds with a shaggy head and all-white plumage. They have a thin neck and a black bill with yellow skin in front of their eyes. Their legs are black with yellow feet. Adults in breeding plumage have long, lacy back plumes, and the skin on their face turns orange. Juveniles look similar to adults except for a paler bill and a **yellow stripe** up the back of each leg.

HABITAT: Cattle egrets hunt insects in dunes and fields, often following large grazing animals or tractors. Snowy egrets stalk and chase fish in the swash zone, runnels, and near inlets. In the spring, both of these egrets nest in branches of brushy trees hanging over water.

DID YOU KNOW? Cattle egrets came to the Americas from Africa in the late 1800s. Snowy egrets use a foot-stirring method to flush small fish and shrimp from the bottom in shallow waters. Their golden toes may help achieve success by either spooking or luring their prey.

Reddish Egret

RELATIVES: Egrets and herons are in the family Ardeidae.

IDENTIFYING FEATURES:

Reddish egrets *(Egretta rufescens),* 25 in (64 cm), are midsize wading birds with gray legs and a long neck. They show two **color forms** (morphs), dark and white, which are not related to age. Dark-form adults are gray with a rusty head and neck, and a pinkish, dark-tipped bill. Dark juveniles are pale gray with a dark bill. White-form birds (least common) have all white plumage. Adults of either color develop shaggy neck plumes. This egret has a unique **fishing dance** during which the bird runs through shallow waters, staggers, leaps, and flails its wings to herd schools of small fish before bill-stabbing at them.

HABITAT: Reddish egrets chase fish in the swash zone and in runnels flooded by the tide. After a short southern migration in spring, reddish egrets nest in multi-species colonies with mangroves, scrubby brush, or cacti to support nests.

DID YOU KNOW? The reddish egret is North America's rarest egret or heron, with a population less than a tenth of what it was in the 1800s. The bird's decline followed intense hunting and has continued as its coastal habitats are developed. The reddish egret's white form makes up just 20 percent of the population. Mates don't discriminate between color forms. Two dark birds can have white chicks, but two white birds can never have dark chicks. This egret's seemingly uncoordinated fishing dance may aid success by startling prey. Shade from the bird's open wings may also lure in sheltering fish and reduce distracting sun glare.

Adult reddish egret, dark form

Juvenile reddish egret, dark form, in its fishing dance

Adult white form fishing a beach runnel

219

Yellow-crowned night-heron with ghost crab

Adult tricolored heron in spring

Adult little blue heron, juvenile white phase (inset)

Herons *(Yellow-crowned Night-, Tricolored, Little Blue)*

Night-heron

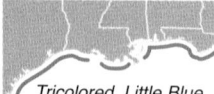

Tricolored, Little Blue

RELATIVES: Tricolor and little blue herons are closer to reddish and snowy egrets than to night herons, but all share the family Ardeidae.

IDENTIFYING FEATURES:

Yellow-crowned night-herons *(Nyctanassa violacea)*, 21 in (53 cm), are short wading birds with stout black bills. Adults have a black head with a yellow crown and a white streak below a red eye. Their body plumage is blue-gray. Juveniles are brownish with light streaks and have amber eyes.

Tricolored herons *(Egretta tricolor)*, 21 in (53 cm), are slender with a dark, slate-blue neck and back, white and rust throat stripe, light, ponytail plumes, and a white belly. The bill is blue with a dark tip. Juveniles are rusty reddish on their neck and sides.

Little blue herons *(Egretta caerulea)*, 24 in (60 cm), as adults are slate grayish with pale green legs and a pale, dark-tipped bill. In good sunlight, adult plumage appears blue-gray on the body and maroon on the neck and head. Juveniles have white plumage. As immatures molt to assume the adult pattern, young herons display a patchwork of white and blue.

HABITAT: Yellow-crowned night-herons lurk on the upper beach between dusk and dawn. Most arrive to breed in early spring and spend winters in Central and South America. Tricolored and little blue herons fish in shallow protected waters, and nest locally in shrubs near water.

DID YOU KNOW? To catch fish, tricolored herons stalk and strike; little blues stand and wait. Night-herons hunt crabs on the shoreline in the dark.

220

Great Blue Heron

RELATIVES: Great blue herons and great egrets share the genus *Ardea*, and are allied within the family Ardeidae with other herons and egrets.

IDENTIFYING FEATURES:

Great blue herons *(Ardea herodias)*, 38 in (97 cm), are tall, grayish wading birds with a long neck and a thick, pointed bill. Adults have a shaggy foreneck, rusty nape, and a white face. Adults in spring have stylish black streaks that trail thin, pony-tail plumes. Juveniles have a subdued cast to the adult colors.

HABITAT: Although these herons do most of their fishing in calm shallows, they also stalk the swash zone. Nesting is in trees when available. Otherwise, shrubs, cacti, or artificial platforms will suffice. Local herons stay year-round, but in winter, the Gulf Coast sees additional northern birds.

DID YOU KNOW? These are America's largest herons, but as impressive as they are, these tall birds weigh only about five pounds (2.3 kg) thanks to their hollow bones. Great blue herons specialize in fish but also eat snakes and rodents. They hunt both during the day and under moonlight, having excellent night vision from an abundance of rod-type photoreceptors. Like some other herons, great blue herons have dual patches of special feathers hidden beneath the contour feathers on their upper breast. These "powder down" feathers continually grow and fray, releasing a fine waxy powder that may help keep plumage free of oily fish residues. When a heron preens, it uses the powder-down patches almost like a washcloth to remove fish slime from its feathers.

A flying great blue heron tucks in its long neck

Adult great blue heron

Great blue herons have good night vision

221

In flight, an egret's neck is held in an S-shape

Great egret with mullet

Breeding plumage shows long plumes past the tail

Great Egret

RELATIVES: Great egrets share the family Ardeidae with other herons and are more closely related to great blue herons than to other "egrets."

IDENTIFYING FEATURES:

Great egrets *(Ardea alba),* 32 in (81 cm), are tall, white herons with a yellowish bill and long, dark legs. In breeding adults, the bill is orange, the skin in front of the eyes (the lores) is lime green, and long, wispy plumes extend from the back beyond the tail (both sexes).

HABITAT: Great egrets stalk the surf and other coastal waters for fish but are more common inland. They nest in trees May–August, and share island colonies with other herons and ibises.

DID YOU KNOW? By the early 1900s, plume hunters had wiped out all but a tiny fraction of the Gulf region's great egrets. Their breeding plumes were in high demand as part of the millinery trade, which used the gossamer feathers to decorate women's hats. In 1905, the newly formed National Audubon Society began a mission to protect these birds from hunting, which resulted in the US Migratory Bird Treaty Act (MBTA) of 1918. Populations began to quickly recover following their protection. The great egret is now the symbol of the Audubon Society. The Gulf's population of great egrets increases slightly in winter due to an influx of migrants from the north. The word "egret" comes from the French word *aigrette,* for a grooming brush. The bird's long, frilly breeding plumes are also called "egrets."

Vultures

RELATIVES: These vultures (family Cathartidae) are distantly related to hawks, ospreys, and eagles, all sharing the order Accipitriformes.

IDENTIFYING FEATURES: Vultures are dark, broad-winged, soaring birds.

Black vultures *(Coragyps atratus)*, 22 in (56 cm), are black with gray wingtips and a stubby, squared tail. Adults have a gray, wrinkled, featherless face, and juvenile faces have a less-wrinkled, youthful look. Black vultures soar with straight wings and give quick flaps between glides.

Turkey vultures *(Cathartes aura)*, 25 in (64 cm), are blackish-brown with pale flight feathers and a long, rounded tail. They have a bare head (red in adults, dark in juveniles) with large nostrils. These vultures soar with wings held in a V, rocking erratically as if trying to maintain a drunken balance.

HABITAT: Black and turkey vultures catch dune updrafts and use the beach as a travel corridor. Vultures on the beach are there because of dead fish or other stranded delicacies. Both species may roost together on poles and snags near water. They may also circle together high within slowly whirling thermal updrafts, which help the birds gain altitude before they spread out to search for carrion. These vultures lay their eggs on the ground in spring, often in thick brush.

DID YOU KNOW? Turkey vultures find carrion by smell and are often first to arrive at a carcass. Black vultures find carrion by sight and in groups they are tougher competitors that may chase away turkey vultures and caracaras.

Black vulture soaring

Adult black vulture

Turkey vulture soaring

Adult turkey vulture

223

An osprey hovers over a target fish

Taking flight after a feet-first catch

Fish are carried as aerodynamic cargo

Osprey

RELATIVES: Ospreys (family Pandioni-dae) are with new-world vultures, hawks, and eagles in the order Accipitriformes.

IDENTIFYING FEATURES:

Ospreys *(Pandion haliaetus),* 22 in (56 cm), are raptors that, in flight, hold their wings with a characteristic bend at the wrist (halfway out the leading edge of each wing). Ospreys have a white head and a dark stripe behind each eye. Their upper-parts are dark brown, and their breast, belly, and leg feathers are mostly white. Females are slightly larger than males and have a more streaked breast. Juveniles look similar to adults but have a streaked breast and a lighter back.

HABITAT: Ospreys live and breed near open water. They are most likely to nest in the tallest local dead trees. But where tall trees aren't available, ospreys nest atop towers, electrical poles, and channel markers. In the northern Gulf, ospreys occur year-round, but in most of Texas, these birds are only seen in winter.

DID YOU KNOW? Ospreys hover over water to target fish, then plunge head-and-feet first to grab their prey. In addition to oversized curved talons, an osprey's feet have spiked pads for gripping slippery fish. After the plunge and grab, and a few take-off flaps, the fisher shivers to shake off water-weight. The fish may be heavier than the bird. In flight, prey are held aero-dynamically, as if they were a bomb ready to be dropped. This fore-and-aft grip is possible due to the osprey's reversible outer toes. Ospreys are excellent fishers. They average a fish every four dives, and their time spent searching for each fish is less than 15 minutes.

Crested Caracara

RELATIVES: Caracaras share the order Falconiformes and the family Falconidae with Falcons.

IDENTIFYING FEATURES:

Crested caracaras *(Caracara cheriway),* 21 in (53 cm), have a regal-eagle profile, but are more closely related to falcons. Adult birds stand tall on yellow or orange legs and have a white neck, black cap, and an orange face. Their body is dark, but in flight, they show white flight feathers and undertail. Juveniles are brown and white with drab legs and a pinkish face. Caracaras soar like eagles, but also fly near the ground, and often walk. They frequently join vultures at a tasty carcass.

HABITAT: A crested caracara's quest for carrion spans open fields, dunes, and the beach. They like to perch on bare trees or fence posts, and their nests (late winter through spring) are atop tall trees, shrubs, or cacti.

DID YOU KNOW? Caracaras eat mostly dead animals, but will take live prey such as insects, reptiles, and small mammals. Unlike other raptors that strike at their prey with an aerial ambush, caracaras often use their long legs to run down their quarry. Crested caracaras watch carefully for feeding opportunities. They will pounce on fleeing animals ahead of an advancing fire, pluck unlucky critters flushed by mowers, and follow vultures to large fresh carcasses, which they cannot open on their own. This bird is commonly called the "Mexican eagle," perhaps because it resembles the regal bird on the Mexican national flag, although the bird on the flag's emblem is actually a golden eagle.

Crested caracara in flight

A crested caracara's catfish carryout

The handsome profile of an adult caracara

Adult in winter plumage showing a white rump

Adult before spring breeding migration, hind toe (arrow)

Adult in winter

Black-bellied Plover

RELATIVES: Plovers are in the family Charadriidae and are allied within the order Charadriiformes, which includes other shorebirds, gulls, and terns.

IDENTIFYING FEATURES: Plovers of all stripes have a habit of running in straight lines and stopping abruptly in a still, head-up posture.

Black-bellied plovers (*Pluvialis squatarola*), 9.5 in (24 cm) tail to bill tip, are medium-size, stocky shorebirds that have gray legs and a dark, thick bill. Most beach birds have a pale, mottled back and breast with white underparts and a white rump. In spring and late summer, many show either the beginnings or vestiges of **breeding plumage**, which is black from face to belly, with a thick, white border around the face and throat. Sexes appear similar, although the female's breeding plumage is less vibrant than the male's. Juveniles look like winter adults but have more contrast to their upperparts.

HABITAT: Breeding takes place in Arctic tundra. Gulf beaches provide important winter foraging locations. These birds are seen as individuals or in small groups as they probe for worms, coquina clams, and mole crabs in the swash zone.

DID YOU KNOW? Black-bellied plovers are sensitive to disturbance and serve as sentinels for other shorebird species. They are the only American plover to have a **hind toe** (albeit tiny) on its foot. This plover's large eyes help it spot signs of small invertebrate prey under the sand, even at night. Their foraging schedule is driven strongly by tides. As soon as a falling tide begins to expose sandy or muddy flats, these plovers leave their roost to feed.

Plovers *(Snowy, Piping)*

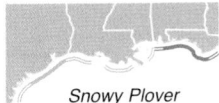
Snowy Plover

Piping Plover

RELATIVES: These plovers share the family Charadriidae with other plovers.

IDENTIFYING FEATURES:

Snowy plovers *(Charadrius nivosus),* 5.25 in (13 cm) tail to bill tip, are dainty shorebirds the color of pale sand. They have white underparts, a black bill, and gray legs. In the spring, males have a black patch behind each eye.

Piping plovers *(Charadrius melodus),* 5.5 in (14 cm) tail to bill tip, are similar to snowy plovers but have a white collar and orange legs. In breeding plumage, these plovers sport a black bar across the forehead, a dark breastband, and a dark orange bill with a black tip. Juveniles look like winter adults.

HABITAT: These plovers forage for small invertebrates in the wrack, swash, and nearby flats. They need undisturbed, barren, dry sand for nesting. Snowy plovers breed on isolated Gulf beaches March–September. Piping plovers wintering along the Gulf breed between Nebraska and southern Canada, April–August.

DID YOU KNOW? Snowy plover chicks leave the nest within hours of hatching and find their own food near watchful parents. Threats from a predator will prompt parents to call, which causes the puffy chicks to flatten against the ground. Many of these birds are **banded** (third image) to study their survival and breeding success. Alabama and Florida respectively list snowy plovers as Endangered and Threatened. The piping plover has a federal Threatened designation. Both species have lost undisturbed areas for breeding, feeding, and resting.

Adult male snowy plover in spring plumage

A female snowy plover on her sand-colored eggs

Adult piping plover in colors ready to breed

Adult piping plover in winter

227

Adult Wilson's plover in winter

A female on eggs within a "scrape" nest

Immature Wilson's plover with a mole crab

Wilson's Plover

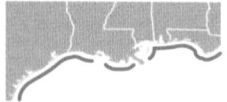

RELATIVES: Other plovers, family Charadriidae.

IDENTIFYING FEATURES:

Wilson's plovers *(Charadrius wilsonia)*, 6.25 in (16 cm), are small shorebirds with grayish-brown upperparts and white underparts. They have a white collar, and a white patch between their eyes. The large black bill and tan legs of Wilson's plover distinguish it from other small plovers. In the spring, the male has a black chest band, and the breeding female's band is brown. Juveniles look like winter adults but with scaly-patterned backs.

HABITAT: Wilson's plovers nest April through June in open spaces between dunes. They feed on fiddler crabs, mole crabs, and other invertebrates in the swash zone and on tidal mudflats.

DID YOU KNOW? A male Wilson's plover makes a **"scrape" nest** in the sand, often near a small plant, and lines it with shell bits. The female lays 2–4 speckled eggs that blend in well. Incubation is about 26 days. The chicks are precocial, fleeing the nest immediately after hatching. These and all nesting plovers have dramatic "distraction displays." A threatened parent will flutter quickly away from their nest, pretending to have an injured wing. The diversion is meant to draw predators away from eggs and chicks. It is also a signal to you that you are too close to a nest. Like other beach-nesting birds along the Gulf Coast, Wilson's plover has lost undisturbed areas for breeding, feeding, and resting. There is concern that this species has declined to the point of needing special protection, but it has yet to be federally listed as Endangered.

Semipalmated Plover and Killdeer

Semipalmated Plover

Killdeer

RELATIVES: Other plovers, family Charadriidae.

IDENTIFYING FEATURES:

Semipalmated plovers *(Charadrius semipalmatus)*, 6 in (15 cm), have brown upperparts, a white collar, and white underparts. Their bill is short, and their legs are yellow-orange. In flight, white stripes are visible on the wings, and the dark tail has a fringe of white. In spring and summer breeding plumage, semipalmated plovers have a thick, black, breast band and an orange base to their bill. The sexes look similar, although the female is larger and duller in color. Juveniles resemble adults but have yellower legs and scaly plumage on the upperparts.

Killdeer *(Charadrius vociferus)*, 9 in (23 cm), are tall plovers with a long tail and wings, show two black breast bands, and have a bright red eye ring.

HABITAT: Semipalmated plovers breed during summer in open areas near the lakes of northern Canada. During winter along the Gulf Coast, they forage on small invertebrates found in the wrack and swash zone of varied beaches. Killdeer forage and breed in open fields and dunes near the coast and inland.

DID YOU KNOW? In their northern breeding areas, semipalmated plover parents swim with their chicks across water to forage on islands. They are aided by the partial webbing between their "semipalmated" toes. Killdeer were named for their shrill *DEE-DEE* … call. If you hear this, it probably means you are too close. The bird's scientific name literally means "noisy plover."

A semipalmated plover shows its wingstripes

Adult semipalmated plover in spring colors

Adult semipalmated plover in winter plumage

Killdeer are noisy plovers

229

Juvenile with blackish bill tip

Adult American oystercatcher

Adults have a bright orange bill and eye ring

American Oystercatcher

RELATIVES: Oystercatchers are alone in the family Haematopodidae and are distantly related to plovers, sandpipers, gulls, and terns.

IDENTIFYING FEATURES:

American oystercatchers *(Haematopus palliatus),* 16 in (41 cm), are large, boldly colored, long-billed shorebirds. Their back is a dark brown, their neck and head are black, and their belly is stark white. They have an unmistakable red-orange bill, orange rings encircling their yellow eyes, and robust, pinkish legs. Summer and winter plumages are almost identical and the sexes look similar. In comparison to adults, juveniles have a darker end to their bill, a darker eye with a less conspicuous eye ring, and lighter upperparts. Oystercatchers are often seen in pairs.

HABITAT: These skittish shorebirds forage for oysters, clams, crabs, worms, and mole crabs along beaches and estuarine shores. They are solitary nesters on undisturbed beaches, exposed shell and sandbars, and dredge-spoil islands between March and July.

DID YOU KNOW? As their name suggests, oystercatchers use their chisel-like bill to open oysters and other bivalves. They are the only local birds able to open and eat these large mollusks. Oystercatchers are monogamous and may maintain a pair bond for life. The birds are unusual in that after a breeding season, young oystercatchers don't follow their parents to winter foraging grounds, and may even migrate in the opposite direction. Some adults forage near their previous breeding area, and some migrate hundreds of miles to distant foraging grounds.

Willet

RELATIVES: Willets are sandpipers in the family Scolopacidae and are allied with plovers, gulls, and terns in the order Charadriiformes.

IDENTIFYING FEATURES:

Willets *(Tringa semipalmata)*, 14 in (35 cm), are long-legged, long-billed, drab-looking shorebirds. A willet in winter is gray-brown above and white below with a gray chest. Their long, straight bill is dark, and their legs are bluish-gray. In flight, willets show a striking wing pattern with a white band through black wingtips. In spring and summer, adult willets are mottled on the upperwings, back, neck, and head. The sexes look alike with the female being slightly larger. Juveniles resemble adults but are browner with white edges on their back feathers.

HABITAT: Breeding **eastern willets** make nests in grassy dunes and saltmarsh along the Gulf Coast from April through mid-June. These birds are the only sandpiper that breeds on Gulf beaches, and they remain here year-round. **Western willets** are from a population that breeds on the prairies of the northern US and southern Canada, and are on the Gulf Coast only in winter. Willets from either group probe the swash zone for coquina clams and mole crabs.

DID YOU KNOW? The willet gets its name from its loud and repetitious *pill-will-willet* alarm call. Western willets are slightly larger and paler gray compared to eastern willets, which have stouter bills and more barring on their breast and back. The two "races" also differ in the pitch of their calls, and in how each responds to the others voice.

Adult in winter showing contrasting wing pattern

Western willet in winter

Eastern willet in summer

231

Ruddy turnstones reveal bold patterns in flight

Winter adult. Juvenile (inset)

Adult before spring breeding migration

Ruddy Turnstone

RELATIVES: Ruddy turnstones are sandpipers in the family Scolopacidae.

IDENTIFYING FEATURES:

Ruddy turnstones *(Arenaria interpres)*, 7 in (18 cm), are stocky shorebirds with orange legs and a dark, wedge-shape bill. In flight, they show a bold set of white wing stripes and a white rump. Juveniles and winter adults are white below with a brownish head, bib, and back. In spring, turnstones develop a black-and-white head, white belly, black bib, and a rusty red back and wings. The sexes look similar, and juveniles resemble winter adults.

HABITAT: Turnstones forage on both the lower and upper beach but favor the wrack line. They also frequent piers and jetties. These birds migrate to nest on islands in the Canadian Arctic and return to winter on Gulf beaches.

DID YOU KNOW? Ruddy turnstones earn their name by flicking aside stones, shells, and beach wrack that may hide amphipods and other tasty invertebrates. These birds often allow close observations by cautious beachcombers and are a joy to watch. Like other shorebirds that are marathon migrants, ruddy turnstones face the critical challenge of gaining fat for their breeding journey. This fat comes from winter feeding and is the fuel needed to power their 3,500-mile (5,600 km) migration. Turnstones that are not plump must continue feeding and depart late, and thin birds may never make it back to Gulf wintering beaches. To cover their flight distance in advance of seasonal weather changes, these little birds must average up to 45 mph (72 kph).

Red Knot

RELATIVES: Red knots are sandpipers in the family Scolopacidae.

IDENTIFYING FEATURES:

Red knots *(Calidris canutus),* 10 in (25 cm), are stout, robin-size sandpipers with greenish legs and a straight dark bill. Red knots in flight show a pale, mottled-gray rump as their key identifier. Birds on Gulf beaches are generally pale gray above and light below, but some may show hints of breeding plumage during spring and late summer—a brick-red head, neck, and breast, and a gray back with rusty spots.

Red knots have mottled rumps

HABITAT: Red knots prefer the lower beach for foraging and roost on upper beaches that are broad, flat, and undisturbed. In the spring, these shorebirds migrate to nest on the high, open tundra of the central Canadian Arctic.

DID YOU KNOW? During spring and fall, many red knots seen on Gulf beaches are just passing through. Some of these belong to a group that winters in southern South America and flies over 20,000 miles (32,000 km) on each migration circuit. Keep this in mind when you see a flock of birds on the upper beach being "lazy." Chances are, the birds are taking some critical downtime between connecting flights. Some of the most important staging areas for these birds' spring migration lie north in Delaware Bay where formerly massive flocks fed on eggs from spawning horseshoe crabs (p. 177). Reduction in this food source has caused severe declines in the subspecies, rufa red knot *(C. c. rufa),* which recently received federal protection as a Threatened species under the Endangered Species Act.

Late summer adult, with colored leg bands

Winter adult

Adult sanderling in flight, late summer

Adult before spring breeding migration

Adults foraging in winter

Sanderling

RELATIVES: Sanderlings are sandpipers in the family Scolopacidae.

IDENTIFYING FEATURES:

Sanderlings *(Calidris alba),* 7 in (18 cm), are frantic, swash-dodging little sandpipers with black legs and a straight, black bill the same length as their head. In winter, adult plumage is pale gray above and white below. Just before and after their spring migration, a breeding adult is rusty and speckled on the back, head, and breast. Juveniles are black-white-gray checkered above, and white below.

HABITAT: Sanderlings probe like tiny sewing machines in the wet sand briefly exposed between swash and backwash. Most adults migrate in spring to nest on the Arctic tundra. Gulf sightings become uncommon in summer.

DID YOU KNOW? Although sanderlings are mostly monogamous, some females feeling sufficiently plump will stray from their mate and lay eggs in the nest of another male. At the opposite extreme, birds that do not feel plump will remain in winter foraging areas while other birds migrate to breed. Sanderlings dodge waves to catch worms, mole crabs and other small invertebrates that burrow within swash-zone sands. They may either swash-run in groups, or go it alone. Lone birds may be defending a patch of beach with lots of prey. These territorial birds can be seen chasing others around in a hunched-over run. A flock of sanderlings will leave a scattering of regurgitated pellets (p. 256) in their wake. The pellets contain invertebrate shell fragments and sand. Regurgitating is a way to dispense with ingested items that are too tough to digest.

Sandpipers *(Semipalmated, Western)*

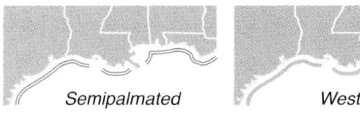

Semipalmated   Western

RELATIVES: These sandpipers are in the family Scolopacidae.

IDENTIFYING FEATURES: Both birds are small, rust-mottled to grayish-brown shorebirds with a black bill and dark legs. They walk slowly, rapidly probing tidal flats for tiny prey. These "peeps" (small sandpipers) are tough to tell apart.

Semipalmated sandpipers *(Calidris pusilla),* 6 in (14 cm), have a bill that is shorter than head length, stout, and straight. Like the western sandpiper, adults in winter are gray-brown above, and in late summer are rufous above with dark mottling and a streaked breast. Semipalmated sandpipers taking flight give out a rapid, laugh-like alarm, and high-pitched *cherk* flight call.

Western sandpipers *(Calidris mauri),* 6 in (14 cm), have a bill that is longer than the head and has a slight droop. Otherwise, they look similar to semipalmated sandpipers and are most reliably distinguished by their vibrato alarm call and high-pitched *cheep* flight call.

HABITAT: Both of these peeps breed in Arctic tundra. Semipalms fly to central Canada, and westerns travel to western Alaska. Although western sandpipers spend cold months on the Gulf Coast, semipalms seen on Gulf beaches in fall and spring are just passing through on flights to and from central America.

DID YOU KNOW? Semipalmated sandpipers get their name from the webs between their toes. Although this trait is rare, western sandpipers have it too. The webbing helps the birds swim at the water's surface.

Juvenile semipalmated sandpiper, fall

Adult semipalmated sandpiper, winter

Western sandpiper, late summer

235

Dunlins in flight, winter

Adult in winter plumage

Adult in spring

Dunlin

RELATIVES: Dunlins are sandpipers in the family Scolopacidae.

IDENTIFYING FEATURES:

Dunlins *(Calidris alpina)*, 7.5 in (19 cm), are medium-size sandpipers with dark legs and a long, slightly droopy, black bill. Flying dunlins show their white rump with a dark central line. Adults with rusty-red backs and black belly patches are seen in near Gulf beaches before and after their spring breeding migrations. The sexes look similar, although the female is slightly larger.

HABITAT: Dunlins prefer to probe for invertebrates on low-wave-energy tidal flats and are uncommon on beaches with rough surf. These birds winter on the Gulf Coast and migrate to nest in the Arctic coastal tundra, May through July.

DID YOU KNOW? Although they are occasionally found with sanderlings, dunlins are much slower and more methodical in their feeding. In the late fall and winter, beach flocks of dunlins may number in the thousands. Bill length between individual dunlins can differ 30 percent or more, which may reflect connections between the world's dunlin populations. Some American dunlins have breeding areas that overlap with breeding sites of birds that winter in Asia.

Least Sandpiper
and **Marbled Godwit**

Least Sandpiper   Godwit

RELATIVES: These birds are sandpipers in the family Scolopacidae.

IDENTIFYING FEATURES:

Least sandpipers *(Calidris minutilla),* 5 in (13 cm), are the smallest sandpiper, about the the size of a sparrow. Otherwise, they are similar in appearance to other "peeps," but least sandpipers differ in having yellow-green legs. Other peeps on Gulf beaches have longer bills and dark legs.

Marbled godwits *(Limosa fedoa),* 20 in (50 cm), are large shorebirds with a long, slightly upturned bill, which has a pink base and a dark tip. Their plumage is marbled buff-brown with cinnamon wing linings. Adult birds in spring and summer have chest barring; winter and immature birds have a plain breast.

HABITAT: As the tiniest peep on the beach, least sandpipers rarely enter the wave wash, preferring to probe for invertebrates in runnels, on tidal flats, and near rock jetties. Marbled godwits probe deeply for invertebrates within muddy tidal flats. These birds winter on the Texas coast and most migrate in spring to nest in short-grass prairies near wetlands in central Canada. Least sandpipers breed across subarctic North America.

DID YOU KNOW? Marbled godwit nests are well hidden. A parent's tactic is to remain still when a predator intrudes, to the extent that a biologist can pick a bird up off its nest to check the eggs. Although the least sandpiper is the smallest shorebird in the world, they are marathon flyers. This tiny bird makes nonstop flights over the ocean spanning 2500 miles.

Adult least sandpiper in spring foraging on a jetty

A marbled godwit shows its cinnamon wing linings

A trio of marbled godwits, late spring

237

Long-billed curlew in flight

Long-billed curlews are our largest shorebird

A short-billed dowitcher in winter plumage

Adult short-billed dowitcher in spring

Long-billed Curlew
and **Short-billed Dowitcher**

Curlew

Dowitcher

RELATIVES: These long-nosed sandpipers are in the family Scolopacidae.

IDENTIFYING FEATURES:

Long-billed curlews (*Numenius americanus*), 23 in (58 cm), are the Gulf's largest shorebird. They have an exquisitely long curved bill and are speckled with dark and light over cinnamon undertones, with a plain cinnamon belly. The whimbrel (*Numenius phaeopus*), 17 in (44 cm), is similar and related, but less common on beaches. Whimbrels have a much shorter curved bill, are smaller birds, and show a striped crown.

Short-billed dowitchers (*Limnodromus griseus*), 10 in (25 cm), are medium-size shorebirds with a plump body, greenish-yellow legs, and a long, straight, dark bill. Winter birds are grayish, and birds ready to breed become flecked with cinnamon and black.

HABITAT: In winter, curlews and dowitchers probe wet sand and mud for small invertebrates along Gulf beaches and tidal flats. Long-billed curlews migrate in spring to breed on grasslands of the Great Plains. Short-billed dowitchers nest near water in central Canada.

DID YOU KNOW? Although long-billed curlews are uncommon, the lower Texas coast has more of these elegant birds than anywhere else in eastern North America. These sandpipers use both visual and tactile cues to locate worms, shrimps, and small crabs hidden in sediment. More so than other sandpipers, dowitchers have a flexible, sensitive bill tip that allows them to grasp deeply buried prey.

Laughing Gull

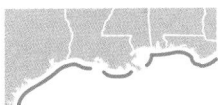

RELATIVES: Gulls are allied with terns and skimmers in the family Laridae and are distantly related to plovers and sandpipers.

IDENTIFYING FEATURES:

Laughing gulls *(Leucophaeus atricilla),* 16 in (40 cm), are slender, long-winged gulls. Adults have a smooth gray back and dark legs. Juveniles are brownish with a scaly back. Like most gulls, their bill tip droops. In the spring, adult laughing gulls have a black head and a deep red bill and legs.

HABITAT: These gulls are common on beaches and throughout other coastal areas. They nest in May and June in saltmarsh and on partially bare islands within bays and lagoons. Most laughing gulls along the Gulf Coast remain all year round.

DID YOU KNOW? Laughing gulls are the beach bird most likely to steal food from your hand, and their call will make you think that they've thoroughly enjoyed the prank—*ha-ha-ha-hah-haah-haah.* Their forward nature is promoted by beachgoers who keep these birds addicted to junk food. Their natural forage is opportunity seafood, including relatively fresh wrackline treats (sea carrion). Spring breeding colonies in Texas number more than 25,000 pairs. Their nests range from simple hollows in the sand to elaborate structures of interwoven saltmarsh grasses.

Adult winter plumage

Juvenile in its first summer (L), adult (R)

Juvenile before its first winter

239

Adult Franklin's gull in spring

Adult ring-billed gull in late winter

A juvenile ring-billed gull's first winter plumage

Gulls *(Franklin's, Ring-billed)*

Franklin's   Ring-billed

RELATIVES: These gulls share the family Laridae with terns and skimmers.

IDENTIFYING FEATURES:

Franklin's gulls *(Leucophaeus pipixcan)*, 14 in (36 cm), are small, gray-backed gulls with a slim, short bill. Gulf Coast birds are passing through, mostly April and November. Spring adults have a black head with white crescents above and below the eye, and the legs and bill are reddish. Fall migrants have a dark hood or mask, and the bill and legs are dark.

Ring-billed gulls *(Larus delawarensis)*, 18 in (45 cm), are medium-size, large-headed gulls with pale gray backs. Their light bill has a distinct black ring at its tip. Juveniles have a black band on their grayish tail, pink legs, and a wide ring at the end of their pinkish bill. Adults are lighter, have a yellow bill with a narrow band, and sport yellowish legs.

HABITAT: Franklin's gulls nest in freshwater marshes of central Canada and winter along coastal South America. Ring-billed gulls seen in winter along the Gulf breed between the Great Lakes and central Canada. Nests are in colonies on sparse terrain near water.

DID YOU KNOW? In spring, Franklin's gulls gain a pink blush to their breast. The color comes from carotenoid pigments in the birds' food. Ring-billed gulls are resourceful scavengers with the reputation of being the "fast food gull." The notoriety comes from their habit of hanging out where french fries are served. The healthier part of their diet comes from fish, dipped on-the-fly from surface waters, and from treats gleaned from wracklines and tidal flats.

Gulls *(Bonaparte's, Lesser Black-backed)*

Bonaparte's

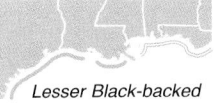
Lesser Black-backed

RELATIVES: These gulls share the family Laridae with terns and skimmers.

IDENTIFYING FEATURES:

Bonaparte's gulls *(Chroicocephalus philadelphia),* 11 in (28 cm), are dainty, ternlike gulls with a gray back and wings, pinkish-orange legs, and a thin, black bill. Their head is white with a distinct dark spot behind the eye. Juveniles are brown above with dark markings on the head and paler legs. First-year birds look similar to juveniles but have gray, dark-patterned upperparts. Breeding adults up north have a jet-black head with white eye crescents.

Lesser black-backed gulls *(Larus fuscus),* 20 in (52 cm), are large gulls with a thick bill. Adults have a dark gray back, yellow legs, and a yellow bill with red spot near the lower tip. Juveniles are dusky brown with a pale face, light rump, dull pink legs, and wings barred with dark and light. Rarer, great black-backed gulls *(Larus marinus)* are much larger, have darker backs, and show pink legs as adults.

HABITAT: Bonaparte's gulls pluck small fish from the water, either on-the-fly or while paddling in the surf. They breed during summer in central Canada where they nest in fir and spruce trees near freshwater marshes. Lesser black-backed gulls nest on the Icelandic tundra.

DID YOU KNOW? Unlike other gulls, Bonaparte's gulls shun table scraps and carrion. Its principle diet comprises live insects and small fish. The bird is named for Charles Lucien Bonaparte, ornithologist and nephew of French emperor, Napoleon. Lesser black-backed gulls are much more common in Europe.

Adult Bonaparte's gull in winter

Bonaparte's gull in first-year plumage

Adult lesser black-backed gull in winter

Juvenile lesser black-backed gull

241

Late summer adult in flight

Adult winter plumage

Juvenile in first winter. Gull in second winter (inset)

Herring Gull

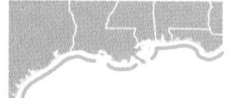

RELATIVES: These gulls share the family Laridae with terns and skimmers.

IDENTIFYING FEATURES:

Herring gulls *(Larus argentatus),* 23 in (59 cm), are our most commonly seen large gull. Juveniles and first-year birds are brownish and have a dark bill with a pale base. Second-year birds become paler with a hint of gray and have a pinkish, black-tipped bill. Adults have a light gray back and a yellow bill with a red spot on the lower tip. Although the adult's head is streaked in winter, summer birds have a head that is immaculately white. All ages show dull pink legs. These gulls take four years to reach adult plumage. The sexes look alike, although the male is slightly larger.

HABITAT: These gulls are colonial nesters, breeding in summer on islands across the northern US and Canada. They feed most commonly near water but also frequent garbage dumps. The birds frequently loaf on open beaches and on the asphalt of beach-side parking lots.

DID YOU KNOW? Herring gulls are moving south. Some of their northernmost breeding areas have been taken by great black-backed gulls, and their southern shift has displaced some laughing gulls. Most herring gulls seen on Gulf beaches are immatures and winter adults. Breeding birds tend to stay near nesting areas. These birds inspired Richard Bach's novel *Jonathan Livingston Seagull.* Despite the title there are no "seagulls," only gulls. Herring gulls can live to almost 30 years.

Caspian Tern

RELATIVES: Terns share the family Laridae with gulls and skimmers, and are distantly related to plovers, oystercatchers, and sandpipers (order Charadriiformes).

IDENTIFYING FEATURES:

Caspian terns *(Hydroprogne caspia),* 20 in (49 cm), are large terns (the world's largest). They have black legs, a black cap, and a thick, pointed, reddish bill with a dark tip. Juveniles have black edging to their back feathers, but adults are silvery gray and white. In winter, their cap fades into speckles, beginning on their forehead. The sexes look alike.

HABITAT: These terns loaf on Gulf beaches and sandbars as singles or in small groups where they are often outnumbered by royal terns. They feed on fish by plunge-diving into coastal waters. In summer, Caspian tern nesting colonies return to open areas near water around the Great Lakes, in eastern Newfoundland, and on artificial islands in Tampa Bay, Florida.

DID YOU KNOW? Because their numbers have declined, the Caspian tern is classified as Vulnerable in Canada, and as Endangered, Threatened, or Species of Special Concern in US states throughout the bird's range. If they avoid threats, these birds can live a long life. One wild Caspian tern was re-sighted over a 32-year period. Like some other terns, young Caspian terns require a lengthy apprenticeship period to learn the art of fish catching. They often remain with their parents during the migration south, and the young gulls mooch food during their first winter.

Adult in winter

Adult in spring

Caspian terns are our largest tern species

Immature royal tern in winter. Note dark wingtips

In late summer, a juvenile begs from an older tern

Adult with a sargassum frogfish plucked from wrack

Adult in winter

Royal Tern

RELATIVES: Terns share the family Laridae with gulls and skimmers.

IDENTIFYING FEATURES:

Royal terns *(Thalasseus maximus),* 18 in (45 cm), are large terns with a dark cap and a relatively slender orange bill. They are similar to Caspian terns but differ in that royal terns are smaller with a lighter bill. In winter, the royal tern's black cap recedes, trailing shaggy feathers in the style of wind-blown male-pattern baldness. Juveniles have dark wing tips and orange legs. Older immatures look similar to winter adults, but have dark wingtips. The sexes look alike.

HABITAT: Royal terns feed on fish from coastal and offshore waters. Crowded, seasonal breeding colonies are scattered in the Gulf from Texas through Florida. Most nesting is April–July. Royal terns make a scrape-nest on sandy or shelly islands. The mated pair surrounds their egg (often only one) with a circular nest rim that is cemented with their own guano.

DID YOU KNOW? After hatching, young royal terns from many nests hang out together in a group known as a crèche, which may eventually accept every chick in the colony. Parents find their own chick among hundreds by recognizing its call. Juveniles are fed by parents even after migrating away from the breeding colony. Young birds may still **beg** parents for food several months after leaving the nest. Royal terns remain social birds outside the breeding colony, often joining large mixed-species flocks on the beach.

Sandwich Tern

Winter adult. Note the aluminum leg band

RELATIVES: Terns share the family Laridae with gulls and skimmers.

IDENTIFYING FEATURES:

Sandwich terns (*Thalasseus sandvicensis*), 16 in (40 cm), are medium-size terns with a black cap, black legs, and a long, thin, black bill, which in adults has a pale yellow tip. Their black cap is complete during summer breeding, but in winter it recedes on their head to the back and sides. Immatures look like winter adults, but lack the yellow bill tip. Males and females look alike.

A juvenile (left) begs a parent for food

HABITAT: These terns plunge-dive for fish in the surf and other shallow coastal waters. They nest May–July on coastal islands from Texas to Virginia. Most Gulf nesting occurs with royal terns on islands in Texas' Laguna Madre, off Louisiana near the Mississippi Delta, and in Tampa Bay. Young sandwich terns group in cooperative crèches with royal terns. Immature terns migrate with their parents and continue **begging** for food through their first winter.

Mating pair of adults in spring

DID YOU KNOW? Remember this bird by its mustard-tipped bill—as in sandwich mustard. The tern is named for the town of Sandwich in Kent, England, origin of the specimen for which the species was described. Also named for this town— John Montagu, 4th Earl of Sandwich, who was a fan of meals between bread slices. Ponder that full circle as you eat your picnic lunch and watch the terns dive for theirs. Sandwich terns have a mixed relationship with people. They have lost some breeding habitat due to disturbance and human-induced sea level rise, but they also benefit from artificial nesting habitat such as dredge-spoil islands.

Adult in winter

Adult Forster's tern in spring showing forked tail

Forster's tern, early-winter adult

Adult gull-billed tern in spring

Gull-billed terns have a straight, stout, black bill

Terns (*Forster's, Gull-billed*)

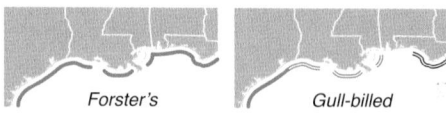
Forster's   Gull-billed

RELATIVES: Terns share the family Laridae with gulls and skimmers.

IDENTIFYING FEATURES:

Forster's terns (*Sterna forsteri*), 13 in (33 cm), are medium-small terns with a long forked tail and orange legs. Winter adults have a black eye mask, which turns into a cap by the spring breeding season. The bill is dark, and has an orange base in breeding adults. Immatures resemble winter adults, and the sexes look alike.

Gull-billed terns (*Gelochelidon nilotica*), 14 in (35 cm), are stocky, mid-size terns with wide wings, a stout black bill, and black legs. Breeding adults have a black cap, and in winter, the head is white with a dark smudge behind each eye. Wings are pale with dark tips. Immature birds resemble winter adults.

HABITAT: These terns feed on small fish in shallow coastal waters. Forster's terns nest on floating grass mats in marshes of Texas and Louisiana, and up north in the Great Lakes region. They nest singly or in loose colonies. In winter, both northern and Gulf-nesting Forster's terns are on Gulf beaches. Gull-billed terns are uncommon. They nest on coastal islands, spoil banks, and gravel rooftops.

DID YOU KNOW? Although the travels of a Forster's tern can span the US from north to south, this bird has one of the smallest ranges of our tern species. They breed in the same areas as black terns where cross-species feeding of chicks has been reported. Johann Reinhold Forster was a pastor and naturalist who made many ornithological discoveries on Captain Cook's world voyage in 1772.

Least Tern

RELATIVES: Terns are with gulls and skimmers in the family Laridae.

IDENTIFYING FEATURES:

Least terns (*Sternula antillarum*), 9 in (23 cm), are tiny terns with short yellow legs. Their sharp yellow bill has a black tip. During spring and summer, adults have a black cap with a white forehead. Winter adults (not seen in the Gulf) have a black eyestripe, a white cap, and a dark bill. Beach colonies have **chicks** visible, which begin life as sand-camouflaged puffballs. Juveniles have a buffy crown and scaly back. Older immature birds resemble winter adults, and the sexes look alike.

HABITAT: These terns feed on small fish in coastal waters. Gulf breeding least terns winter in South America from Venezuela to Brazil. Their nesting here once occurred mostly on beaches, but due to human disturbance, the majority of nesting now takes place on the gravel rooftops of large buildings built near water. Breeding takes place from late April to early August.

DID YOU KNOW? Breeding least terns are sensitive to disturbance and will dive-bomb nest-colony intruders. With undisturbed beaches rare, and with gravel rooftops being replaced by more modern roofing, the future of the world's smallest tern is uncertain. A few beaches remain where signs and string protect tern colonies from human visitors, who are able to enjoy the bird's aerial feats and family life from a distance.

Adult hovering above the surf before a plunge

A male courts a female with a gift anchovy

A female succumbs to her fisherman's charm

A cryptic, least tern chick, juvenile (inset)

247

A bridled tern searches a sargassum line out at sea

Adult bridled tern on flotsam. Juvenile (inset)

Adult black tern in spring

Molting adult black tern in late summer

Terns *(Bridled, Black)*

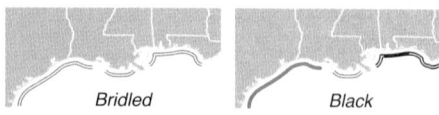

Bridled   Black

RELATIVES: Terns share the family Laridae with gulls and skimmers.

IDENTIFYING FEATURES:

Bridled terns *(Onychoprion anaethetus)*, 15 in (38 cm), are medium-size terns with dark upperparts, a deeply forked tail, black legs, a black bill, a white forehead, and a bridle-like cowl. Immature birds have a white head streaked with black. Bridled terns and the similar sooty tern *(Onychoprion fuscatus)* are seen on beaches only after hurricanes. Sooty terns have a darker black back, darker undersides of flight feathers, and white on the forehead that does not extend past the eye.

Black terns *(Chlidonias niger)*, 10 in (25 cm), are small, dark terns with a thin black bill. Spring adults have a dark head and chest and gray wings. Winter birds are white in front with a dusky crown and nape. **Molting adults** have a messy pattern of black and white.

HABITAT: Bridled terns pluck fish and crustaceans from offshore sargassum, and are seen near beaches only after storms. These tropical terns nest on isolated rock-islands in the West Indies and Central America. Black terns breed in freshwater marshes of the Midwest and Canada. While inland, these small terns eat insects. On the coast and at sea, they dip up small fish. Migrating black terns are often seen loafing in flocks on Gulf beaches.

DID YOU KNOW? Black terns are social birds that forage, roost, and migrate in flocks. Many other pelagic birds show up on Gulf beaches following hurricanes, including shearwaters (family Procellariidae), storm-petrels (family Hydrobatidae), and boobies (family Sulidae).

Black Skimmer

RELATIVES: Skimmers are in the family Laridae with gulls and terns.

IDENTIFYING FEATURES:

Black skimmers *(Rynchops niger)*, 16 in (41 cm), are medium-size, short-legged seabirds decked out in Halloween colors. They have a long, unmistakable, scissor-like bill with a red-orange base and dark tip. Their lower bill is knife-thin and much longer than the upper bill. Adults have a black cap and upperparts, and immatures are darkly mottled above. Sexes look alike, although males are slightly larger and have a longer bill.

HABITAT: Skimmers feed by gracefully slicing their lower bill through surface waters where small fish are caught unaware. They are year-round Gulf residents. In April, skimmer breeding colonies form on remote beaches, exposed sandbars, and gravel rooftops, often with nesting least terns. By the end of August, most chicks have fledged, and become juveniles at a little over a month old. In the fall and winter, the Gulf's skimmer population increases with migrants from Georgia and the Carolinas.

DID YOU KNOW? Skimmers fish by feel before dawn and at dusk. Birds resting during the day have probably returned from an early morning fishing trip. They appreciate each other's company, but not yours. If disturbed, a flock will yip like a pack of excited chihuahuas. Our black skimmer population has declined almost 90 percent since the mid 1960s, with the principal cause being nesting habitat loss and disturbance. Research has shown that simple signs and strings between sticks work well to keep nesting colonies undisturbed.

A black skimmer skimming (adult in winter plumage)

Black skimmers nest in colonies

Skimmers loafing on the lower beach

A knife-thin lower bill allows skim-fishing

An adult American avocet in April

Adult avocet wades in the swash zone

Black-necked stilt adult in May

Avocet and Stilt

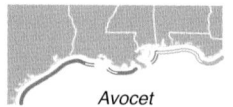

Avocet

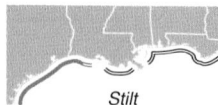

Stilt

RELATIVES: Avocets and stilts (family Recurvirostridae) are in the order Charadriiformes with distantly related plovers, sandpipers, oystercatchers, gulls, and terns.

IDENTIFYING FEATURES:

American avocets *(Recurvirostra americana),* 18 in (46 cm), are tall shorebirds with a long, thin, upturned bill. Adults in spring have a rusty head and neck that turns to gray after the breeding season. Their body is mostly white except for black on their back and wings. Their long legs are bluish gray.

Black-necked stilts *(Himantopus mexicanus),* 15 in (38 cm), have extremely long, bright-pink legs and a thin, straight, black bill. Adults are dark on top and white below, and immatures are brownish above with a duller leg color.

HABITAT: Avocets and stilts tweeze small invertebrates from low-energy beaches, wetlands, and tidal flats. Their nesting takes place May through August in loose colonies. On the coast, both species nest in open areas near water, but many avocets wintering on the Gulf Coast migrate to breed on barren dikes and islands in the central US and Canada. Most local stilts breed in coastal Texas, Louisiana, and southern Florida.

DID YOU KNOW? In proportion to their bodies, stilts have the second-longest legs in the bird world, bested only by flamingos. Black-necked stilts and American avocets are so closely related that they occasionally hybridize, producing an offspring known as an "avo-stilt." These hybrids have a black neck, dull-pink legs, and a slightly upturned bill.

Grackles

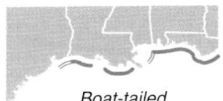
Boat-tailed

Great-tailed

RELATIVES: Grackles are blackbirds (family Icteridae) sharing the order Passeriformes with crows, swallows, and other perching birds.

IDENTIFYING FEATURES:

Boat-tailed grackles *(Quiscalus major)*, 14 in (36 cm), are long-legged, strutting blackbirds with a tapered, downcurved bill and dark eyes. Adult females are about half the male's size and are brown with a dark back, dark wings, and a pale eyebrow. Adult males appear dark from a distance, but are iridescent indigo in the sun. The male's long tail is folded in a V, like a boat keel. This grackle's call ranges from a slow series of *chut*s, to clacks and screams when predators approach.

Great-tailed grackles *(Quiscalus mexicanus)*, 18 in (46 cm), are similar to boat-tailed grackles except for their distribution, eye color, size, and call. The two species overlap only near the Texas-Louisiana border. Great-tailed grackles have much lighter eyes and are larger than boat-tails. Great-tails are also louder, with calls that include an ascending slide whistle, rattling notes, and a rapid-fire *KI-KI-KI …*

HABITAT: These grackles feed in a wide array of open areas, including fast-food parking lots. Boat-tails nest in saltmarsh, and great-tails nest in shrub tops.

DID YOU KNOW? Breeding boat-tailed grackles form defended harems. In this system, males compete for mating access to females that have clustered their nests together. The winning male defends his harem from other males, but sneaky males almost always father some chicks. Great-tailed grackles often form enormous winter roosts with up to half a million birds.

A female boat-tailed grackle feeds on seaoats

A male boat-tailed grackle struts on the beach

Adult female great-tailed grackle

Male great-tailed grackles have loud calls

251

Male red-winged blackbirds have bright epaulets

Red-winged blackbirds, calling male (L) and flock (R)

A female red-winged blackbird snacks on seaoats

Tree swallows (left and top). Barn swallow (bottom)

Red-winged Blackbird and **Swallows**

Blackbird

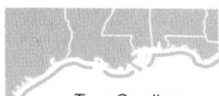
Tree Swallow

Barn Swallow

RELATIVES: Blackbirds (family Icteridae) and swallows (Hirundinidae) are passerines in the order Passeriformes.

IDENTIFYING FEATURES:

Red-winged blackbirds (*Agelaius phoeniceus*), 8 in (20 cm), are stocky blackbirds with a thin, conical bill. Females are dark brown with lighter streaks and a pale eyebrow. Males are glossy black with shoulder patches (epaulets) of red and yellow.

Tree swallows (*Tachycineta bicolor*), 6 in (15 cm), have a darting flight pattern and are shaped like stubby fighter jets. They are white below, shiny blue-green on top, and have a short bill. Similar **barn swallows** (*Hirundo rustica*) have a longer, forked tail, and are rusty-orange below.

HABITAT: Red-winged blackbirds feed in open space near water and nest in marsh reeds. Swallows eat insects in flight and migrate along coastlines. Tree swallows may stay the winter along the Gulf, but barn swallows are just passing through between Central American wintering sites and structures across North America (like barns) where the birds nest.

DID YOU KNOW? Popular male red-winged blackbirds have up to 15 mates. The males fiercely defend their breeding territories, spending about a quarter of their daylight hours chasing rivals away. Before sunset, roosting tree swallows form large, dense, bird clouds that swirl like a living tornado.

Fish Crow, Flycatcher, and **Nighthawk**

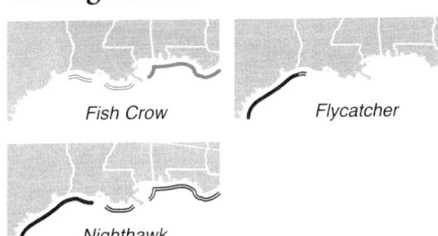

Fish Crow

Flycatcher

Nighthawk

A fish crow in flight

RELATIVES: Crows (family Corvidae) and flycatchers (Tyrannidae) share the order Passeriformes, and are distantly related to nighthawks (order Caprimulgiformes, family Caprimulgidae).

IDENTIFYING FEATURES:

Fish crows *(Corvus ossifragus)*, 15 in (38 cm), are dark, broad-winged birds, similar to the more inland, American crow *(C. brachyrhynchos)*, but are smaller and have a different call—a two-toned, nasal-sounding *UH-uh*, a bit like the negative indication children give when they don't want to do something.

Fish crows feed on a variety of beach scraps

Scissor-tailed flycatchers *(Tyrannus forficatus)*, 14 in (36 cm), are pale gray with dark wings. Males have more intense colors to include salmon-colored sides.

Common nighthawks *(Chordeiles minor)*, 9 in (23 cm), have pointed wings, a forked tail, large eyes, and a short bill. They are camouflaged in mottled white, gray, brown, and black, and have light wing patches. Their erratic flight involves swoops and dives as thay chase insects under low light.

Scissor-tailed flycatcher on driftwood

HABITAT: Crows nest in treetops, and flycatchers in dense brush. Nighthawk nests are on the ground in open areas like beaches.

DID YOU KNOW? Crows are clever, using imagination and prediction to solve problems. They use stick tools, and recognize human faces.

A common nighthawk near her beach nest

253

A mixed-species flock of migrating sandpipers

A travelling savannah sparrow

The journey was too much for this evening grosbeak

A stranded gannet. Healthy birds are offshore

Bird Migrations

Gulf beaches serve as routes and stopover points for birds travelling the Central Americas Flyway during spring and fall migrations. Many of these birds roost or feed near beaches, but others are only seen high above as their flocks pass by in V-formations. Birds that are likely to stop in for a beach visit include large flocks of **sandpipers** and individual perching songbirds (passerine species) including warblers, wrens, thrushes, finches, and sparrows like the **savannah sparrow** (*Passerculus sandwichensis*). Other birds passing through in high-altitude flocks include ducks, geese, and swans. Some birds that are seldom seen in groups, like the great blue heron, can be spotted flying in formation along this well-flown aerial highway.

Birds of prey typically pass though as individuals. Of these, peregrine falcons (*Falco peregrinus*) may swoop in over the dune in attempts to catch unwary shorebirds. To avoid both predation and heat stress, many songbirds are nocturnal migrants, warbling to each other to keep the flock together as they navigate using the stars to maintain their course.

Migrations are difficult for birds. Many songbirds must fly great distances nonstop over open Gulf waters. Those that are not well fed, or that have bad luck with a headwind or poor weather, don't survive. Migration casualties like the **evening grosbeak** (*Coccothraustes vespertinus*, a finch), are common in the spring and fall. In winter, migrant **northern gannets** and common loons wash ashore after depleting energy reserves. Long-distance migrants like red knots can lose more than half their body weight during their 4,300-mile (6,900-km) nonstop flights. Benefits that drive arduous fall trips are prospects for better food availability and weather. In the spring, migrations are directed northward where similar benefits favor raising the next generation of birds.

For the Birds

Birds are beautiful but vulnerable. As much as we appreciate them, they are among the first elements lost from a living beach. Keeping birds part of our beach experience requires accommodating some of their needs and offering occasional assistance.

LEAVE SOME SPACE: In part, birds hang out on beaches to relax (sound familiar?). Birds that seem to be **"loafing"** are probably desperate to get a little rest after an exhausting flight, swim, or run. Give these and other resting birds a wide berth. Enjoy them from a distance (get close with binoculars), and never allow dogs or children to scatter a flock. Breeding birds (March–August) need extra room. When intruders approach, agitated tern parents take flight and plovers feign wing injury to distract predatory attention. These are clues that you are too close. Their **nests** are mere scrapes, **eggs** are cryptically beach-colored, and **chicks** resemble fluffy, sand-colored cotton balls that virtually disappear when still. Both eggs and chicks can be easily stepped-on or run over, and continual harassment may force parents to leave their young to die.

LEND A HAND: Don't feed bony fish carcasses to begging birds (exposed spines can pierce their insides), don't fish where birds will go after your bait, and never miss the opportunity to pick up discarded fishing line. If you hook a bird, never just cut the line. Reel in the bird and toss a shirt or towel over it (to calm the bird and control its bill). To complete the rescue of hooked, entangled, or otherwise troubled birds, phone 411 and ask for your local wildlife hospital.

See a flock of loafing beach birds? Try to let them rest.

Wilson's plover nest and eggs on the upper beach

A cryptic but vulnerable snowy plover chick

A royal tern hooked and entangled by fishing line

255

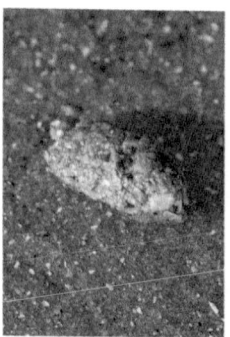 

Sanderling pellets. Left contains mole-crab bits

Pellet from a willet containing coquina clam-shell bits

Bill pokes from a sand-probing sanderling

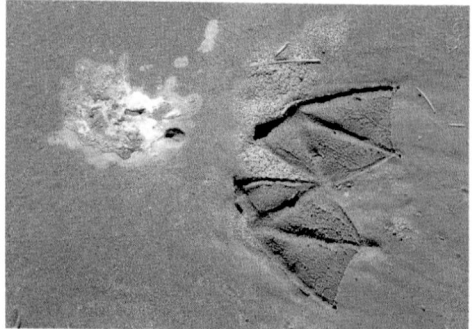

A brown pelican leaves a clocal calling card

Birds Were Here
(Pellets, Pokes, Poops, and Prints)

IDENTIFYING FEATURES:

Shorebird pellets, to 1 in (25 cm), are neatly compact bits of bird barf. They contain undigested parts of a bird's last meal, along with some incidental sand. Shorebirds occasionally regurgitate (cast) pellets while they are out looking for their next menu item. Pellets are formed in the bird's gizzard (muscular stomach) within several hours of a meal. Contents of a shorebird pellet vary widely between bird species and individual pellets, but are pretty uniform within a single pellet. This shows how focused the birds are on their prey. Sandpiper pellets often contain nothing but **mole-crab shell** fragments, or only broken **coquina clam shells**.

Bill pokes are marks in wet sand where a shorebird used its bill to probe for food. Sanderlings and other "peeps" (small sandpipers) leave marks from short bills poked rapidly in tight succession. Plovers run, stop, and make a single selective pluck, leaving a conspicuous mark only when larger prey are pulled from the sand. Dowitchers make repeated bill pokes in muddy sand from their slow, sewing-machine-like movement. Other shorebirds with longer bills, like willets, curlews, and avocets, feed in water and leave no marks.

Seabird poop, is the pasty, white-and-gray excrement from a bird's cloaca, its main exit point. The "poop" actually contains both urine and feces. The urine has crystals of ammonium urate and struvite (magnesium ammonium phosphate), which are products of gut bacteria that break down uric acid before it's excreted. Uric acid is how birds get rid of nitrogenous waste. Unlike urea (how mammals excrete), uric acid does not need much water to flush it out, allowing the bird to conserve water. The feces don't have much undigested material, which was largely cast out as pellets.

Watching Birds

Beach birds are eminently watchable because they flaunt their beauty in the open and do interesting things. But enjoyable, ethical bird watching requires balance. A great beach-birding experience balances a good view with remaining inconsequential to the birds you're viewing. It's easy to get too close, which can cause birds to leave, stress-out, or worse. These hints should help both you and the birds have a delightful experience.

JUST WATCH: Focus on the birds. Trying to watch birds in addition to playing music, walking dogs, or flying kites will leave you little to look at. Do involve kids, but keep them reverent. Watching birds is a good way to learn patience.

KNOW BIRD BODY LANGUAGE: Relaxed birds are relaxing to watch. Relaxation signs include various **"loafing"** postures, including **sitting** and **standing on one leg**, often with the bird's bill turned back to rest under a wing. Contented birds also do maintenance, like feather preening and surf bathing. Shorebird idle time is also an opportunity for yoga. A typical **stretch** involves extending a wing and leg on the same side. But when a shorebird stretches both wings up above its back, it's often because it is perturbed and about to fly away.

LET OPTICS BRING YOU CLOSE: The best way to watch birds while staying safely distant is to use binoculars or a spotting scope. If you want pictures, be sure you have a long-focal-length lens (400 mm or longer).

LIGHT AND TIDE: To see birds in their most vivid detail and colors, view them with sunlight at your back. To catch shorebirds as they feed, low tide is best. But take care not to chase them. Stay low and still, and let the birds walk by. At high tide, those feeding shorebirds will be roosting (loafing) on the upper beach.

"Loafing" shorebirds aren't lazy, they're exhausted

Sanderling and willet in relaxed, one-legged stands

A stretching sanderling in a yoga pose

Watching birds is also knowing when to turn away

257

Armadillos are armored mammals

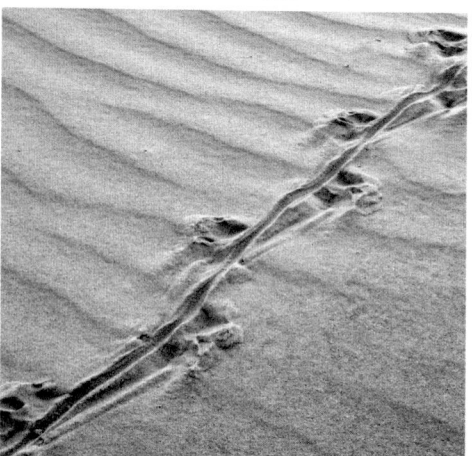

Armadillo tracks, width 4 in (10 cm)

Marsh rice rat on a saltmarsh beach

Armadillo and Marsh Rice Rat

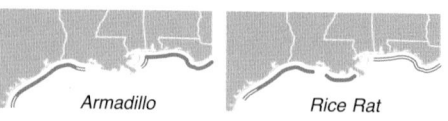

Armadillo   Rice Rat

RELATIVES: These are mammals (class Mammalia). Armadillos are in the order Cingulata. Rice rats are rodents (order Rodentia) in the family Cricetidae with American native mice, and are only distantly related to the dreaded black rat *(Rattus rattus)* from the old world.

IDENTIFYING FEATURES:

Nine-banded armadillos *(Dasypus novemcinctus),* length 28 in (71 cm), are hump-backed, armor-plated critters with stubby legs, a long, tapered tail, and cup-shape ears. This squinty-eyed Mister Magoo of the animal world is likely to be found with its sensitive nose probing the ground in a seemingly oblivious search for small subsurface animals and eggs. They leave **tracks** with a distinct tail mark.

Marsh rice rats *(Oryzomys palustris)* are medium-size rodents, 12 in (30 cm), with gray- to red-brown upperparts that sharply contrast with their whitish belly and feet. Their fur is thick and short with longer, unpigmented guard hairs that trap air underwater and help a swimming rat stay buoyant and semi-dry. Rice rats have ungual tufts (furry toes) as adaptations for swimming.

HABITAT: Armadillos occupy lowland areas and are active on beaches only at night. Rice rats live in wetlands and occasionally forage near the beach where sand intersects with saltmarsh.

DID YOU KNOW? Armadillos invaded the US from Mexico about 150 years ago. They are now extensive predators of reptile eggs, including those of sea turtles. Marsh rice rats make nests of sedge and grass, and eat plants, fungi, and insects.

Beach Mice

1. Alabama
2. Perdido Key
3. Santa Rosa
   1 2 3 4 5 4. Choctawhatchee
5. St. Andrews

RELATIVES: Beach mice are rodents in the family Cricetidae with rice rats.

IDENTIFYING FEATURES:

Beach mice (*Peromyscus polionotus* sspp.), 5.5 in (14 cm), are distinctly pale-furred versions of the inland, oldfield mouse (*P. p. polionotus*). There are remnant populations of six Gulf subspecies named for their home beach locations:

Alabama (*P. p. ammobates*)—Pale brown above including the tail.

Perdido Key (*P. p. trissyllepsis*)—Pale gray-tan above.

Santa Rosa (*P. p. leucocephalus*)—Pale tan on back only with white face and flanks.

Choctawhatchee (*P. p. allophrys*)—Pale orange-brown above with a dorsal stripe on the tail.

St. Andrews (*P. p. peninsularis*)—Pale brown above with a white face.

HABITAT: Beach mice rest, cache seeds, and bear pups within multiple burrows in the dune. As nocturnal herbivores, beach mice feed primarily on the seeds of seaoats (p. 272) and other grasses. The mice also eat insects during the spring and summer. **Burrow openings** are triangular and about two fingers wide. **Tracks** look like tiny, four-fingered hands or lines of dots in paw-clusters of four.

DID YOU KNOW? Beach mouse burrows lead to a main chamber, which has a back-door tunnel ending just below the sand surface. The back door allows escape from front-door predators like snakes. All but the Santa Rosa beach mouse are Endangered. They are declining due to an inability to live in parking lots.

Alabama beach mouse

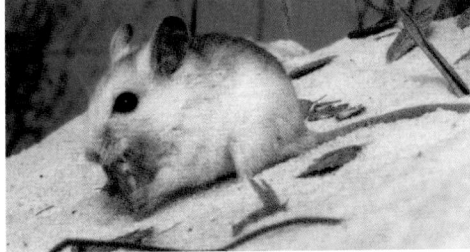

Santa Rosa beach mouse

Choctawhatchee beach mouse

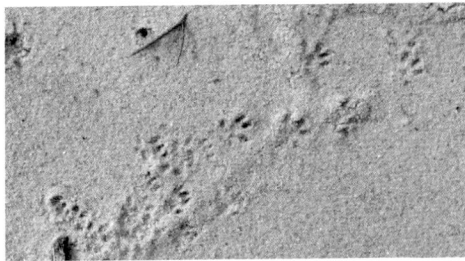

Tracks from nocturnal foraging, width 0.3 in (0.8 cm)

A beach mouse burrow in the dune

259

Gulf Coast kangaroo rat, 8.7 in (22 cm)

Hind-feet hop prints from a Gulf Coast kangaroo rat

Pocket gopher mound in dune, and critter (inset)

Spotted ground squirrel, 10 in (25 cm), at burrow

Kangaroo Rat, Pocket Gopher, and **Ground Squirrel**

Kangaroo Rat, Ground Squirrel

Pocket Gopher

RELATIVES: Kangaroo rats (family Heteromyidae) are more closely related to pocket gophers (Geomyidae) than to ground squirrels (Sciuridae). All are rodents (order Rodentia).

IDENTIFYING FEATURES:

Gulf Coast kangaroo rats *(Dipodomys compactus)* have large eyes, a tufted tail, and big, 5-toed feet with furry soles. Their upperparts range from pale gray to golden brown, and their cheeks are white. Burrows are tennis-ball size with an apron of loose sand. **Tracks** show hop marks from big hairy feet.

Texas pocket gophers *(Geomys personatus)*, 12 in (30 cm), remain hidden, but push up conspicuous, 20-in (51-cm) **mounds** along burrows. The critter is drab gray with beady eyes and enormous front digging claws. They have large upper incisors with two grooves and large, fur-lined shoulder pouches for food transport.

Spotted ground squirrels *(Xerospermophilus spilosoma)* are shy rodents that emerge from burrows early and late in the day. They are sand colored with scattered white spots. The burrow is about 2 in (5 cm) wide with an apron mound.

HABITAT: These burrowing rodents prefer sandy, well drained areas including Gulf dunes.

DID YOU KNOW? Kangaroo rats roam the dunes at night to collect seeds in their cheek pouches. A single Texas pocket gopher may have 100 ft (30 m) of passages in its knee-deep, home burrow system. These gophers eat grass roots and dune sunflower plants that they pull from below into their burrow.

Rabbits and Mole

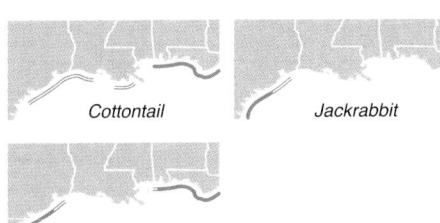

Cottontail

Jackrabbit

Mole

RELATIVES: These rabbits are allied in the order Lagomorpha and family Leporidae. Moles (Talpidae) are with shrews in the order Soricomorpha.

IDENTIFYING FEATURES:

Eastern cottontail rabbits *(Sylvilagus floridanus)*, 17 in (43 cm), have gray to brown fur with frosted tips and a cottony tail. Their pea-size fibrous **droppings** are in piles, often near their conspicuous **tracks**.

Black-tailed jackrabbits *(Lepus californicus)*, 24 in (61 cm), are long-legged rabbits with tall, dark-tipped, naked ears, and a black upper tail.

Eastern moles *(Scalopus aquaticus)*, 7 in (17 cm), are recognized by their surface **ridge tunnels**. An exposed mole shows gray fur, a naked snout, spadelike forelimbs, and tiny eyes covered by fur.

HABITAT: These rabbits are mostly nocturnal, and forage for grasses behind the dune. Jackrabbits also occur in dry desert. Eastern moles search for earthworms and insects hiding within sandy soils.

DID YOU KNOW? The rabbit poop we see may have gone through twice. A special type of poop called a caecotroph contains nutrients extracted from ingested plant material by hind-gut bacteria. These nutrient-filled poops are valuable, so the rabbit eats them. Jackrabbits can run 30 mph (48 kph) and cover 20 ft (6 m) in a leap. Moles dig deep nest-burrows, which are a hub for multiple **ridge tunnels** used only for gathering groceries.

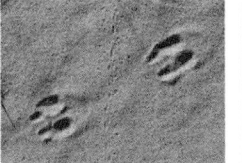

Eastern cottontail, hop-and-stop tracks, droppings

Black-tailed jackrabbit, droppings (inset)

Temporary tunnels (ridges) made by an eastern mole

261

A bandit caught on a nocturnal, sea turtle nest-raid

Wild raccoons are nervous in open daylight

Prints from forepaws (below) and hindpaws (above)

Raccoon

RELATIVES: Raccoons (family Procyonidae) share the order Carnivora with cats and dogs.

IDENTIFYING FEATURES:

Northern (common) raccoons *(Procyon lotor),* 35 in (89 cm), are lumbering, heavy-bodied critters with a dark mask and a thick, ringed tail. Their coat color varies with habitat but most are grayish-red or buff. Raccoons have dexterous forepaws and a reputation for troublesome handi-work. Their **tracks** show marks from spread-toed paws.

HABITAT: Adaptation to suburban life has made the raccoon one of the most common wild mammals found in human populated areas. It is likely that fresh water from impoundments and lawn irrigation, and food subsidies from garbage cans and fishing discards, have allowed raccoons to densely populate some beach areas.

DID YOU KNOW? The northern raccoon is one of seven raccoon species from Mexico and the Caribbean. Suburban raccoons are clever and coordinated enough to lift trashcan lids, climb bird feeders, and open simple latches. Purposefully feeding raccoons is illegal in some states because it encourages all sorts of delinquent behavior. Subsidized raccoons will happily move in with you, overpopulate the local habitat, and decimate nearby populations of birds and reptiles. This crafty mammal is responsible for most of the depredated sea turtle nests in the northern Gulf (p. 209).

Bobcat and Coyote

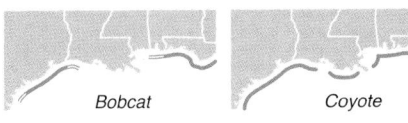

Bobcat   Coyote

RELATIVES: Bobcats (family Felidae, cats) and coyotes (Canidae, dogs and foxes) are in the order Carnivora.

IDENTIFYING FEATURES:

Bobcats *(Lynx rufus)*, 35 in (89 cm), resemble supersize house cats with dark ear tufts and stubby tails (but longer than their northern cousins). Their spotted coats are tawny-gray in winter and reddish-brown in summer. **Tracks** are without claw marks and 1.5 in (3.8 cm) long.

Coyotes *(Canis latrans)*, 51 in (1.3 m), look like a medium-large dog with a narrow muzzle, small nose pad, and large erect ears. Their **tracks** are similar to a domestic dog's tracks, but with notable differences. Coyote tracks are more narrow than most dog tracks, and are about 2.5 in (6.4 cm) long, with parallel toenail marks and crisp edges. The trail left by a coyote is typically straight, with hind foot tracks often over forefoot tracks.

HABITAT: These predators are nocturnal and prefer to hunt small mammals and birds where dense growth meets open space, sometimes on the beach.

DID YOU KNOW? Bobcats are solitary, territorial, and require hundreds of acres of living space. Although coyotes are native to North America, they are new to the southeastern US. This canine began a range expansion in 1900, and over several decades has spread from Texas through all of Florida. Their diet includes opossums, armadillos, raccoons, domestic cats, skunks, rodents, rabbits, birds, lizards, frogs, insects, grasses, and berries. Coyotes have benefitted from the way we've eliminated forests and extirpated major predators like wolves.

Bobcats are seen mostly at dawn and dusk

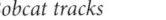

Bobcat tracks   *Coyote tracks*

Coyotes are relatively new to the southeast US

263

Dogs dig the beach

Red fox, 3 ft (0.9 m)

Sea turtle nest dug by a fox. Tracks (inset)

Dog and Fox

Dog   Gray and Red Fox

RELATIVES: Foxes and dogs are carnivores (order Carnivora) that share the family Canidae with coyotes.

IDENTIFYING FEATURES:

Domestic dogs (*Canis lupus familiaris*) vary in size and form among breeds, and hey ... you know what a dog looks like. But can you tell their tracks from coyotes (p. 263) and foxes? These canines leave similar tracks, but often have different trails. Domestic dogs tend to wander in a non-businesslike, meandering fashion, which separates their hind-foot tracks from forefoot tracks. Wild canines travel along straight trails and often have superimposed tracks.

Red foxes (*Vulpes vulpes*) look like dainty dogs with a narrow muzzle and a bushy, white-tipped tail. Their **tracks** are like a small dog with extra fur between the paw-pads. Gray foxes (*Urocyon cinereoargenteus*) have a black tail tip and are rare on beaches.

HABITAT: Domestic dogs are everywhere, including beaches not designated as "dog beaches." Because red foxes prefer open areas, they are the most common fox seen on beaches. Expecting parents dig dens out of old burrows and protect their litter of kits there until they can hunt on their own.

DID YOU KNOW? Free-running dogs are a key cause of shorebird nest abandonment. Please keep dogs away from birds. Dogs share species affiliation with gray wolves. Their cousin, the red wolf (*Canis rufus*) has been reestablished on the Gulf's St. Vincent Island, Florida. The red fox has the widest distribution of any wild member of the canine family, occupying most of the northern hemisphere.

Wild Boar and Deer

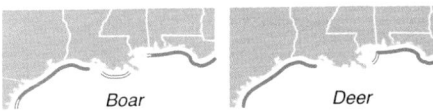

Boar   Deer

RELATIVES: Deer (family Cervidae) and boar (pigs, family Suidae) are even-toed ungulates in the order Artiodactyla.

IDENTIFYING FEATURES:

White-tailed deer *(Odocoileus virginianus)*, 4 ft (1.2 m) shoulder height, are tan or brown in summer and grayish brown in winter. They are generally skittish, and raise their white tail underside when running away. Only males (bucks) have antlers, which start growing as velvety buttons in spring and are shed as multi-prong racks in late winter. **Tracks** show a cloven hoof with pointed tips toward the inside of the track.

Wild boar (hogs) *(Sus scrofa),* are dark, coarse-haired, scruffy wildtypes of the familiar domestic pig. They vary in color from brown to black and often have a noticeable mane. Their dual-toed **tracks** differ from deer in having rounded hoof tips and widely set rear marks from the pig's dewclaws.

HABITAT: Deer and wild boar occupy many terrestrial habitats on the Gulf Coast and inland.

DID YOU KNOW? During warm months, white-tailed deer browse on leaves, and in the fall, eat seeds and fruits, including prickly pear, yaupon, and wolfberry. Deer are ruminants, which means they get additional nutrients from their food by fermenting it in a special stomach compartment. This "cud" is regurgitated and re-chewed. A motivated white-tailed deer can clear a 10-ft (3-m) tall fence. Pigs were brought to Florida by Spanish explorers in the 1500s. Their wild descendents are highly destructive to native wildlife and their habitats.

Young male white-tailed deer, tracks (inset)

Wild boar, 28 in (70 cm) shoulder height

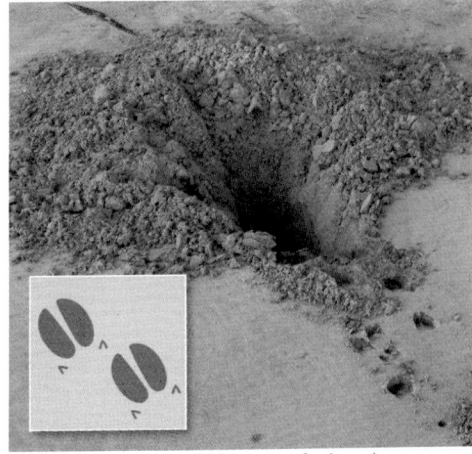

Beach rooting by a wild boar. Tracks (inset)

265

Bottlenose dolphins play off Gulf beaches

Gulf groups taking calls on live and dead strandings

Keep animals wet and cool until help arrives

Cetaceans

RELATIVES: Cetaceans are marine mammals in the order Cetacea, and are divided between the toothed whales (Odontoceti, like dolphins) and baleen whales (Mysticeti, like humpback whales).

IDENTIFYING FEATURES:

Bottlenose dolphins (*Tursiops truncatus*), 8.5 ft (2.6 m), are small, sleek, toothed whales that are dark gray above and lighter on their flanks, with no spots. Their dorsal fin is rounded back and their head is melonlike with a distinct snout.

Strandings occur when marine mammals are sick, injured, or disoriented. All beached marine mammals should be reported to hotlines in respective states:

> **Texas:** 800-962-6625
> **Louisiana:** 504-235-3005
> **Mississippi:** 888-767-3657
> **Alabama:** 877-942-5343
> **Florida:** 888-404-FWCC (3922)

While waiting for help for live animals, drape a wet cloth over them and keep the blowhole clear. Do not attempt to push a stranded animal back into the water. They are likely to strand again after losing additional energy.

HABITAT: Bottlenose dolphins inhabit coastal and inshore waters, and often swim and feed off beaches. Other dolphin and whale species live farther offshore.

DID YOU KNOW? Each year, hundreds of marine mammals strand on the Gulf Coast from Texas through Florida's Panhandle. The vast majority are bottlenose dolphins. Other species include short-finned pilot whales (*Globicephala macrorhynchus*) and rough-toothed dolphins (*Steno bredanensis*).

266

Verte-bits *(Fish Parts)*

Vertebrates are animals with backbones. These critters commonly leave behind bits of their skeletons and other hard parts.

Fish bones are often thin and membranous, but jaws, skulls, vertebrae, and some fin spines can be thick and tough. Skulls come apart and are hard to identify, but a catfish cranium (**A**), with its underlying crucifix, and black drum cranium (**B**), are common and recognizable. A fish's cleithrum bone (**C**) anchors its pectoral fin to the cranium. Jaw bones of bluefish (dentary—lower jaw with teeth, **D**) and grouper (premaxilla—upper jaw sans teeth, **E**) often persist on the beach. A fish's preopercular (gill bone, **F**) is also robust. But perhaps the sturdiest parts of a fish are its otoliths (earstones, **G**), which are balance structures located behind the brain of bony fishes. To chew their crustacean food, black drum possess crusher plates (**H**) on the roof and sides of their throat that are studded with molar-like teeth (**I**). Eagle rays process their food with boomerang-shape teeth plates (**J**). Vertebrae (backbone segments) from bony fishes (**K**, **L**) occasionally have long spines, whereas those from sharks (**M**) are simple discs, but all fishes have vertebrae that are deeply cupped at each end. Osteoderm (skin-bone) spines from porcupine pufferfish (**N**) are distinctive T-shape bones. Fish fin bones include fin-ray bones from many species (**O**), barbed spines from catfish (**P**), and their pterygiophores (**Q**), which are dorsal spine's support bones. In large black drum, these dorsal fin-support bones have dense, inflated knobs (**R**). Stingrays defend themselves with a long, gently curved, barbed spine (**S**). Fish ribs (**T**) are conspicuous from large fish, and occasionally, fish swim bladders (**U**) float ashore, and dry in the sun.

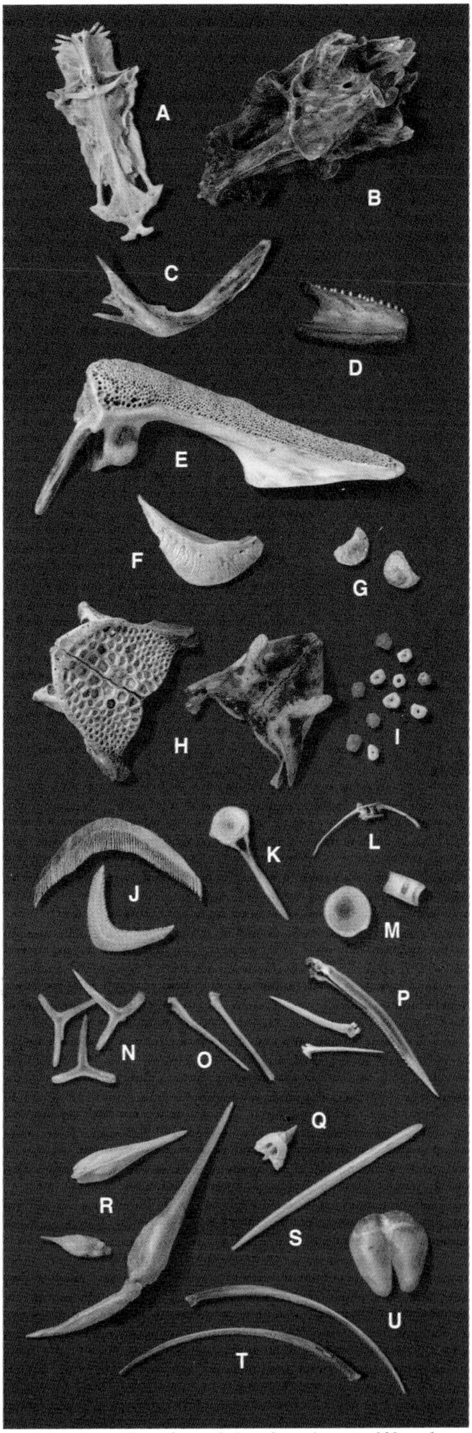

Widely varied bits from fishes found on Gulf beaches

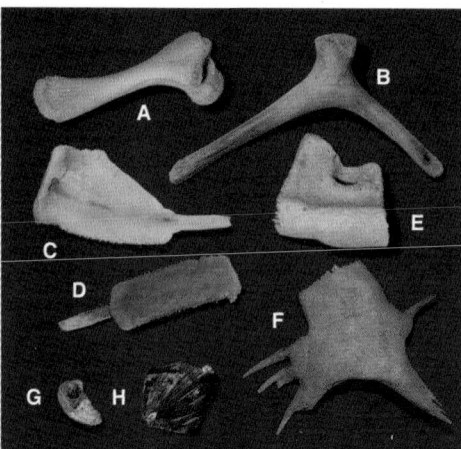

Sea turtle parts

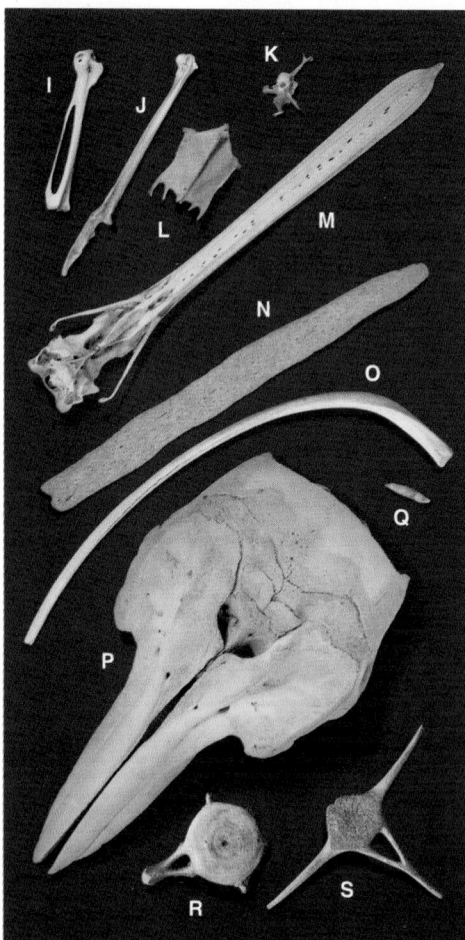

Bird and marine mammal parts

Verte-bits
(Sea Turtles, Birds, Marine Mammals)

Sea turtles have persistent, large scales (scutes) covering their shells, and have dense bones, some of which have strange shapes. Their limb bones, like the femur (**A**), are stout and heavy. Supporting their powerful front flippers is a pectoral girdle containing a Y-shape scapula (**B**). Their flat, carapace (upper-shell) bones (**C**, **D**) are fused with their ribs or spine, and are jagged where they join each other. The edge of the carapace is rimmed with marginal bones (**E**). The turtle's lower shell (plastron) is made of nine, flat, spiked elements like the hyoplastron (**F**) bones, which are paired in the middle lower shell and anchored by spikes into surrounding cartilage. Although a sea turtle's finger bones are within a webbed flipper, they do have a claw with a nail (**G**) at the leading edge. Shell scutes (**H**) are often scattered by wind and waves away from where a stranding occurred.

Birds have some bones light enough to float in water. Easily recognized bird bits include wing bones like the carpometacarpus (**I**), which is the fusion of carpal and metacarpal bones between wrist and knuckles. Leg bones are typically lightweight, but this tibia from a loon (**J**) is more solid to aid the bird's diving. Bird neck vertebrae (**K**) often have delicate projections, and the breastbone or sternum (**L**) has a keel that anchors the bird's large flight muscles. A pelican skull (**M**) shows a long premaxilla (beak) bone.

Marine mammals have large, dense bones. Their ribs (**N**, **O**) are often long and curved. Dolphin skulls (**P**) have a long rostrum with uniform, conical teeth (**Q**). A concavity in front of the nostrils (blowhole) is where the dolphin's fatty melon was cradled. The melon is an echolocation lens to focus the dolphin's clicking noises. Dolphin vertebrae (lumbar, **R**; thoracic, **S**) have three, flat processes.

BEACH PLANTS
AND FUNGI

What are Beach Plants?

Plants use sunshine to turn carbon dioxide and water into sugar, starch, fiber, and wood. But pulling this off at the beach can be difficult. Sure, the beach has sunshine, but it also has toxic salt, desiccating sands, and earth-altering sea-storms to endure. These tough conditions cull the list of plants that can thrive near the beach. Yet, the list represents a wide array of plant groups, including Gymnosperms (for example, pines), and Angiosperms (flowering plants). Flowering plants near the beach include monocots like grasses, palms, and yuccas, and eudicots such as sunflowers, cacti, legumes, and many other familiar herbaceous or woody plants. Although diverse, these beach plants share traits that make them exquisitely adapted to the beach and dune. Many are so closely tied to this habitat that they live no place else but a narrow strip of coastal land.

Not all plants found near the beach have their roots anchored in sand. Some lived elsewhere, such as the *Marine Plants*—algae and seagrasses. These are what most might call "seaweed." At sea, these plants are fundamental pillars of marine food chains, and on the beach they are essential elements of the wrack. The energy gathered and food they make out at sea are put to good use within the beach community. Many beach plants and animals are dependent upon the regular arrival of this gift from the sea.

Some of the most intriguing beach plants are those we never see except for their ocean-drifting pieces and parts. These plants may live many hundreds of miles away in places far from a sandy beach. But because their parts persist and float, they are able to make grand ocean journeys and herald their presence by arriving on a beach. The seeds, nuts, fruits, and pods that make these voyages are collectively known as *Seabeans*. Other drifting plant parts include stems, corky bark, and entire tree trunks—the sea-borne stuff generically categorized as *Driftwood*. Note that despite its woody origin, we've placed lumber in the section called, *Hand of Man.*

The flamboyant flowers and spiny stems of a Texas pricklypear, a plant adapted to seaside sand and salt

Plant Anatomy and Leaf Shape

Plant species can be identified by their characteristic patterns of vegetative structures—leaves and stems. Here are images defining the principal terms used in this section.

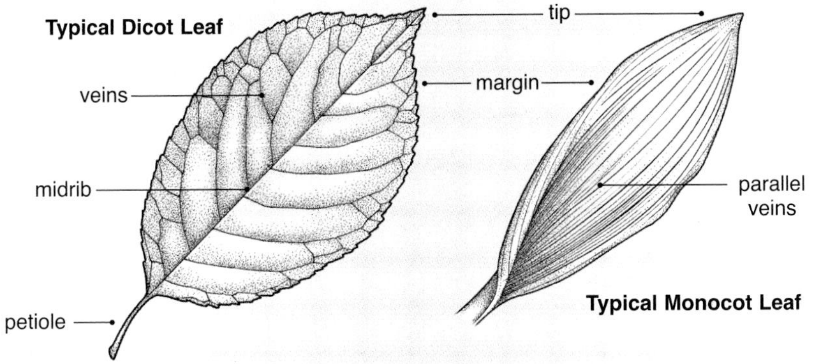

Typical Dicot Leaf

tip

margin

veins

midrib

parallel veins

petiole

Typical Monocot Leaf

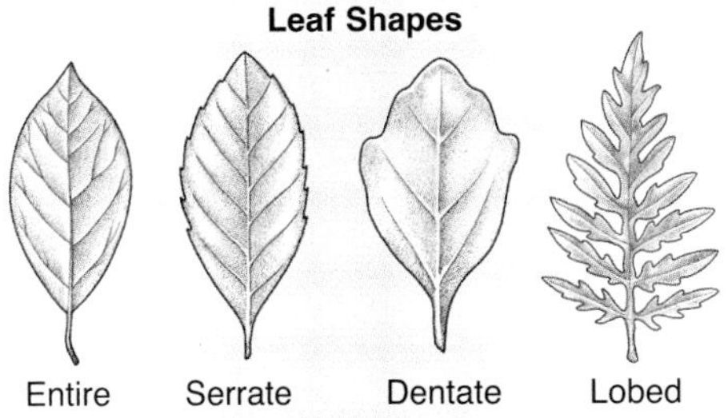

Leaf Shapes

Entire   Serrate   Dentate   Lobed

Types of Compound Leaves

Pinnate   Trifoliate   Palmate

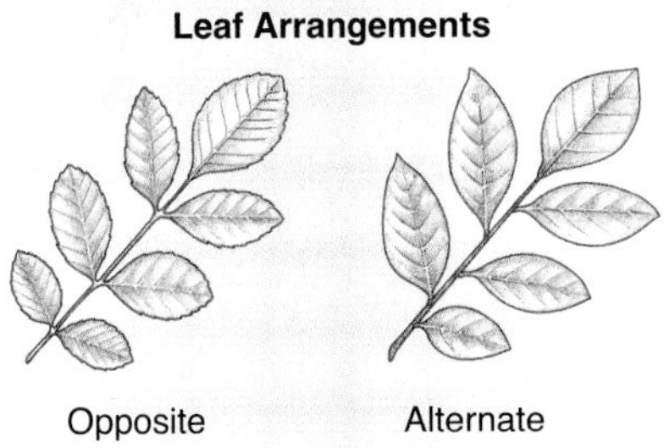

Leaf Arrangements

Opposite   Alternate

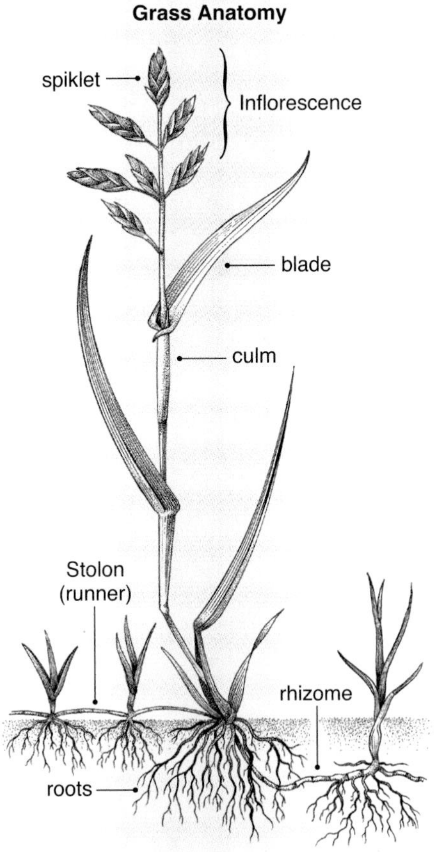

Grass Anatomy

spiklet

Inflorescence

blade

culm

Stolon (runner)

rhizome

roots

271

Fall seaoats flag on a Florida Panhandle dune

An immature clump of seaoats in late summer

Seaoat panicles, August

Seaoats

RELATIVES: Seaoats are flowering plants (angiosperms) in a group called the Lilianae (monocots), with other grasses in the family Poaceae.

IDENTIFYING FEATURES:

Seaoats (*Uniola paniculata*) grow in 24 in (61 cm) high clumps that dominate most Gulf dune faces. This perennial grass has curl-edged blades growing from clumps that spread by underground rhizomes. The gracefully flagging clusters (**panicles**) of golden oatlike seeds mature in summer and reach 6 ft (1.8 m) tall.

HABITAT: Seaoats grow on the dune and out onto the open, upper beach. The plant prefers coarse to medium-grain sands, and grows poorly in fine sediments.

DID YOU KNOW? Seaoats require three summers of growth before they set seed. Their greenish, flowering spikelets appear in July and are fertilized by wind-blown pollen. By late summer, winds also disperse the mature seeds, which remain dormant through winter and germinate with spring rains. Seaoat seeds are a critical component in the diet of endangered beach mice. The grass lives in partnership with nitrogen-fixing bacteria and water-absorbing fungi, which reside within the plant's massive root system. The root-held fungi (mycorrhizae) send out microscopic tendrils (mycelia), thinner than the tiniest rootlet. This vastly increases the surface area for absorbing water and nutrients, allowing the plant to thrive in dry, nutrient-poor beach sands. Seaoats grow best in areas with blowing sand. Partial burial of the plant stimulates growth and promotes spread by rhizomes. Seaoats play a critical role in building and maintaining dunes.

Bitter Panicgrass and Sandburs

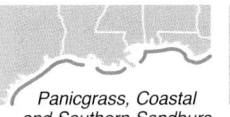

Panicgrass, Coastal and Southern Sandburs

Dune Sandbur

RELATIVES: Panicgrass and sandburs are with other grasses in the family Poaceae.

IDENTIFYING FEATURES:

Bitter panicgrass *(Panicum amarum)* is a perennial grass with waxy, bluish-green, broad blades in clumps that spread by rhizomes. Their pale **panicles** with small seeds mature in summer.

Coastal sandbur *(Cenchrus spinifex)*, 6 in (15 cm) high and sprawling, is a perennial grass most conspicuous when the barbed prickles of its burs (seed capsules) penetrate tender feet. Winter through early summer, the plants may be without burs and look like lawn grass. Coastal sandbur seed capsules mature in fall, are slightly pubescent, and age from green to tan.

Southern sandbur (sandspur) *(C. echinatus)* is an annual grass similar to coastal sandbur, but has less crowded burs on a visible rachis (stalk) and spines that are purplish toward the tip.

Dune sandbur *(C. tribuloides)* is an annual grass resembling these other sandburs, but has densely woolly bur capsules that are crowded on the rachis.

HABITAT: Bitter panicgrass grows on the upper beach and dune. Coastal sandbur is found in disturbed areas, especially dune paths. Southern and dune sandburs occur in sandy dune areas.

DID YOU KNOW? The "panic" in panicgrass refers to the plant's seed-bearing panicles. Like seaoats, bitter panicgrass has a symbiotic relationship with mycorrhizae. To unstick sandburs, spit on the fingers you use to pull them out and don't squeeze. This keeps the micro-barbs on the bur-spines from clinging anew.

Bitter panicgrass clump, typically 3 ft (0.9 m) high

Bitter panicgrass panicle with seeds, max 6 ft (1.8 m)

Coastal sandbur

Southern sandbur

Dune sandbur, woolly seed capsule (inset)

273

Saltmeadow cordgrass at the crest of the dune

Saltmeadow cordgrass clump, seed heads (inset)

Smooth cordgrass, inflorescence (inset)

Crowsfoot grass, inflorescence (inset)

Cordgrasses and Crowsfoot Grass

RELATIVES: These monocots are related to other grasses in the family Poaceae.

IDENTIFYING FEATURES:

Saltmeadow cordgrass *(Spartina patens),* 3 ft (0.9 m) high, is a perennial grass that grows in tight bunches. The leaf blades have their edges rolled inward and appear wiry and stiff. The flower stem (inflorescence) has several alternating spikes at 45-degree angles. The seed heads turn tan in fall.

Smooth cordgrass *(Spartina alterniflora),* 3.5 ft (1.1 m), has tough, 1/4-in-wide (6-mm-wide) leaf blades that are ribbed above and smooth below. Flower spikes appear in spring. Although the grass is a perennial, its stems die back in late fall.

Egyptian crowsfoot grass *(Dactyloctenium aegyptium),* 8 in (20 cm) and sprawling, is an annual grass with blades that are hairy at the margins and midrib. It has no runners, and soon after sprouting in the spring, it produces a green inflorescence arranged like a bird's foot. The foot turns brown with seeds in fall.

HABITAT: Saltmeadow cordgrass grows in dune areas, washover flats, and marsh. Smooth cordgrass is emergent from saline waters and estuarine intertidal zones. Crowsfoot grass grows on disturbed dunes and within mowed turf near the beach.

DID YOU KNOW? Saltmeadow cordgrass was an important, coastal forage for cattle in pioneer days. Smooth cordgrass is the dominant plant in most Gulf saltmarsh ecosystems. Crowsfoot grass is an alien invader from Africa. It grows well in the alkaline sediments of many bulldozed, artificial dunes.

274

Paspalum, Bluestem, and **Common Reed**

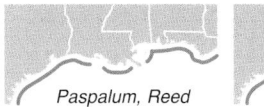
Paspalum, Reed

Bluestem

RELATIVES: These monocots are related to other grasses in the family Poaceae.

IDENTIFYING FEATURES:

Paspalum grasses (*Paspalum* spp.) near the beach are perennial in dense mats with long runners (stolons) and have 2–5 forked flower/seed spikelets up to 16 in (40 cm) tall. Species near the beach include gulfdune paspalum (*Paspalum monostachyum*), and **seashore paspalum** (*Paspalum vaginatum*).

Seacoast bluestem (*Schizachyrium littorale,* pronounced skits-ah-KEER-ee-um lit-or-AL-ay) is a perennial grass in 18 in (46 cm) tall clumps with blue-green stems and leaf blades that turn coppery in winter. In late summer, the grass sends up a 5-ft (1.5 m) tall inflorescence that holds onto lush, feathery seeds through fall. The closely similar, Gulf bluestem (*S. maritimum*) is limited to dunes between eastern Louisiana and the Florida Panhandle.

Common reed (*Phragmites australis*) is a tall, perennial, canelike grass that forms dense stands persisting through winter. The hollow shoots and leaves are stiff and sharp. Purple flowers appear in late summer, and wispy seeds disperse in fall.

HABITAT: Paspalum grasses cover low dune areas. Seacoast bluestem and muhly grass are most common behind dunes. Common reeds emerge from estuarine and freshwater wetlands.

DID YOU KNOW? Common reed is one of the most widespread plants on Earth. Although the Gulf region has a native lineage of this plant, our coast is dominated by a nonnative lineage accidentally introduced from ship ballast in the late 1700s.

Seashore paspalum with long stolons

Seashore paspalum showing forked spikelets

Seacoast bluestem (L), and with fall inflorescence (R)

Common reed grows to 16 ft (4.9 m) tall

275

Shore grass

Saltgrass, inflorescence (inset)

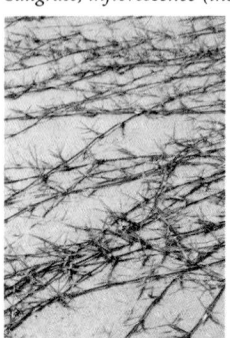 

Seashore dropseed stolons (L) and seed head (R)

Seashore dropseed on a Gulf dune

Shore Grass, Saltgrass, and **Seashore Dropseed**

Shore Grass

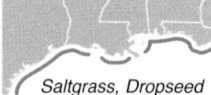

Saltgrass, Dropseed

RELATIVES: These monocots are related to other grasses in the family Poaceae.

IDENTIFYING FEATURES:

Shore grass *(Distichlis littoralis),* to 10 in (25 cm) tall, is a perennial, short-bladed grass that grows in scraggly mats. Its branches sprout along wiry runners with long internodes. In summer, a tiny inflorescence forms at the branch ends.

Saltgrass *(Distichlis spicata)* is a perennial grass that forms dense areas of dark-green, leafy stems, up to 3 ft (0.9 m) tall. The inflorescence and seed head are a cluster of short spikelets, seen summer and fall.

Seashore dropseed *(Sporobolus virginicus),* 18 in (46 cm) tall, is a perennial grass with long stolons above or below the sand. On the beach, this grass grows upright with long blades, but in the dune it may spread densely and have shorter, spiky blades. The seed head is a single spike evident in summer and fall.

HABITAT: Shore grass and saltgrass occupy upper saltmarsh and wet washover areas. Seashore dropseed inhabits the dune but may also spread out onto the open beach.

DID YOU KNOW? Seashore dropseed was important forage for grazing cattle in dry coastal areas. The grass has an amazing ability to grow in salty sand, with a tolerance to salt at concentrations three times that of seawater. The grass has tissues that are tolerant of toxic salt, and has an ability to expel salt through secretions of leaf-blade salt glands.

Lovegrass, Muhly Grass, and **Needlegrass Rush**

Lovegrass, Muhly

Needlegrass Rush

RELATIVES: All are monocots. Lovegrass and muhly grass are with other grasses in the family Poaceae. Needlegrass is with other rushes in the family Juncaceae, and is closer to sedges than to grasses.

IDENTIFYING FEATURES:

Red lovegrass *(Eragrostis secundiflora)*, 12 in (30 cm) tall, is a perennial grass with branching, spindly stems and blue-green leaf blades. In late summer, inflorescences protrude to 18 in (46 cm) and have clusters of flat spikelets tinged with red, becoming straw yellow in the fall.

Pink (hairawn) muhly grass *(Muhlenbergia capillaris)*, 3 ft (0.9 m) tall, grows in perennial, dense, wiry clumps without runners. In late summer and fall, wispy, purplish-pink flower heads appear.

Needlegrass rush *(Juncus roemerianus)*, 18 in (46 cm) tall near the beach, grows in perennial tufts. The plant's dark, grayish-green, stiff leaves are rolled tightly to form pointed tubes. Its inflorescence has flowers at the tips of branches and branchlets, and seed capsules in clusters.

HABITAT: Red lovegrass grows in back dunes and disturbed sites. Muhly grass frequents back dunes and sandy uplands. Needlegrass rush grows in saltmarsh and intertidally on low, protected beaches.

DID YOU KNOW? Muhly grass makes native gardens beautiful. This tough, easily maintained grass was voted *2012 Plant of the Year* by the Garden Club of America. Needlegrass rush is more tolerant of petroleum than many other saltmarsh plants and has been used to remediate wetlands impacted by coastal oil spills.

Red lovegrass, seed heads (inset)

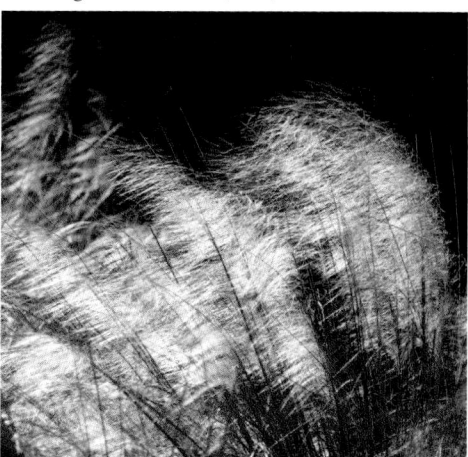
Pink muhly grass in late-summer bloom

Needlegrass rush showing seed capsules

277

Marsh fimbry in late summer, with spikelets (inset)

Yellow nutsedge, with floral spikelets (inset)

Spikesedge showing its terminal spikes (inset)

Common threesquare, with spikelets (inset)

Fimbry, Nutsedge, Spikesedge, and **Threesquare**

RELATIVES: These plants are sedges, which are monocots in the family Cyperaceae, closer to rushes than to grasses.

IDENTIFYING FEATURES: Sedges have edges (stems triangular in cross section).

Marsh fimbry (*Fimbristylis spadicea*), 3 ft (0.9 m) tall, is a densely clumped, perennial, grasslike sedge with narrow, inrolled, dark green leaf blades. In summer, brown stems end in cone-shape terminal spikelets covered with brown scales.

Yellow nutsedge (*Cyperus esculentus*), 8.5 in (21 cm), is a perennial sedge with an erect central stem, mostly covered by sheaths of light-green leaf blades. The stem ends in clusters of yellowish floral spikelets, which appear in summer.

Spikesedges (*Eleocharis* spp.), 24 in (61 cm) tall, are perennial, grasslike, clumped sedges with a small, terminal, cylindrical spike that matures from light green to dark brown through summer.

Common threesquare (*Schoenoplectus pungens*), 3 ft (0.9 m) tall, is a perennial sedge with an erect stem and 1–5 arching leaves. In summer, the inflorescence is a cluster of 1–5 bristly spikelets that turn orange-brown at maturity. The similar seacoast bulrush (*Bolboschoenus robustus*) is taller and has more than five spikelets.

HABITAT: Marsh fimbry grows in overwashed dune swales. Yellow nutsedge grows in scattered areas of the dune and landward. Spikesedges and threesquare are found in sandy wetlands.

DID YOU KNOW? Yellow nutsedge grows from an edible tuber that has a sweet, nutty flavor. Its species name, *esculentus,* is Latin for delicious.

Yuccas

Texas Spanish Dagger

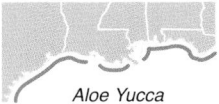

Aloe Yucca

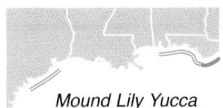

Mound Lily Yucca

RELATIVES: Yuccas are monocots allied with other agaves and yuccas in the family Asparagaceae, along with asparagus.

IDENTIFYING FEATURES: These plants have swordlike leaves.

Texas Spanish dagger *(Yucca treculeana)* is a treelike yucca that often has branches high on the trunk. Its stiff, yellow- or blue-green leaves have wickedly sharp tips and reach 4 ft (1.2 m) in length. The leaves are cupped above and have smooth margins lined with curly filaments toward the plant center. Flowers are cream-colored and often tinged with purple.

Aloe yucca (Spanish bayonet) *(Yucca aloifolia)* is similar to Texas Spanish dagger, but seldom reaches 10 ft (3 m) tall and has shorter, 30-in (76-cm) leaves. Aloe yucca's deep-green leaves are only slightly cupped above, and have finely serrated edges that will cut skin. Stalks with white, purplish-tinged flowers appear in spring.

Mound-lily yucca *(Yucca gloriosa)*, 6.5 ft (2 m) high, has bluish-green leaves with smooth margins and gently pointed tips. Leaves are flexible enough for some to bend toward the ground. Stalks with white flowers appear during the summer.

HABITAT: All grow on the dune crest and landward.

DID YOU KNOW? Yucca roots contain saponins, which are soapy, bitter-tasting, compounds that have medicinal effects including control of infections from protozoans like giardia. Mound-lily yucca is a protected, Endangered Florida plant.

Texas Spanish dagger, blooming, to 20 ft (6.1 m)

Aloe yucca. Flower stalk (inset)

Mound-lily yucca. Flower stalk (inset)

279

A sabal palm on a Florida Gulf-coast dune

Mexican fan palms near a Texas beach

Saw palmetto, max 10 ft (3 m) tall

Palms

Sabal Palm

Mexican Fan Palm

Saw Palmetto

RELATIVES: Palms are monocots in the order Arecales, family Arecaceae.

IDENTIFYING FEATURES: These palms have large, fan-shape fronds.

Sabal (cabbage) palms *(Sabal palmetto),* 50 ft (15 m) tall, have a vertical trunk that may retain the boots (sheathes and petiole stubs) from old fronds. A conspicuous arcing midrib gives fronds a three-dimensional shape. Clusters of small white flowers bloom in spring, and dark, pea-size drupes (stone fruits) appear in summer.

Mexican (Washington) fan palms *(Washingtonia robusta)* grow to 90 ft (27 m) tall and have a slender trunk with a wide base. Shorter palms retain a cloak of old fronds down the trunk. The fronds are relatively flat and have spiny petioles. Flower clusters are pale orange-pink, and the fruits are dark, pea-size drupes.

Saw palmetto *(Serenoa repens)* has a sprawling, branching, fibrous trunk, and fronds with a serrated leafstalk (petiole) and no obvious midrib. Spring flower clusters result in summer, olive-size drupes.

HABITAT: These palms grow on the dune crest and landward.

DID YOU KNOW? The sabal palm is Florida's state tree, which is interesting, because the plant is more closely related to grasses than to the woody plants that most would call "trees." Their other name, cabbage palm, generously describes the taste of their "heart" (frond bud). Mexican fan palms are native to northwestern Mexico, but have been "naturalized" in Texas.

Blue-eyed Grass, Spiderwort, and **Dayflower**

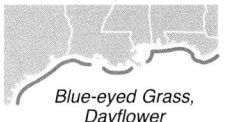

Blue-eyed Grass, Dayflower

Spiderwort

RELATIVES: Blue-eyed grass shares the family Iridaceae with irises, and is in the order Asparagales with yuccas and asparagus. Spiderworts and dayflowers are in a separate monocot order, Commelinales, family Commelinaceae.

IDENTIFYING FEATURES:

Blue-eyed grass (*Sisyrinchium angustifolium*) is a clump-forming perennial with narrow grass-like leaves. In spring and summer, its branched, flowering stems have violet-blue flowers with six, point-tip petals (technically, tepals, which combine petals and sepals) and a yellow eye.

Ohio (bluejacket) spiderwort (*Tradescantia ohiensis*), 24 in (61 cm), is a clumped perennial with narrow leaves and branched, erect stems. In spring and summer, stems are topped with clusters of delicate, blue, three-petal flowers that open in the morning.

Whitemouth dayflower (*Commelina erecta*), 24 in (61 cm), is a perennial with soft stems bearing alternate, lance-shape leaves. Its flowers bloom spring through mid-fall and have two upper petals of sky blue, a tiny, white lower petal, and bright yellow anthers. Blooms last a day, but several buds may bloom on subsequent days.

HABITAT: Blue-eyed grass grows in dune swales and other wet areas. Spiderworts and dayflowers occupy dry, back dunes and other sandy places.

DID YOU KNOW? Bees and other pollinators love to visit spiderworts and dayflowers, but not for the nectar. These flowers have none, offering only pollen as a reward.

Blue-eyed grass flower, max 12 in (30 cm)

Ohio spiderwort flowers and buds

Whitemouth dayflower

281

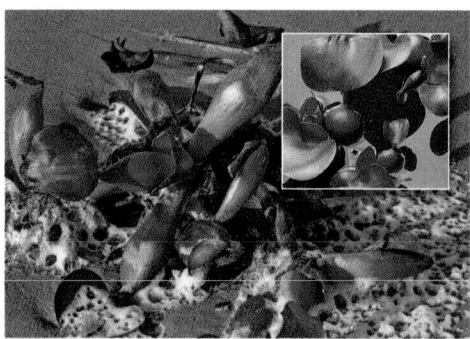

Moribund water hyacinth, greener and fresher (inset)

Earleaf greenbriar, fruits and tendrils (inset)

Saw greenbriar, typically 6.5 ft (2 m)

Wild taro, stranded, then rooted on the upper beach

Water Hyacinth, Greenbriars, and Wild Taro

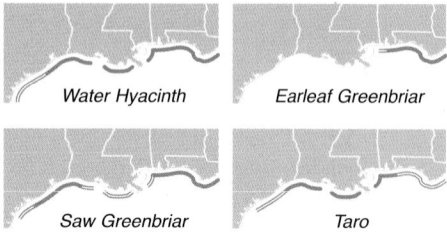

Water Hyacinth

Earleaf Greenbriar

Saw Greenbriar

Taro

RELATIVES: Water hyacinth shares the monocot order Commelinales with spiderworts and dayflowers, but is in the separate family, Pontederiaceae. Greenbriars are in the order Liliales, family Smilacaceae. Taro is categorized in the order Alismatales, and family Araceae.

IDENTIFYING FEATURES:

Water hyacinths *(Eichhornia crassipes)* on the beach are limp, moribund clumps of bulbous, spongy stalks with fleshy, 5-in (13-cm) leaves and dark, feathery roots. In life, the plant clogs waterways and has clusters of lavender flowers.

Earleaf (dune) greenbriar *(Smilax auriculata),* 6.5 ft (2 m) in entwining clumps, is a tough vine with prickles, tendrils, and dark-green, smooth-margined leaves.

Saw greenbriar *(Smilax bona-nox)* is similar to earleaf greenbriar, but seldom climbs where it is exposed to salt spray. It has leaves that are frequently variegated (dual colored) with prickled margins.

Wild taro *(Colocasia esculenta),* to 3 ft (0.9 m) tall, is a fleshy plant with large, triangular leaves that are dark green above and lighter beneath. Stranded plants survive on the beach only through the wet season.

HABITAT: Water hyacinth and wild taro are freshwater plants that arrive on Gulf beaches after being swept down rivers. Both are invasive, alien invaders. Greenbriars live throughout the dune. Earleaf greenbriar frequently covers shrubs on the salt-pruned dune face.

Gregg's Amaranth, Crested Saltbush, and Seepweed

Gregg's Amaranth

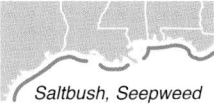
Saltbush, Seepweed

RELATIVES: These dicots share the order Caryophyllales, which includes cacti and carnations, and are more specifically allied within the family Amaranthaceae, containing spinach, beets, and quinoa.

IDENTIFYING FEATURES:

Gregg's amaranth *(Amaranthus greggii)*, 3 ft (0.9 m) tall, is a large annual with leggy, linear-striped stems and fleshy, deeply creased leaves. Its small light-green flowers bloom summer and fall on branched spikes.

Crested saltbush *(Atriplex cristata)*, 18 in (46 cm) tall, is an annual (north) or perennial (south) that forms multi-branched clumps. Leaves are alternate or opposite, oval to slender, and tend to curl upward. Stems are often reddish. Yellowish flowers are on terminal spikes summer and fall.

Seepweed (sea blite) *(Suaeda linearis)*, 3 ft (0.9 m) tall, is an annual in the northern Gulf and a perennial in the south. Its slightly woody, branching stems are striped toward the ends, and bear alternating, fleshy, lance-shape leaves. Tiny, pale or reddish flowers bloom August–September, and fruits ripen September–October. Leaves and stems vary from green to red.

HABITAT: All grow on the dune face and open beach. Seepweed is also common in other salty places like lagoon shorelines.

DID YOU KNOW? All are pre-salted, edible potherbs. Seepweed is a little bitter raw, but delightful when cooked. It's nickname, "blite," originates from the Latin word for spinach, to which seepweed is closely related.

Gregg's amaranth leaves

Gregg's amaranth flowers

Crested saltbush, with flowers (L)

Seepweed with fruits (L) and with reddish growth (R)

283

Virginia glasswort

Late-season dwarf glasswort (L), with flowers (R)

Woolly tidestromia clump, with flowers (inset)

Silverhead, with flowers

Glassworts, Woolly Tidestromia, and **Silverhead**

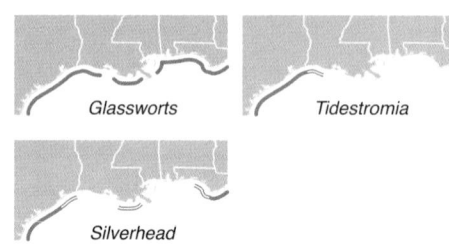

Glassworts

Tidestromia

Silverhead

RELATIVES: These dicots share the family Amaranthaceae with ornamental amaranths, and are kin to beets and spinach.

IDENTIFYING FEATURES:

Glassworts (pickleweeds) (*Salicornia* spp.), 12 in (30 cm) tall, are succulent, fleshy perennials with jointed stems. The leaves are mere scales, paired in opposite fashion along the newest stems. The flowers (August–October) are tiny yellow nubs. The plants turn wine-red during cold months. **Dwarf glasswort** (*S. bigelovii*) has its terminal, scaled, and flowering portion wider than the supporting stem, which is about the same width in **Virginia glasswort** (*S. depressa*).

Woolly Tidestromia (*Tidestromia lanuginosa*), 18 in (46 cm) clumps, is an annual with reddish, yellow, or greenish stems and rounded leaves that are gray-green due to dense fuzz. Where the fuzz has worn, the leaves are greener. Yellow flowers appear in leaf axils (July–October).

Silverhead (samphire or saltweed) (*Blutaparon vermiculare*), 24 in (61 cm) tall, is a sprawling, succulent perennial (or occasionally annual) with thick, shiny, elongate leaves. Its flower heads are white or light pink, and dry to a silvery white. Blooms are seen all year.

HABITAT: These plants are on the upper beach, dune, and salty overwash areas.

DID YOU KNOW? Glassworts are used as a crunchy, salty garnish in Michelin-starred restaurants.

Russian Thistle, Jointweed, Squareflower, and Sea-lavender

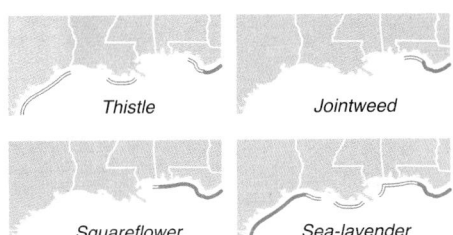

Thistle

Jointweed

Squareflower

Sea-lavender

Russian thistle, "tumbleweed" in winter (inset)

RELATIVES: These dicots share the order Caryophyllales, but are in the separate families, Amaranthaceae (Russian thistle), Polygonaceae (jointweed), Caryophyllaceae (squareflower), and Plumbaginaceae (sea-lavender).

IDENTIFYING FEATURES:

Russian thistle *(Salsola kali),* 3.3 ft (1 m) high, is a bushy annual with short, sharptipped, spiky leaves. Its lower stems and leaves may be red. Dried bushes in winter take on the familiar look of tumbleweeds.

Largeflower jointweed (Sandhill wireweed) *(Polygonella robusta),* is a perennial, 18 in (46 cm) tall, that has stiff, jointed, tangled stems and abundant, pinkish flowers that appear in fall.

Largeflower jointweed

Sand squareflower *(Paronychia erecta),* 18 in (46 cm) tall, is a perennial with purplish stems and small, opposite leaves. Showy, whitish inflorescences appear March to November, and are branched in a regular pattern giving a square outline.

Carolina sea-lavender *(Limonium carolinianum)* is a perennial with large, ovate, basal leaves and smooth stems that in summer, bear small purple flowers along one side of the terminal branches.

Sand squareflower

HABITAT: Russian thistle grows on the upper beach. Jointweed and squareflower occupy dry dune, and sea-lavender prefers salty, wet areas including saltmarsh.

DID YOU KNOW? Sea-lavender is harvested for dried floral arrangements, but this pressure harms the plant's abundance.

Carolina sea-lavender, max 12 in (30 cm) tall

285

Sea purslane is a salty, succulent, seaside herb

Sea purslane leaves (L) and 5-sepaled flower (R)

Kiss-me-quick, with flowers

Cockspur pricklypear

Sea Purslane, Kiss-me-quick, and **Pricklypear Cactus** (*Cockspur*)

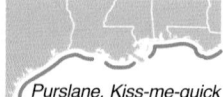

Purslane, Kiss-me-quick   Cockspur Pricklypear

RELATIVES: These dicots share the order Caryophyllales, but are in the separate families, Aizoaceae (sea purslane), Portulacaceae (kiss-me-quick), and Cactaceae (pricklypear cacti).

IDENTIFYING FEATURES:

Sea purslane (*Sesuvium portulacastrum*), 24 in (61 cm) tall, is a sprawling, fleshy, perennial herb with inflated, green or red leaves arranged oppositely on the stem. Its starlike, purple-pink flowers bloom all year, and have five sepals (no petals).

Kiss-me-quick (pink purslane) (*Portulaca pilosa*), 3 in (8 cm) tall, is a fleshy-stemmed annual arranged with alternating, fingerlike leaves, many with hairs at their base. Its pink flowers have five petals and are slightly smaller than a dime. They bloom spring through fall.

Cockspur pricklypear (*Opuntia pusilla*), 12 in (30 cm) tall, is a perennial cactus with inflated barrel-shape pads (stems), and gray spines as long as the pads are wide. Large, yellow flowers bloom in the spring.

HABITAT: Sea purslane tolerates salty conditions on the upper beach. Kiss-me-quick grows on the dune crest and landward, including between the cracks in sidewalks. Pricklypears grow throughout the dune.

DID YOU KNOW? Sea purslane has edible stems and leaves that taste like a salty green bean. In many parts of Asia, this plant is sold in vegetable markets and is believed to treat kidney trouble and scurvy. Sea purslane can be propagated simply by poking a cut stem into moist soil and ignoring it.

Pricklypear Cacti
(Texas, Erect, Devil's Tongue)

Texas Pricklypear

Erect, Devil's Tongue

Texas pricklypear

RELATIVES: Pricklypears are dicots in the order Caryophyllales, and with other cacti in the family Cactaceae.

IDENTIFYING FEATURES: Pricklypears are perennial cacti with succulent, branching, oval pads (stems), and occasional sharp spines (modified leaves). The spines protrude from areoles that also have tufts of tiny barbed bristles called glochids. Their large flowers bloom in spring.

Texas pricklypear *(Opuntia engelmannii),* 6 ft (2 m) tall and sprawling, has a short trunk and yellow-green to blue-green pads, each up to 16 in (40 cm) long. Flowers are yellow, orange, or pink-red with greenish stigma lobes (the center, female flower parts).

The Texas pricklypear flower has a green stigma

Erect pricklypear *(Opuntia stricta)* reaches 6 ft (2 m) tall and may be sprawling or erect. Its pads are dull green and up to 6 in (15 cm) long, with margins scalloped between raised areoles. The flowers are light yellow throughout.

Devil's tongue (eastern) pricklypear *(Opuntia humifusa)* grows in sprawling clumps, typically no more than 2 ft (0.6 cm) or two pads tall. The compressed pads range from glossy green to dark green and are 6.5 in (17 cm) long. The flowers are yellow with white stigma lobes.

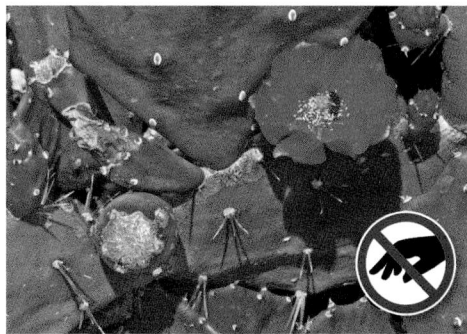

Erect pricklypear with yellow flower and purple fruit

HABITAT: These cacti grow throughout the dune.

DID YOU KNOW? Erect pricklypear is declining due to habitat loss and is a Threatened species in Florida. Pricklypears of several species are being lost to a cactus moth from Argentina *(Cactoblastis cactorum)*, which was introduced into Florida and has spread to Texas.

Devil's tongue pricklypear with apricot-color fruit

287

Gulf searocket, racemes with flowers and new fruits

Southern searocket, raceme with new fruits

Gulf Coast searocket flowers

American searocket flowers (L) and dried fruits (R)

Searockets

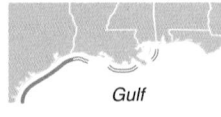

Gulf

Southern

Gulf Coast, American

RELATIVES: Searockets share the dicot family, Brassicaceae, with mustard, broccoli, cabbage, kale, turnips, and radishes.

IDENTIFYING FEATURES: Searockets are occasionally bush-size, 24 in (61 cm), succulent annuals with alternate leaves and white or pale lavender flowers (spring, summer). Green fruits in summer turn yellow in fall, and dry to be brown and corky. One fruit section remains attached, and the end section drops.

Gulf searocket (*Cakile geniculata*) has fleshy, broad-tipped leaves with irregular margins, and fruit-bearing stems (racemes) that zigzag between alternating fruits.

Southern (coastal) searocket (*Cakile lanceolata*) has long, leggy branches with compressed (not very fleshy), elongate, blunt-tipped leaves. Racemes are straight and bear fruits with narrow tips.

Gulf Coast searocket (*Cakile constricta*) has fleshy leaves with wavy margins. Its racemes are long and straight, with point-tipped fruits, four-angled in cross-section.

American searocket (*Cakile edentula*) has fleshy leaves that are serrated or with outward-pointing lobes. Its racemes are mostly straight, and bear eight-ribbed or four-angled fruits with a conical or blunt tip.

HABITAT: Upper beach

DID YOU KNOW? Searocket tastes like its salad-cousin, arugula. "Rocket" comes by way of *eruca* (ancient Rome), *ruchetta* (northern Italy), *roquette* (France), and *rocket* (British Isles).

Pepperweed, Turtleweed, and **Sandmats**

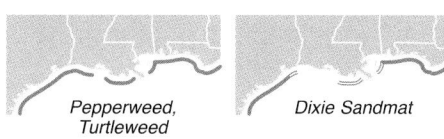

Pepperweed, Turtleweed

Dixie Sandmat

Heartleaf Sandmat

Virginia pepperweed fruits, to 24 in (61 cm) tall

RELATIVES: Pepperweed is in the family Brassicaceae with searockets, and shares the order Brassicales with turtleweed (family Bataceae). Sandmats (order Malpighiales) share the family Euphorbiaceae with other spurges.

IDENTIFYING FEATURES:

Virginia pepperweed *(Lepidium virginicum),* is an erect annual with its leaves largest at the plant's base. Racemes bear tiny white flowers in spring and summer, followed by flake-like, green fruits that turn light orange then brown. The leaves and fruits have a peppery taste.

Turtleweed with yellow fruits

Turtleweed (seaside saltwort) *(Batis maritima)* is a succulent, sprawling, woody-stem perennial as tall as 3 ft (0.9 m). The leaves look like tiny pickles. Fruits look like lumpy green peas, which turn yellow when ripe.

Dixie sandmat *(Euphorbia bombensis),* 12 in (30 cm) tall, is a sprawling annual herb with reddish stems and opposite, oblong leaves. Small white flowers are seen spring–fall.

Heartleaf sandmat *(Euphorbia cordifolia),* 6 in (15 cm) tall, is similar to Dixie sandmat, but has rounder leaves and more branched stems.

Dixie sandmat

HABITAT: Throughout the dune

DID YOU KNOW? *Euphorbia* stems contain a milky latex, which is irritating to our skin but serves as antifungal and antibacterial protection for the plant.

Heartleaf sandmat

289

Touch-me-not, max 12 in (30 cm) tall

Touch-me-not, flowers and stinging prickles

Beach croton, max 3 ft (0.9 m) tall

Winged flax, max 16 in (40 cm) tall

Touch-me-not, Beach Croton, and **Winged Flax**

Touch-me-not

Beach Croton

Winged Flax

RELATIVES: These plants share the order Malpighiales. Touch-me-not and beach croton (Euphorbiaceae) are distantly related to winged flax (family Linaceae).

IDENTIFYING FEATURES:

Touch-me-not *(Cnidoscolus urens)* is an erect perennial with three- or five-lobed leaves. Nearly the entire plant is covered with stinging prickles that cause a burning rash when brushed against sensitive skin. Bright-white flowers bloom spring through fall. West of the Mississippi, the related Texas bullnettle *(C. texanus)* forms clumps to 31 in (80 cm) tall, has large leaves, and is uncommon near the beach.

Beach croton *(Croton punctatus)* is a perennial shrub covered by fuzz. Its rusty branches bear gray-green leaves. The fruit is a three-lobed capsule seen all year.

Winged flax *(Linum alatum)* is an annual wildflower with simple, lanceolate leaves. Its yellow, five-petaled flower has a reddish center and blooms spring–summer.

HABITAT: Touch-me-not is often tucked between grasses in the dune. Beach croton grows on the dune face and upper beach. Winged flax is scattered in back dunes.

DID YOU KNOW? The stinging prickles of touch-me-not are an adaptation to protect the plant from grazing. The prickles are sharp, brittle, and seated on a gland that secretes a caustic fluid comprising a variety or irritants. Dominating this fluid is histamine, which is related to acute inflammatory effects in mammals.

Evening-primrose and Pennywort

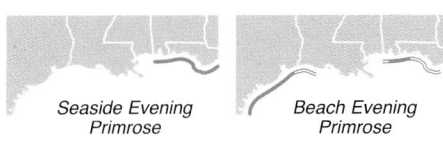

Seaside Evening Primrose

Beach Evening Primrose

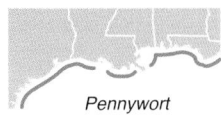

Pennywort

Seaside evening-primrose, wilted flower (inset)

RELATIVES: Evening-primroses (order Myrtales, family Onagraceae) are not directly related to pennyworts (order Apiales, family Araliaceae), which are allied with ivies and ginseng.

IDENTIFYING FEATURES:

Seaside evening-primrose *(Oenothera humifusa),* 12 in (30 cm) tall, is an upright or sprawling biennial with thick, fuzzy leaves. In late spring and summer, its yellow flowers have half-inch (1.3 cm) petals and turn orange-pink as they wilt. The fruit is an elongate, fuzzy capsule.

Beach evening-primrose *(Oenothera drummondii)* is similar to seaside evening-primrose, but grows bushy, to 20 in (50 cm), and has larger flowers.

Largeleaf (beach) pennywort *(Hydrocotyle bonariensis),* 8 in (20 cm) tall, is a perennial herb with creeping lateral stems buried in the sand, and circular, scalloped-edged leaves that are peltate, which means that stems attach to the leaf center. Its white or pale-yellow flower clusters appear spring through summer.

HABITAT: These plants occupy the dune face and between dunes.

DID YOU KNOW? Evening-primrose flowers open at dusk to attract moths and other night-flying pollinators. The plant's genus is derived from the Greek word for wine, *oinos.* The herb is believed to effectively treat hangovers that result from drinking too much fermented grape juice.

Beach evening-primrose, flower (inset)

Beach evening-primrose is common on Texas dunes

Largeleaf pennywort with flowers

291

Beach sunflower, leaf (inset)

A tall stand of silverleaf sunflower

Silverleaf sunflower leaves, flower (inset)

Sunflowers

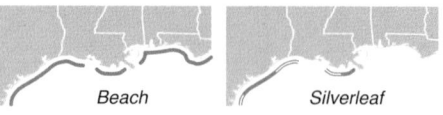

Beach   Silverleaf

RELATIVES: These dicots share the order Asterales and family Asteraceae with asters, daisies, and the tall flower of sunflower-seed fame.

IDENTIFYING FEATURES: Sunflowers have composite flowers made of florets.

Beach (cucumberleaf) sunflower *(Helianthus debilis)*, 24 in (61 cm) tall, is a sprawling perennial (annual, where freezes occur) that shows long-stalked, yellow flowers with brown centers. The flower-base leaflets (bracts) are hairy, as are the green, alternating, stalked, triangular leaves. The plant flowers all year with some die-back in winter.

Silverleaf sunflower *(Helianthus argophyllus)* is a tall-standing annual, to 7 ft (2.1 m), with silky, silver-green leaves that drop as the plant grows. Its large yellow blooms appear in late summer and fall.

HABITAT: Sunny areas throughout the dune and at sandy inland sites. Beach sunflowers also grow on the upper beach.

DID YOU KNOW? Plants in the sunflower family (Asteraceae) have composite flowers, which are actually flower-like inflorescences that comprise a flattened disc bearing many individual florets. Encircling the disc are more than a dozen florets, each with a single, petal-like ligule. The florets within the disc have a tubular shape composed of fused petals. Sunflowers present an example of plant heliotropism, which is tracking of the Sun across the sky. It's the flower buds that follow the Sun, and when the flowers open in the morning, this tracking stops, which generally fixes the flowers on Gulf beaches facing east-southeast toward the sunrise.

Camphorweed, Black-eyed Susan, and **Seaside Oxeye Daisy**

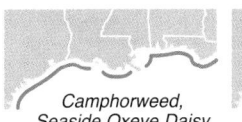

Camphorweed,
Seaside Oxeye Daisy

Black-eyed Susan

Camphorweed

RELATIVES: These dicots share the family Asteraceae with asters, daisies, and sunflowers.

IDENTIFYING FEATURES:

Camphorweed *(Heterotheca subaxillaris),* 18 in (46 cm) tall, is a biennial with thick, roughened, wavy-edge leaves and all-yellow flowers on long stems. Although most flowering is during summer and fall, flowers may be present all year. Its crushed leaves have a distinct camphor-like aroma.

Black-eyed susan *(Rudbeckia hirta),* 30 in (76 cm), is a biennial or short-lived perennial that has fuzzy, lanceolate leaves and stems. Its daisy-like flowers are yellow with a domed, chocolate-brown center.

Seaside oxeye daisy *(Borrichia frutescens),* 3 ft (0.9 m) tall, is an upright, shrublike perennial with fleshy leaves covered by dense, gray fuzz. The flowers have short, yellow petals (ligules) and a raised, brownish-yellow center. Blooms can be seen during all warm months.

HABITAT: Camphorweed lives in dry, sandy dunes and landward. Black-eyed susans prefer damp swales. Seaside oxeye grows in sunny, damp, salty areas of the dune and saltmarsh.

DID YOU KNOW? To deter leaf grazers, a damaged camphorweed's glands emit a pungent concoction of camphor, pinene, and other volatile components of turpentine. The fuzz (pubescence) covering the leaves of seaside oxeye and other coastal sunflowers traps humidity and allows the plants to conserve water.

Black-eyed susan

Bushy seaside oxeye daisy, flower and seed heads (R)

Indian blanketflower, max 18 in (46 cm) tall

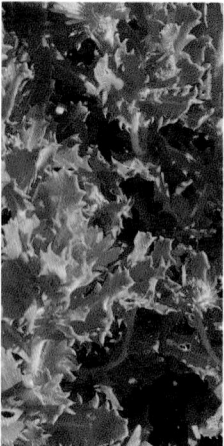

Camphor daisy, max 24 in (61 cm) tall

Lazy daisy, max 18 in (46 cm) tall

Blanketflower, Camphor Daisy, and **Lazy Daisy**

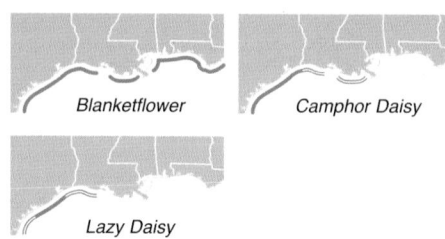

Blanketflower   Camphor Daisy

Lazy Daisy

RELATIVES: These sunflowers share the family Asteraceae with asters, daisies, and ragweeds.

IDENTIFYING FEATURES:

Indian blanketflower (fire-wheel) *(Gaillardia pulchella)* is an upright annual (north) or biennial (south) with alternating fuzzy leaves. Its long-stalked flowers come in yellow, orange, red, and two-tone combinations, blooming summer and fall.

Camphor daisy *(Rayjacksonia phyllocephala)* is an erect or sprawling perennial with thick, succulent, pubescent leaves that often have a sticky residue. Most leaves have deeply serrated margins, often with fleshy spines. The yellow flowers are seen all year. Fruits are within domed tufts and have gossamer hairs that help them disperse in the wind. The plant's crushed leaves smell like turpentine.

Lazy daisy *(Aphanostephus skirrhobasis)* is an annual wildflower with soft, fuzzy, alternate leaves and stems. The flower is the size of a quarter, with 20–45 white rays (ligules) and a yellow center. Spring and summer, blooms open late in the day, giving the plant its common name.

HABITAT: Each lives in sunny areas with sandy soils on the dune, with camphor daisy found most seaward.

DID YOU KNOW? Indian blanketflower is the state flower of Oklahoma. Lazy daisies are favored by dune-grazing white-tailed deer (p. 265).

Horseweed, Plain's Fleabane, Greenthread Aster, and Ragweed

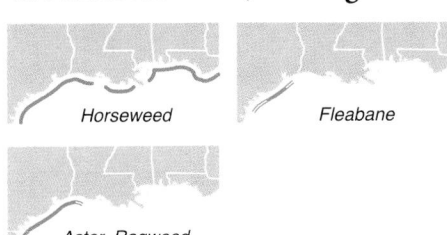

Horseweed

Fleabane

Aster, Ragweed

Canadian horseweed grows to 5 ft (1.5 m) tall

RELATIVES: These plants share the family Asteraceae with sunflowers, daisies, goldenrods, and palafoxias.

IDENTIFYING FEATURES:

Canadian horseweed *(Conyza canadensis)* is an annual that begins as a rosette with lobed leaves and matures to produce branched stems bearing small, thin leaves without petioles. Tiny white flowers (June–October) wilt to pink-orange, and result in white-bristled seeds.

Plains fleabane *(Erigeron modestus)* is a short-lived perennial that grows from a fuzzy-leaved clump. Spring through fall, thin, leafless stalks support blooms with about a hundred, thin, white rays (ligules).

Stiff greenthread aster *(Thelesperma filifolium)* is a short-lived perennial with slender, branched stems that bear threadlike leaves. Flowers (spring–summer) have eight yellow rays around a brown center.

Western ragweed *(Ambrosia psilostachya)* is an erect perennial with a fuzzy stem and leaves, which are deeply lobed. Male (pollen-producing) flowers rise on a terminal spike, summer–fall, and fruits are brownish burs. The similar common ragweed *(A. artemisiifolia)* is uncommon near the beach.

HABITAT: Landward of the dune crest

DID YOU KNOW? Beachcombers with allergies take care; some get a reaction from handling horseweed and ragweeds. Grazing animals ignore these plants due to their bitter taste.

Plains fleabane, max 12 in (30 cm)

Stiff greenthread aster, max 24 in (61 cm)

Western ragweed, max 6 ft (1.8 m)

295

Young seaside goldenrod begins as a rosette of leaves

Seaside goldenrod flowers

Texas palafox flowers and leaves

Goldenrod and Palafox

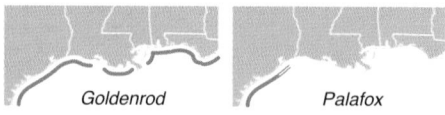

Goldenrod   *Palafox*

RELATIVES: These members of the sunflower family (Asteraceae) are allied with asters, daisies, ragweeds, and seacoast marsh-elder.

IDENTIFYING FEATURES:

Seaside goldenrod (*Solidago sempervirens*), 3.5 ft (1.1 m) tall, is a perennial herb with a tight clump of narrow, evergreen leaves at its base (called a **rosette**), topped by erect leafy stalks that may arch due to sea breezes. The flowering heads (August–November) are dense with deep-yellow flowers.

Texas palafox (*Palafoxia texana*) is an annual herb with an erect, slender stem bearing upper branching that holds most of the plant's alternate, rough-textured, lanceolate leaves. Upper branches reach to about 3 ft (0.9 m), and in warm months bear pinkish, starburst, flower heads without ray florets. The tubular disc florets are the most noticeable flower parts and include purple stigmas and curved, pale-pink styles.

HABITAT: Both species grow in dune swales and other sandy coastal areas.

DID YOU KNOW? Goldenrod is wrongly accused of causing "hay fever" because it blooms with local ragweed, which is a pernicious allergen. Although ragweed sheds pollen in the wind, goldenrod pollen is transfered only by insects. Ironically, a tea made from goldenrod leaves provides an excellent sinus remedy. The plant's genus comes from *solido*, Latin for "to heal." Botanists also name plants for their heroes. The palafox flower honors José de Palafox, who as captain-general attempted to defend Spain against Napoleon's armies during the 1807–1814 war.

Beach Morning-glory and **Railroad Vine**

RELATIVES: These dicots are distantly related to others in the order Solanales (ground cherries, tomatoes, nightshades), and share the family Convolvulaceae with other morning-glory flowers and the sweet potato.

IDENTIFYING FEATURES:

Beach morning-glory *(Ipomoea imperati)* is a ground-level, perennial vine with long runners (stolons) that are frequently buried, showing only the plant's leathery, variably elongate or three-lobed leaves. The plant's funnel-shape flowers (summer–fall) are white with a yellow center.

Beach morning-glory, 6 in (15 cm) tall

Railroad vine *(Ipomoea pes-caprae)* is a perennial vine with leaves at ankle height, but with stolons that often span the beach width between dune and high-tide mark. The long, trailing stolons root at their nodes and bear shiny, thick, circular leaves that tend to fold at the midrib. Flowers are purple to pink and bloom spring through fall. Dark, papery seed capsules contain fuzzy seeds (p. 326) that disperse in the wave-wash.

Railroad vine, 6 in (15 cm) tall

HABITAT: Beach morning-glory and railroad vines grow on the dune face and upper beach. By late summer, railroad vine may have its "tracks" stretched to the tide line.

DID YOU KNOW? Extracts from these plants have been shown to reduce inflammation, with railroad vine demonstrating a particular effectiveness for reducing dermatitis caused by jellyfish stings. The floating seeds from each of these plants are spread widely by ocean currents. Railroad vine is found on nearly all of the world's warm, sandy beaches.

Railroad vines have long stolons

297

Smallflower groundcherry, fruit sack (inset)

Carolina wolfberry, max 6 ft (1.8 m) tall, fruit (inset)

Partridge pea, max 24 in (61 cm) tall

Beach pea can climb dune plants to head height

Groundcherry, Wolfberry, Partridge Pea, and Beach Pea

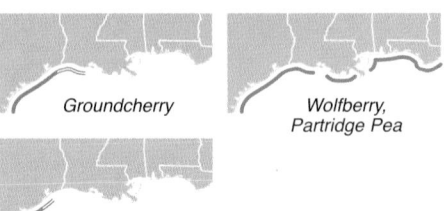

RELATIVES: Groundcherries and wolfberry are with nightshades, potatoes, and tomatoes in the family Solanaceae. Partridge and beach peas are in the family Fabaceae with other legumes.

IDENTIFYING FEATURES:

Smallflower groundcherry *(Physalis cinerascens)*, 18 in (46 cm) tall, is a perennial herb with rough, hairy leaves. Its yellow-green flower (March–November) is a downward-hanging bell with a burgundy star at its center. The fruit is a papery, five-sided sack enveloping a green berry.

Carolina wolfberry *(Lycium carolinianum)* is a perennial shrub with woody branches bearing small, succulent leaves. Four-petaled, lavender flowers are followed by fleshy, red berries.

Partridge pea *(Chamaecrista fasciculata)* is an annual herb with alternate, compound leaves, each with about a dozen leaflet pairs. Its yellow, five-part flowers bloom summer to mid-fall.

Beach pea (bay bean) *(Canavalia rosea)* is a perennial vine with leathery leaves in compound threes. Its pink flowers bloom mostly in summer, and hotdog-size seedpods are evident in the fall.

HABITAT: Groundcherry and partridge pea grow in back dunes. Wolfberry prefers salty, wet areas. Beach pea grows on the dune face.

DID YOU KNOW? Partridge pea flowers are either right- or left-handed based on the direction that their stamens bend.

Sensitive Briar and Cowpea

Sensitive Briar

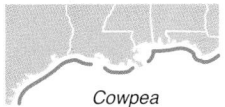

Cowpea

Littleleaf sensitive briar, new (L) and old (R) flowers

RELATIVES: These legumes are in the family Fabaceae with peas, beans, and lupines.

IDENTIFYING FEATURES:

Littleleaf sensitive briar *(Mimosa microphylla)* is a sprawling vine with prickly stems bearing compound leaves with 4–8 pairs of small leaflets. These leaves are sensitive, and fold immediately after being touched. Flower buds look like lumpy spheres, and flowers are purple powder puffs that bloom June to September. The fruits are small, prickly, elongate pods. The plant's woody stems often cover the dune face in southern Texas.

Littleleaf sensitive briar, flower buds

Yellow cowpea *(Vigna luteola)* is a fuzzy, short-lived, perennial vine that both spreads and climbs. It has compound leaves of three and oval leaflets with acute tips. Cowpea flowers are yellow and composed of one large petal, two smaller "wing petals," and two lower "keel" petals. Its fruits are thin, fuzzy pods that turn from green, to dark brown, and twists open to release its black seeds.

HABITAT: Throughout the dune

Littleleaf sensitive briar, leaves, and prickly stems

DID YOU KNOW? Yellow cowpea is yet another edible dune plant. The flowers and youngest seed pods, raw or cooked, taste like green-beans. But examine your harvest. The plant has a special mutualistic relationship with ants. The insects receive sweet rewards from the cowpea's extrafloral nectaries, which are glands throughout the plant. In return, the cowpea gets defended from grazers who would munch on the plant. If you enjoy foraging in the wild, this could mean you. No matter, simply brush off the ants before you nibble.

Yellow cowpea compound leaf, flowers, and bean pods

299

Trailing fuzzy-bean

Scarlet pea

Gulf Coast lupine, fuzzy fruits (inset)

Trailing Bean, Scarlet Pea, and **Lupine**

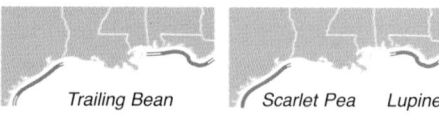

Trailing Bean   Scarlet Pea   Lupine

RELATIVES: These legumes are in the family Fabaceae with peas, beans, and sensitive briar.

IDENTIFYING FEATURES:

Trailing fuzzy-bean *(Strophostyles helvula)*, 3 ft (0.9 m) tall, is a sprawling, annual, pubescent, herbaceous vine with compound leaves of three. Its winged flowers are pinkish and have a dark-purple, upturned, sickle-like keel petal. Flowers and seed pods often occur in pairs.

Scarlet pea (coastal indigo) *(Indigofera miniata)*, 8 in (20 cm), is a multi-stemmed perennial with pinnately compound leaves of 5–9 leaflets bearing flush, pale hairs. Pinkish-red flowers (April–October) have the classic "pea" shape, with a top petal, two side "wing" petals, and two lower petals fused into a "keel." Seed pods are half the length of a pinky finger.

Gulf Coast lupine *(Lupinus westianus)*, 3 ft (0.9 m), is a short-lived perennial that is herbaceous when young but becomes woody as it ages into a clump. Its leaves are densely fuzzy. Showy inflorescences with abundant purple and blue flowers are only seen in April and May.

HABITAT: Trailing bean and scarlet pea grow in the back dune and disturbed areas. Gulf Coast lupine does well on sandy dune tops.

DID YOU KNOW? Trailing beans are wild relatives of cultivated "green beans." They were used as both food and medicine by Native American peoples. Gulf Coast lupine is found almost exclusively on the Gulf dunes of Florida Panhandle beaches and is a State Threatened species.

Milkweed, Texas Bluebell, and **Sand Rose Gentian**

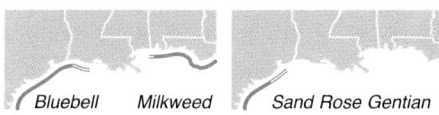

Bluebell   Milkweed   Sand Rose Gentian

Texas bluebell

RELATIVES: These plants are in the order Gentianales. Milkweeds (family Apocynaceae, dogbanes) are only distantly related to bluebells and other gentians (family Gentianaceae).

IDENTIFYING FEATURES:

Texas bluebell *(Eustoma exaltatum)* is a knee-high, erect, annual herb with smooth stems and waxy, bluish green leaves. Its large, bell-shape flowers bloom May–October, and show 5–7 lavender, white, and deep purple petals.

Sandhill milkweed *(Asclepias humistrata)* is a perennial herb (dormant in winter) in leaning clusters about 18 in (46 cm) tall. Smooth unbranched stems bear 5–10 pairs of purple-green leaves. Its flowers (March–June) are cream and pale-purple, and fruits are erect, finger-length pods.

Sandhill milkweed, flowers (inset)

Sand rose gentian *(Sabatia arenicola)*, 12 in (30 cm) tall, is an annual with succulent leaves that shows pink or white, coin-size, five-petal flowers, spring–summer.

HABITAT: Texas bluebell and sand rose gentians grow in dune swales and other wet areas. Milkweed is found in dry, back dunes.

DID YOU KNOW? Sandhill milkweed is an important larval host plant for monarch butterflies *(Danaus plexippus)*. There is coincidence between when this milkweed is available and when the butterflies make their spring migration from Mexico. Caterpillars feeding on this plant take in cardenolides, which are compounds with a specific toxic effect on the part of the brain that causes vomiting. Most caterpillars gain several times the cardenolide concentration required to make an insect-eating bird throw up.

Sand rose gentian

Sand rose gentian flower colors

301

Herb-of-grace, flower (inset)

Woolly stemodia foliage and flower

Turkey tangle frogfruit, flower cluster (inset)

Spotted horsemint flowers above lavendar bracts

Herb-of-Grace, Woolly Stemodia, Frogfruit, and Horsemint

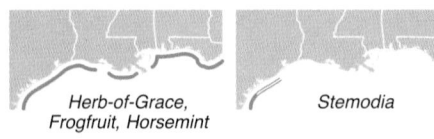

Herb-of-Grace,
Frogfruit, Horsemint

Stemodia

RELATIVES: These plants share the order Lamiales, but are in the separate families, Plantaginaceae (herb-of-grace and woolly stemodia), Verbenaceae (frogfruit and other verbenas), and Lamiaceae (horsemint and other mints).

IDENTIFYING FEATURES:

Herb-of-grace (*Bacopa monnieri*) is a shin-high, perennial, creeping herb with succulent, oppositely arranged, oblong, leaves. Its small, pale, bell-shape flowers (spring–fall) have 4–5 petals.

Woolly stemodia (*Stemodia lanata*), 12 in (30 cm) tall, is a perennial, silver-gray herb with velvety leaves and tiny, pale lavender flowers that bloom April–November.

Turkey tangle frogfruit (*Phyla nodiflora*) has small, ankle-high, white flower clusters with a purple center (May–October). This perennial herb may cover wide areas with its serrate-leaved, creeping foliage, which turns purplish in winter.

Spotted (beebalm) horsemint (*Monarda punctata*) forms short-lived, perennial, herbaceous clumps to 3 ft (0.9 m) tall. Its leaves are opposite, lanceolate, and densely pubescent with serrate margins. Flowers bloom in summer at multiple levels on the upper plant, and are cream with purple spots above large, lavender bracts.

HABITAT: Herb-of-grace occupies dune swales and wet areas inland. Woolly stemodia, frogfruit, and horsemint grow in back-dune areas.

DID YOU KNOW? Herb-of-grace has been used medicinally since before the sixth century for brain function, diabetes, inflammation, ulcers, and cancer.

Broomrape, Seaside Heliotrope, and Southern Dewberry

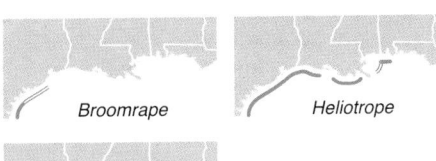

Broomrape   Heliotrope

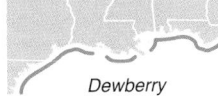

Dewberry

RELATIVES: Broomrapes (order Lamiales, family Orobanchaceae), heliotropes (order Boraginales, family Heliotropiaceae), and dewberries (order Rosales, family Rosaceae, roses), are remotely related.

Louisiana broomrape flower spike

IDENTIFYING FEATURES:

Louisiana broomrape *(Orobanche ludoviciana)* is an annual, parasitic plant that lacks chlorophyll and depends on a host-plant's roots for sustenance. The visible plant is a scaly, spike with densely clustered violet flowers, to 8 in (20 cm) tall.

Seaside heliotrope *(Heliotropium curassavicum)*, 18 in (46 cm) tall, is a mounding or creeping perennial herb with fleshy foliage and thick, oval leaves. Its inflorescences are curled, double rows of white flowers with a purple or yellowish throat.

Southern dewberry *(Rubus trivialis)* is a sprawling, trailing, shrubby herb that can grow head-high, leaning on adjacent plants. Its stems and leaves are covered with clawlike prickles. The palmately compound leaves have 3–5 leaflets with serrated margins. In spring, white flowers are followed by red fruits that ripen black.

Seaside heliotrope, leaves and arching inflorescence

HABITAT: Broomrape and dewberry live behind the dune crest, and heliotropes grow throughout the dune and dune face.

DID YOU KNOW? Broomrape is an obligate parasite of plants in the sunflower family. They connect with the host's roots and steal both water and nutrients. Dewberries are juicy, sweet, and are used to make jams and preserves.

Southern dewberry, flower (inset)

303

Mature slash pine from a former inland forest

A young slash pine in the dune

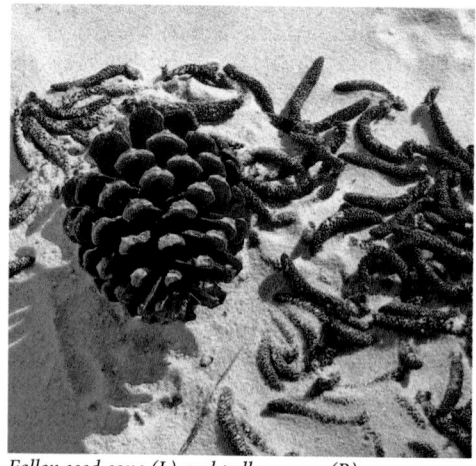

Fallen seed cone (L) and pollen cones (R)

Pines

RELATIVES: Although pines are trees, they are unrelated to the angiosperms (flowering plants) on the following and preceding pages. Pines are conifer (cone-bearing) gymnosperms in the family Pinaceae. As gymnosperms, they have no flowers, and their seeds develop in cones, not fruits.

IDENTIFYING FEATURES:

Slash pines *(Pinus elliottii)*, 60 ft (18 m) tall on the coast, have 4–10 in (10–25 cm) needles, two or three per bundle (fascicle), and bear purplish pollen cones that brown after they drop. The mature seed cones are a lustrous chestnut brown, bear short, stout prickles, and are attached by distinct stalks. Rarer near the beach are sand pines *(Pinus clausa)*, 20 ft (6.1 m), which have 2–3 in (5–8 cm) needles in fascicles of two.

HABITAT: Mostly back-dune areas and landward, although many living pines may be stranded on a retreating beach.

DID YOU KNOW? Established slash pines are more salt-tolerant than other pine species. Slash pines are common in pine plantations throughout the Florida Panhandle, where the trees are harvested for lumber after about 30 years of growth. Younger trees are harvested for pulpwood to make paper. Most of the pines on Florida beaches are from forests being overtaken by the sea, and are soon to be the skeletal remnants of boneyard beaches (p. 31). A mature slash pine forest occupies much of Mississippi's remote, Petit Bois Island, although its barrier sands are eroding fast, and the island moves westward at about 200 ft (60 m) per year. In French, *petit bois* means "little woods."

Magnolia and Saltcedar

RELATIVES: Magnolias (order Magnoliales) are primitive Angiosperms (flowering plants) with conelike flower parts that resemble those of gymnosperms. Saltcedars (family Tamaricaceae) are in the order Caryophyllales with cacti and amaranths, and are not true cedars.

IDENTIFYING FEATURES:

Southern magnolia (*Magnolia grandiflora*) trees near the beach are more like compact shrubs, less than 6 ft (1.8 m) tall. The plant has large leaves that are glossy green above and rusty below. Large, white, fragrant flowers bloom May through July.

Saltcedar (*Tamarix ramosissima*) forms thickets to 12 ft (3.7 m) tall near the beach. It has reddish stems and long, slender branches with feathery leaves. Plumes of deep pink flowers bloom August–October. The seedpods are pinkish to yellow.

HABITAT: More common in inland forests, magnolias are occasionally in backdune areas. Saltcedar grows on the open beach and throughout the dune.

DID YOU KNOW? Southern magnolia is the state tree of Mississippi. The plant was featured on the original Mississippi state flag (1861–1865), and has its blossom on the new flag (2020–present). Saltcedar is an invader from Asia. It's named for its ability to tolerate salty water. The plant concentrates salt in its leaves, which drop to make the surrounding ground saltier, thereby restricting growth of other plants. In the 19th century, saltcedar was introduced to the US as an ornamental shrub that would grow under tough conditions. It did. The species is now considered to be a major threat to habitats throughout the southern half of the country.

Southern magnolia grows shrublike in the dunes

Saltcedar thicket on a Texas dune

Saltcedar in bloom

305

Black mangrove with pneumatophores

Black mangrove seed capsules and flowers

A female yaupon bears ripened fruits

Black Mangrove and Yaupon Holly

Black Mangrove

Yaupon Holly

RELATIVES: These dicot angiosperms are only indirectly related. Black mangroves are in the order Lamiales, family Acanthaceae, and hollies are in the order Aquifoliales, family Aquifoliaceae.

IDENTIFYING FEATURES:

Black mangroves (*Avicennia germinans*), to 20 ft (6.1 m) tall, are trees with oppositely arranged, smooth-edged leaves that are shiny above and fuzzy below. White flowers in summer produce green capsules that look like a split lima bean. This tree almost always grows in wet, salty areas. Its roots "breathe" under water through vertical, shin-high "snorkels" called **pneumatophores.**

Yaupon holly (*Ilex vomitoria*), to 10 ft (3 m) tall, is a shrubby tree with small, alternating, elliptical, dark-green leaves with wavy-toothed margins. Its branches are typically at 45–90 degree angles. Female plants have pea-size holly berries that mature from green to red in the fall.

HABITAT: Black mangroves live in flooded saltmarsh and some lower, wetter, saltier, beach areas. Yaupon may form a low canopy on the dune and live in a severely salt-pruned form, but most are in back-dune areas, and inland.

DID YOU KNOW? Although yaupon berries are toxic, its leaves make a nice tea. It seems to be the only locally native plant from which one might make such a caffeinated drink. The plant is a close cousin to yerba maté, but has more of a caffeine punch—as much as some energy drinks. Partly due to its species name, *vomitoria*, a myth persists that this holly will make you barf. In fact, the tea was preferred by yerba maté drinkers in a blind taste test.

Marsh-elder and Groundsel Tree

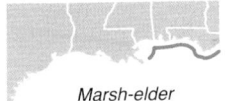

Marsh-elder

Groundsel Tree

RELATIVES: Marsh-elder and groundsel trees share the family Asteraceae with sunflowers, asters, daisies, and ragweeds.

IDENTIFYING FEATURES:

Seacoast marsh-elder *(Iva imbricata)* grows in clumps to 3 ft (0.9 m) tall. It's a perennial, often woody-stemmed shrub with fleshy, alternating leaves. Some lower leaves may be opposite. During intense salt-pruning, the plant is shaped into compact mounds. Late-summer plants are festooned at their tips with green, pealike flowers, then with fruits that turn brown in the fall along with outer branch tips.

Groundsel tree (sea-myrtle, salt bush) *(Baccharis halimifolia)*, to 10 ft (3 m) tall, has alternate, thumb-size, thickened, dentate-margin, silver-green leaves. Yellow male flowers and white female flowers appear on separate plants in early fall. The feathery pappus of the female flower is retained by the seeds, which disperse in the winds of late autumn.

HABITAT: Seacoast marsh-elder grows throughout the dune and out onto the open beach where its clumps gather blowing sand. Groundsel trees are found in low dunes, swales, and higher saltmarsh areas.

DID YOU KNOW? Seacoast marsh-elder is extremely tolerant of salt spray and overwash from sea waves, and is often the closest perennial to the swash zone. This tough plant has fostered many dunes and helps to stabilize them as they mature. The plants can be easily started from transplanted cuttings. Groundsel trees in the toughest growing locations are most likely to be males, with females doing best in wetter, less crowded areas.

Seacoast marsh-elder, leaves (inset)

The pealike flowers of seacoast marsh-elder

Groundsel tree. Unique leaf shape (inset)

Female groundsel tree with fruits, pappus seeds (inset)

Sand live oak, salt-pruned in the dune

Sand live oak leaves and acorns

Wax myrtle, leaves (inset)

Waxy berries of the female wax myrtle

Sand Live Oak and Wax Myrtle

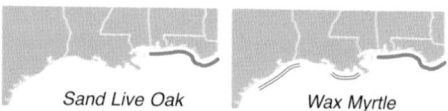

Sand Live Oak   Wax Myrtle

RELATIVES: These trees share the order Fagales. Oaks are in the family Fagaceae with beech trees, and wax myrtle is in the family Myricaceae with eucalyptus and guavas.

IDENTIFYING FEATURES: These plants are woody-stemmed evergreens that grow into large bushes or trees.

Sand live oak *(Quercus geminata),* to 10 ft (3 m) near the beach, is a woody tree with tough, corky bark on its trunk. Most of its thick, older leaves have curled margins and are shaped like upturned canoes with a pale fuzz on their concave underside. This oak's acorns mature in fall and are often set in twins, thus inspiring the plant's species name.

Wax myrtle (southern bayberry) *(Morella cerifera),* 10 ft (3 m) tall, is a shrubby tree with leathery, serrate-margined, olive-green leaves that have a spicy fragrance when crushed. Shrubs are highly branched, especially near salt spray. Waxy, pale-blue berries occur on female plants in winter.

HABITAT: Sand live oaks grow among other salt-spray-shielding plants on the dune. These trees offer some of the most picturesque, salt-pruned, beach-bonsai tree-forms (p. 11). Wax myrtle occupies low dunes and swales.

DID YOU KNOW? American colonists boiled wax myrtle berries to render their paraffin for fragrant "bayberry" candles, which are still used in old-fashioned Christmas decorations.

Hercules Club, Peppertree, and **Coralbean**

Hercules Club

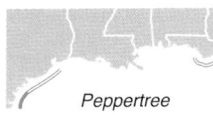

Peppertree

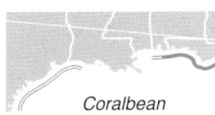

Coralbean

Hercules club leaves (L) and prickly stem (R)

RELATIVES: Hercules club and Brazilian peppertree share the order Sapindales, but Hercules club (family Rutaceae) is allied with citrus, and Brazilian peppertrees (Anacardiaceae) are closer to cashews, mangos, and poison ivy. Coralbean is in the order Fabales, family Fabaceae, with peas, beans, and other legumes.

IDENTIFYING FEATURES:

Hercules club *(Zanthoxylum clava-herculis)*, 10 ft (3 m) tall, is a prickly tree with alternate, compound leaves of about nine asymmetrical, leaflets with wavy margins.

Brazilian peppertree *(Schinus terebinthifolius)*, 20 ft (6.1 m) tall, often grows as multiple arcing trunks within a tangled thicket. Its pinnately compound leaves typically have seven or nine leaflets. Clusters of shiny red berries are obvious in winter. When its branches break they emit the pungent odor of peppery turpentine.

Brazilian peppertree foliage

Eastern coralbean *(Erythrina herbacea)*, 10 ft (3 m) tall, is prickly and shrublike with compound leaves of three that drop in winter. Tubular, scarlet flowers protrude from an otherwise bare plant in early spring.

HABITAT: All grow behind salt-spray-shielding plants on the dune.

DID YOU KNOW? Leaves of the Hercules club have a citrus smell from an aromatic bitter oil called xanthoxylin, which has many folk-medicinal uses. Also known as the toothache tree, chewing this plant's fruit and leaves can numb mouth pain.

Eastern coralbean flowers, compound leaf (inset)

Freshly stranded pelagic sargassum

Sargassum fluitans

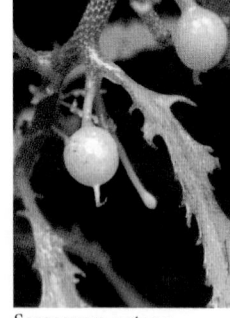

Sargassum natans

Pelagic sargassum as habitat in the open Gulf

Attached sargassum

Brown Algae *(Sargasso Weed)*

Pelagic Sargassum

Attached Sargassum

RELATIVES: Brown algae are heterokonts in the class Phaeophyceae (fay-o-FYE-cee-ee), unrelated to green algae and red algae. Sargasso weeds are in the family Sargassaceae.

IDENTIFYING FEATURES:

Pelagic sargassum (sargasso weed, gulf-weed) (*Sargassum* spp.) forms golden bunches as large as 16 in (40 cm), and turns brown, then black, after stranding. There is no central stem or holdfast. Clumps have numerous branches (stipes) bearing tooth-edged leaves and gas-filled bladders (pneumatocysts). Two species are most common. *Sargassum fluitans* has smooth, oblong bladders and pencil-width leaves, and *Sargassum natans* has spherical, spur-tipped bladders and narrower leaves. The brownish, BB-size bladders are common in the tide line (p. 23).

Attached sargassum (*Sargassum filipendula)*, to 3 ft (0.9 m) long, has brown, serrate-margin, elongate leaves with uneven rows of faint dots. Branches connect to a central stem and may have a single, tough holdfast disc. The longest branches have pea-size gas bladders.

HABITAT: Pelagic sargassum drifts with Gulf and Atlantic currents. Attached sargassum grows anchored to hardbottom in shallow waters, but drifts when rough weather breaks its holdfasts free.

DID YOU KNOW? The mid-Atlantic Sargasso Sea is an end-point for pelagic sargassum, rather than an origin. Circulating currents, collectively called the North Atlantic Gyre, bring the floating algae to its namesake sea. As sargassum drifts, it provides a habitat life-raft for hundreds of species of animals.

Brown Algae *(Three-cornered Hat, Balloon, Padina, Dictyota)*

Three- cornered Hat   Balloon, Dictyota

Padina

Three-cornered hat algae

RELATIVES: All are in the class Phaeophyceae. Three-cornered hat algae are with sargasso weed in the family Sargassaceae. Balloon seaweed (family Scytosiphonaceae) is distantly related to Padina and Dictyota algae (family Dictyotaceae).

IDENTIFYING FEATURES:

Three-cornered hat algae *(Turbinaria turbinata)*, 6 in (15 cm), beach as golden bouquets of pyramid-shape fronds. The algae live attached and float when free due to the air in their fronds.

Balloon seaweed *(Colpomenia* spp.), 1.5 in (3.8 cm), beaches as individual or clustered, green-brown bubble-balls that appear partially flattened.

Padina *(Padina gymnospora)*, 6 in (15 cm), is formed of flat, splitting, green-brown blades with concentric bands and an in-rolled outer margin. Other *Padina* species are less common.

Dictyota *(Dictyota dichotoma)*, 12 in (30 cm), has flat, olive-brown fronds with smooth margins and blunt tips that fork into even branches of two. This alga largely disappears in cold months.

HABITAT: All live attached to hardbottom in shallow water.

DID YOU KNOW? Many brown algae species have bitter chemicals to discourage grazing by fish. In response to a wound, dictyota releases trimethylamine and dimethylsulphide in a distasteful cocktail for any potential nibbler.

Balloon seaweed

Padina

Dictyota

311

Broadleaf sea lettuce on a jetty boulder

Detached broadleaf sea lettuce

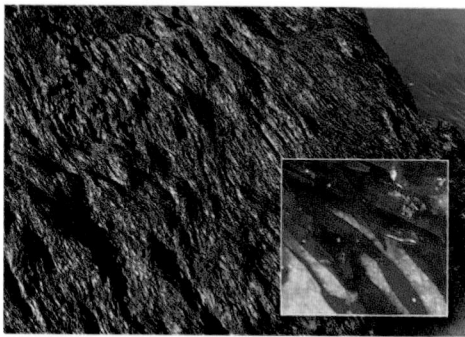

Gutweed on swash rocks, closeup (inset)

Gutweed anchored to a live coquina clam

Green Algae *(Sea Lettuce, Gutweed)*

RELATIVES: Green algae (phylum Chlorophyta) are closer to land plants than to brown or red algae. These species are in the order Ulvales, family Ulvaceae.

IDENTIFYING FEATURES: *Ulva* species are green in life, but bleach white in the sun after stranding.

Broadleaf sea lettuce *(Ulva lactuca)*, 2 in (5 cm) or wider straps, is a bright, apple-green alga with wavy, ruffled edges. Its slick, translucent sheets begin attached but are easily torn free.

Gutweed *(Ulva intestinalis,* formerly *Enteromorpha* spp.) is hairlike when short and forms unbranched, narrow, limp, tubular, intestine-like strands up to 6 in (15 cm) long.

HABITAT: These algae often grow in dense patches covering entire jetty boulders below the high-tide line. They also cover hardbottom during calm summer months, and are common in fresh wrack after rough weather tears them free. Gutweed frequently attaches to surf-zone shell bits and live shells, like **coquina clams**.

DID YOU KNOW? Farm chickens fed broadleaf sea lettuce had higher "dressing weights" and more breast muscle. But you can skip the middleman and enjoy sea lettuce on your own. Also known as "green laver," this alga is tasty and nutritious. Wash it well, toast it, add it raw to salads, boil it in soups, or use it as a substitute for sushi nori. Sea lettuce proliferates in areas where there is runoff of high-nitrogen waters, which may be polluted with bad bacteria, so assess water quality where you forage for algal treats.

Green Algae
(Hair, Caulerpa, Green Fleece)

Hair Algae   *Caulerpa, Green Fleece*

RELATIVES: Hair algae are in the order Cladophorales and family Cladophoraceae. Caulerpa and green fleece share the order Bryopsidales, but are separated in the respective families, Caulerpaceae and Codiaceae.

IDENTIFYING FEATURES:

Hair algae (*Cladophora* spp.), to 12 in (30 cm) long, vary in growth form but are always composed of green, thin, branched, hairlike filaments. A frequent form is a wispy, attached clump. These clumps may also drift along the sea bottom, especially during massibe blooms of "June grass" in the northeastern Gulf. Another growth form is an unattached, densely branched, cottony ball (aegagropila) up to fist size, often associated with pelagic sargassum. Each of these forms may brown, or sun-bleach pale green or white after stranding.

Common caulerpa (*Caulerpa prolifera*), 4 in (10 cm) long fronds, has slightly twisted, dark-green straps with slender stalks attached to rhizomes.

Green fleece (dead man's fingers) (*Codium* spp.), 12 in (30 cm), has forked, spongy, bright- to dirty-green fingers joined at a single holdfast. In the water it forms a domed clump.

HABITAT: Hair algae grow in shallow waters or at the surface. Caulerpa and green fleece are in calm shallows.

DID YOU KNOW? Individual plants in the order Bryopsidales are composed of a single giant cell with lots of nuclei and chloroplasts. To make plant food from light, the chloroplasts move where the light is. If areas are covered by sand, chloroplasts move to sunnier plant parts.

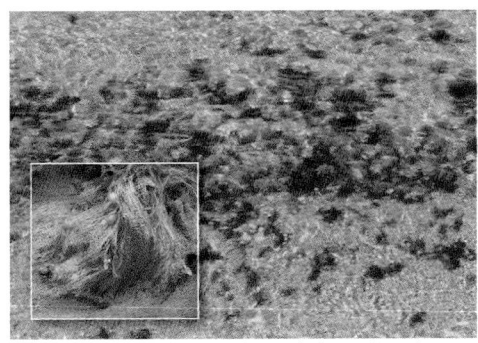

"June grass" hair algae in the surf, clump (inset)

Unattached, cottony, hair-algae balls

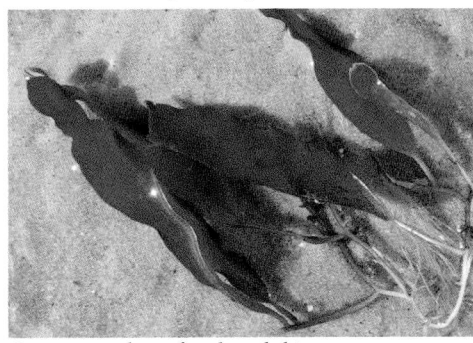

Common caulerpa fronds and rhizomes

Green fleece algae

313

Graceful redweed, branching closeup (inset)

Graceful redweed anchored to a living coquina clam

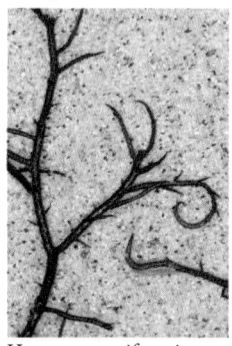

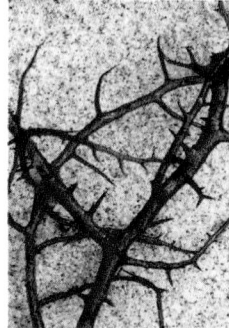

Hypnea musciformis   Hypnea cervicornis

A tuft of Gelidium *on a Texas jetty*

Red Algae *(Graceful Redweed, Hypnea, Turf Algae)*

RELATIVES: Red algae (Rhodophyta) are unrelated to brown algae and somewhat related to green algae. *Gelidium* turf algae (order Gelidiales) are only distantly related to graceful redweed and hypnea (both in the order Gigartinales).

IDENTIFYING FEATURES:

Graceful redweed *(Gracilaria tikvahiae),* 12 in (30 cm), has greenish- or yellowish-red to bright-red, forked thalli (branches) like sinuous clumps of spaghetti. Small, anchored plants often dislodge and drift as larger clumps. This alga quickly bleaches white.

Hypnea *(Hypnea musciformis),* 12 in (30 cm), grows in wiry, tangled yellowish-brown clumps. Thallus tips frequently end in sickle hooks. *Hypnea cervicornis* is similar in color and form, but has numerous, pointy branchlets and no hooks.

Turf algae grow as a dense, shaggy carpet on surf-zone rocks. Much of the turf is *Gelidium crinale,* which is also common as individual tufts. This alga's stubby tangles often bleach pale at the tips. In winter, *Bangia fusco-purpurea* (order Bangiales) occurs in algal turf, looking like fine, purple, wet-matted hair.

HABITAT: All grow in shallow water. Graceful redweed and hypnia anchor in sediment or live as unattached drift algae.

DID YOU KNOW? Red algae have a photosynthetic pigment that absorbs blue light, allowing the plants to live at greater depths than green plants. Many red algae are farmed as seafood and are an important part of the Asian-food lover's diet. Hydrogel (agar) food additives from *Gracilaria* provide texture to many foods.

Red Algae *(Furry Red, Triangular Red, Dragon's Tongue, Fern Weed)*

Furry Red, Triangular Red | Fern Weed | Dragon's Tongue

Furry red alga, max 6 in (15 cm)

RELATIVES: Furry red algae share the order Cermiales with triangular red seaweed, and are only distantly related to dragon's tongue and fern weed (order Halymeniales).

IDENTIFYING FEATURES: These algae were attached by a single, fibrous holdfast.

Furry red alga *(Bryocladia cuspidata)* is brownish to reddish with minimally branched, cylindrical thalli that are densely clothed with short, thin, hairlike branchlets.

Triangular red seaweed, max 8 in (20 cm)

Triangular red seaweed *(Alsidium triquetrum)* has stiff, dark-red branches that are triangular in cross-section with short, spinelike, outward-pointing branchlets lining the branch corners.

Dragon's tongue *(Halymenia floresii)* is bright pink with wide blades that have margins with numerous, narrow branchlets.

Grateloup's fern weed *(Grateloupia filicina)* has moderately flattened, yellow-brown to dark red-brown, fernlike branches that are gradually shorter toward the growing end.

Dragon's tongue, max 20 in (50 cm)

HABITAT: Triangular red seaweed grows sheltered between intertidal rocks and deeper. Furry red, dragon's tongue, and fern weed algae grow in calm areas of shallow rubble or seagrass.

DID YOU KNOW? Dragon's tongue is harvested for food under the more appetizing name, red sea lettuce. Experiments show the alga to change its dominant pigments depending on the color of light exposure. It's thought that this could be a cultivation tool to improve nutrition and appearance.

Grateloup's fern weed grows to 20 in (50 cm)

315

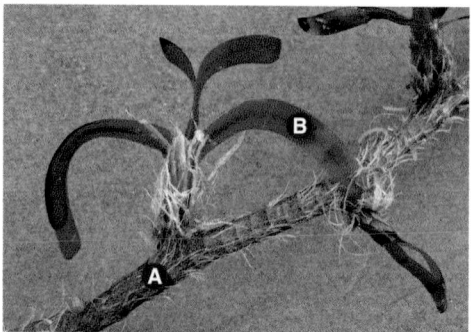

Freshly beached turtle grass, (rhizomes A, blades B)

Turtle grass with sun-bleached blades and rhizomes

Turtle grass rhizomes with fibrous root tufts

Star seagrass leaves and rhizome

Seagrasses *(Turtle Grass, Star Seagrass)*

RELATIVES: Seagrasses are flowering monocots in the order Alismatales, not directly related to true grasses (order Poales). Turtle grass and star seagrass share the family Hydrocharitaceae.

IDENTIFYING FEATURES:

Turtle grass *(Thalassia testudinum)* has ribbonlike blades as long as 24 in (61 cm) with about a dozen parallel veins. Their shoots connect to tough, tubular rhizomes with wiry root-tufts. Blades may be sun-bleached pure white, and rhizomes may strand in tangled masses. Turtle grass fruits are 1/2 in (1.3 cm), green and fleshy with a rough, brown covering. Sprouted seeds are like button-size mushrooms with a tiny protruding blade.

Star seagrass *(Halophila engelmannii)* has unusual leaves for a seagrass—a whorl of four to six oblong leaves at the end of each 4-in (10-cm) stem. These leaves are about an inch (2.5 cm) long and have finely serrated margins.

HABITAT: Relatively calm waters out to 30 ft (9 m) where waters are clear. Turtle grass is most common where salinity is close to seawater. After storms, the leaf blades and corky rhizomes of turtle grass erode out of the sediment, float, and drift to distant beaches.

DID YOU KNOW? Individual plants of these seagrass species are either male or female, with different tiny, underwater flowers. Pollen from male flowers disperses in negatively buoyant slimy strings. Turtle grass is the most common seagrass in the Wider Caribbean region, including the Gulf. This plant is habitat or food for thousands of marine creatures. It's a favored food for the green turtle.

Seagrasses *(Manatee Grass, Shoalgrass, Widgeon Grass)*

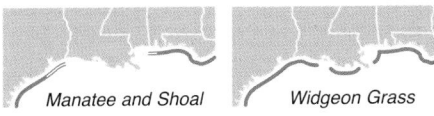

Manatee and Shoal   Widgeon Grass

RELATIVES: Manatee grass and shoal-grass (family Cymodoceaceae) are related to widgeon grass (Ruppiaceae) and other seagrasses, but are only remotely linked to true grasses.

IDENTIFYING FEATURES:

Manatee grass *(Syringodium filiforme)* has thin, arcing, 12-in (30-cm) tubular leaves that are slightly stiffer than cooked spaghetti. Its pale rhizomes are tougher than the fragile leaves, which beach in broken segments. Leaf segments float and frequently drift until they are white from sun-bleaching.

Shoalgrass *(Halodule wrightii)* has leaves of similar length and width to manatee grass, but with flat, tinsel-like blades.

Widgeon grass *(Ruppia maritima)* resembles shoalgrass but has bushy bundles of 4-in (10-cm) threadlike blades that protrude in an alternating fashion along a zig-zagging, yellow-green stem.

HABITAT: Manatee grass and shoalgrass live in calm waters out to 20 ft (6 m). Manatee grass is common in the wrack line because is grows in pastures open to the Gulf. Shoalgrass and widgeon grass tolerate a wide range of light, temperature, and salinity conditions.

DID YOU KNOW? Bleached manatee grass blades may have come from seagrass pastures far away. Shoalgrass is generally the first seagrass species to colonize a disturbed bar of shallow sediment. Its roots and rhizomes stabilize the sediment and allow a succession of dominating species like turtle grass and manatee grass.

Fresh manatee grass blades, closeup (inset)

Stranded, sun-bleached segments of manatee grass

A clump of shoalgrass, tinsel-like blades (inset)

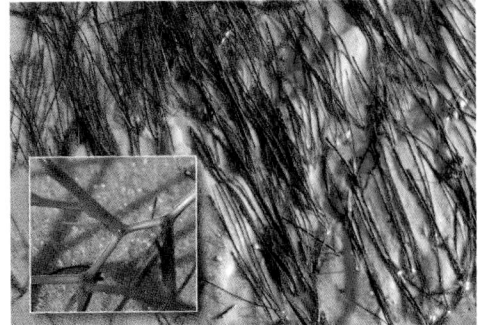

Widgeon grass, zigzag stem (inset)

317

Plant Drifters—Seabeans and Driftwood

A **seabean** is a fruit or seed that has made a sea voyage. Some drift to disperse their genes, and others are lost, having floated far from where they could hope to grow. By design or by luck, these drifters end their journey on Gulf beaches during particular times of the year. Seabean season in the Gulf matches hurricane season (June–October), except in southern Texas, where persistent winds blow drifters out of ocean currents (arrows below) and onto beaches, late March through June. Seabean season, minus travel time, may also correspond to the period when tropical seeds enter the world during the annual Amazonian flurry of flood-season fruiting. Drifting fruits include capsules and nuts (dry fruits with a hard shell and single seed), and seeds include beans and other varied forms. **Driftwood** describes stems, trunks, and other plant parts that also ride currents, and that reach the same beaches as seabeans.

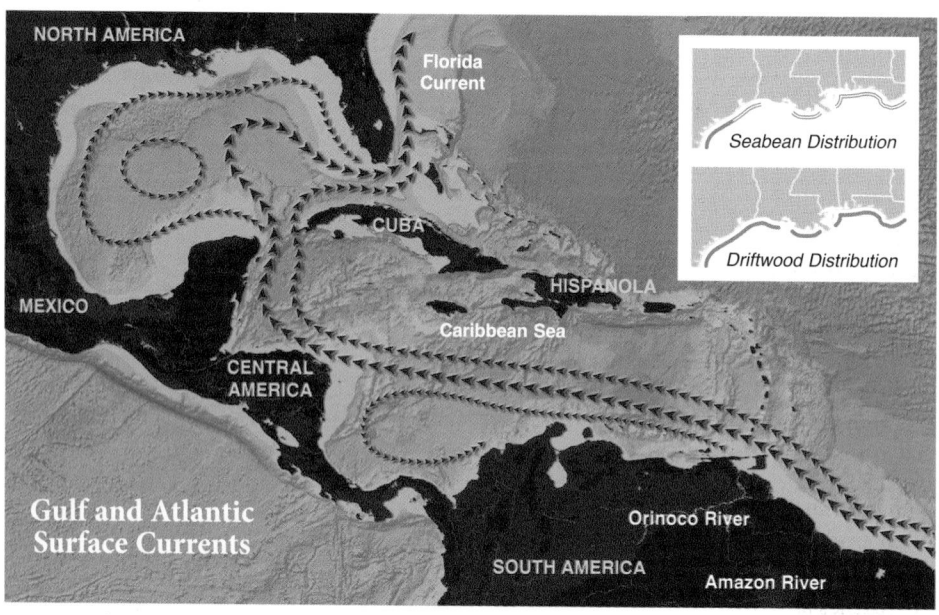

NORTH AMERICA

Florida Current

Seabean Distribution

Driftwood Distribution

CUBA

HISPANOLA

MEXICO

Caribbean Sea

CENTRAL AMERICA

Gulf and Atlantic Surface Currents

Orinoco River

SOUTH AMERICA

Amazon River

Seabean Anatomy

mesocarp (flesh or husk)

endocarp (seed coat)

hilum (attachment scar)

germination pores

exocarp (outer layer)

endocarps

Palm Seeds *(Prickly Palm, Starnut, Coconut, Cohune, Rabbitface)*

RELATIVES: These nuts come from palms (family Arecaceae) in the cocosoid palm group.

IDENTIFYING FEATURES:

Prickly palms (*Acrocomia* spp.) produce fruits with roundish, black, stony nuts. When weathered they may be gray or powdery brown. The outer, hard part (the endocarp) is unique in having three equally spaced pores around its middle.

Prickly palm endocarps, max 1.25 in (3.2 cm)

Starnut palms (*Astrocaryum* spp.) have nuts with hard, black, teardrop-shape endocarps. The rounded end has three pores that may be obscured by fibers. The teardrop shape may vary from sharply pointed to almost spherical. New nuts are oily; worn nuts are chalky brown.

Coconuts (*Cocos nucifera*) strand on beaches as mature, football-size **fruits**, immature fruits, **husks**, and **endocarps**. The three-pored mature endocarp (when fresh) contains a layer of fleshy white meat and is partially filled with clear liquid.

Starnut palm endocarps, max 2 in (5 cm)

Cohune palms (*Attalea cohune*) bear egg-size fruits. Their time at sea leaves only a blackish, 2-in (5-cm) nut worn through to a tan, woody endocarp. Its three pores are often eroded into a ragged hole.

Rabbitface seeds, 1.3 in (3.2 cm), appear on beaches as nonviable drifters from a cocosoid palm yet to be identified. The hard, black, three-pore seed has a rabbit-like face with eyes, nose, and mouth.

Coconut fruit (left), husk (top), endocarp (lower)

ORIGIN: Prickly palms are planted throughout the Caribbean. Starnut palms grow in lowland rainforests of Central and South America. Coconuts grow on warm beaches all over the world. Cohune palms grow in Central American rainforests. Rabbitface seeds come from some myste-rious corner of the Tropics.

DID YOU KNOW? Prickly palms have wickedly long trunk spines.

Cohune palm seed   *Rabbitface seed*

319

Nypa palm fruit, about 4 in (10 cm)

Sea coconuts, max 3 in (7.6 cm)

Sea coconuts with husks

Thatch screw pine fruitlets, max 2 in (5 cm)

Nypa Palm, Sea Coconut, and Screw Pine

RELATIVES: Although all monocots, nypa and sea-coconut palms (order Arecales, family Arecaceae) are not directly related to screw pines (order Pandanales), which are not true pines.

IDENTIFYING FEATURES:

Nypa palm (*Nypa fruticans*) fruits are glossy black, teardrop shape, and lined with about five ridges. The narrow portion ends with a tuft of fibers.

Sea coconuts (*Manicaria saccifera*) grow on the bussu palm in singles, twos, or threes within a brown, lumpy **husk**. The sea often erodes free the spherical seeds, which have a single scar like a bellybutton and are covered by brownish, glossy, endocarp occasionally obscured by flaky layers of stiff, fruit skin. They may be weathered to a tan or gray.

Thatch screw pines (*Pandanus tectorius*) produce fruit masses like a round pineapple composed of dozens of fruitlets. These fruit segments reach distant beaches as brown cones with a fibrous narrow end.

ORIGIN: Nypa palms grow in standing water along river banks in Panama and Trinidad, although the palm is native to China and southeast Asia. Sea coconuts drop from bussu palms growing near rivers of the American Tropics. Screw pines are native to the tropical Indo-Pacific but have been introduced as landscape plants throughout the Wider Caribbean, including Mexico and Florida.

DID YOU KNOW? Bussu palm seeds roll downhill, float, and are dispersed by tropical floodwaters. The palm grows in swampy areas as a slender-trunked palm with enormous fronds to 30 ft (9 m)—the longest leaf known. Locals use the giant leaves as preferred thatching for their houses—rain protection that can last more than a decade.

Mango, Hog Plum, Pond Apple, and **Australian Pine**

RELATIVES: All are dicot flowering plants. Mangos and hog plums (family Anacardiaceae, cashews and poison ivy) are only distantly related to pond apple (Annonaceae) and Australian pine (Casuarinaceae).

IDENTIFYING FEATURES:

Mangos (*Mangifera indica*) contain large, thin, flattened, fibrous seeds that weather at sea to become tan and fuzzy with conspicuous radiating veins.

Hog plums (*Spondias mombin*) arrive on beaches as tannish, oblong fruits with only the fibrous mesocarp showing.

Pond apples (*Annona glabra*) are yellow, pungent fruits that release small, oval seeds. The pond apple seeds making a sea voyage may be shiny brown, or tan if more weathered.

Australian pines (*Casuarina equisetifolia*) drop dime-size, gray-brown fruits that look like tiny, spiny pineapples.

ORIGIN: Mangos are native to India, but have been introduced all over the Tropics. Find a mango seed on a stick? The fruits are sold in Mexico as *paletas de mango* (mango popsicles). Hog plum and pond apple are found throughout tropical America. Australian pines are native to the western Pacific but have made a facilitated invasion of the world's tropical coasts, including Florida and Mexico.

DID YOU KNOW? Mangoes from India indirectly inspired the paisley motif, Europe's oldest weaving pattern, which was copied from Kashmiri shawls designed in the 1400s. The hog plum falls from the tree as a long yellow fruit with a juicy, clear, acidic pulp that is in fact relished by hogs. Pond apple trees often grow with mangroves. The plant is native to Florida and Mexico, but is an invasive pest in tropical Australia.

Mango endocarps, max 4 in (10 cm)

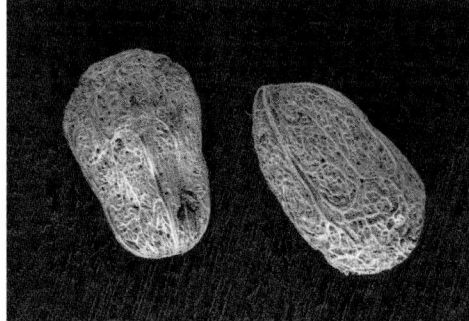

Hog plum mesocarps, max 0.75 in (2 cm)

Pond apple seeds, max 0.5 in (1.3 cm)

Australian pine fruits, max 0.75 in (2 cm)

321

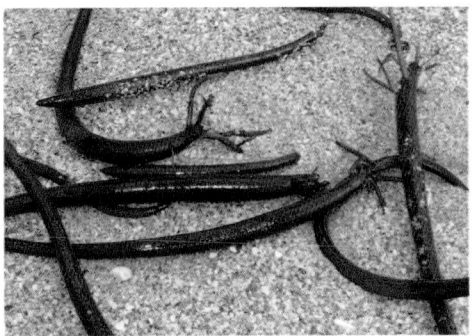

Red mangrove propagules, max 12 in (30 cm)

Black mangrove fruits, max 1.25 in (3 cm)

White mangrove fruits, max 0.75 in (2 cm)

Rangoon creeper fruits, max 1.5 in (4 cm)

Mangroves and Rangoon Creeper

RELATIVES: Mangroves share habitat preferences (marine swamps), but are not direct relatives. Red mangroves are in the order Malpighiales, family Rhizophoraceae, and black mangroves are in the order Lamiales, family Acanthaceae. White mangroves share the order Myrtales and family Combretaceae with Rangoon creeper.

IDENTIFYING FEATURES:

Red mangroves (*Rhizophora mangle*) produce seeds that sprout on the parent tree. They eventually drop as variously curved propagules that look like flamboyantly long, green writing pens. The root-end is brownish and may have begun sending out rootlets before reaching the beach.

Black mangroves (*Avicennia germinans*) have yellow-green fruits like plump, pointed lima beans that have split and sprouted. Their velvety root sprouts may be lengthy or withered.

White mangroves (*Laguncularia racemosa*) have ribbed fruits shaped like spearpoints. New fruits are dark brown and the most weathered fruits are pale tan.

Rangoon creeper (Chinese honeysuckle) (*Combretum indicum*) fruits come ashore as blackish-brown, lightweight, footballshape drupes, with five (sometimes 4–9) longitudinal ribs.

ORIGIN: These mangrove species grow in salty, coastal swamps rimming most of the Gulf and the tropical Americas. Rangoon creeper is native to India and Southeast Asia, but is an ornimental throughout the Wider Caribbean.

DID YOU KNOW? Red mangrove propagules float vertically, root-end down, and can survive a year before arriving home. Black mangrove seeds can root after a four-month drift at sea.

322

Tropical Almond and Calabash

RELATIVES: Tropical almonds share the family Combretaceae with white mangrove and Rangoon creeper, and are not allied with the commercial almond species. Calabash trees are in the bignonia family (Bignoniaceae) and are not gourds.

IDENTIFYING FEATURES:

Tropical almonds *(Terminalia catappa)* beach as corky fruits in various stages of disintegration. The top three images show the range of these beached fruits from newest to oldest. Black or brown fruits with the familiar almond shape may not have drifted very long. More weathered fruits are bleached and fibrous, and the oldest drifters are deeply pitted. Many of the most disintegrated seeds are recognizable only by their deep longitudinal grooves and vague almond shape.

Calabashes *(Crescentia cujete)* from the calabash tree are hard, light, rattling, baseball- to football-size brownish-gray fruits with a protruding scar.

ORIGIN: Tropical almonds fall from a large-leaved tree that grows with vertically separated branches in a pagoda form. The tree is a native of tropical Asia but has been introduced throughout the Tropics and in Florida and Mexico. New-looking seeds may have come from these closer sources, although the tree is most numerous in tropical America and the Caribbean. Calabash fruits drift from Central America and the Caribbean.

DID YOU KNOW? The kernel within a tropical almond tastes like a commercially grown almond and is generally eaten raw. It has been shown to have a sedative or aphrodisiac effect on rats, and nut extracts have promise in treatments for diabetes. The hard shell of the calabash fruit is used for containers, scoops, and artsy decorations, but time at sea makes most beach-found specimens unsuitable for crafts.

Tropical almonds, max 2.75 in (7 cm)

Weathered tropical almonds without outer skin

Pitted tropical almonds have had a long voyage

Calabash fruits, max 9 in (23 cm)

323

A barnacle encrusted box fruit, max 6 in (15 cm)

Anchovy pear fruits, max 4 in (10 cm)

Laurelwood fruits, max 1.5 in (3.8 cm)

Mammee apple, max 2.4 (6 cm)

Box Fruit, Anchovy Pear, Laurelwood, and Mammee Apple

RELATIVES: Box fruit and anchovy pear are allied within the order Ericales, family Lecythidaceae, with the Brazil-nut. Laurelwood and mammee apple are in the distantly related order Malpighiales, and family Calophyllaceae.

IDENTIFYING FEATURES:

Box fruits (sea putat) (*Barringtonia asiatica*) are brownish, lantern-shape fruits with four (rarely five) keels. Weathered fruits become tan and fibrous.

Anchovy pears (*Grias cauliflora*) beach as dried, tan, football-shape fruits with eight veiny ribs and contain a single large seed.

Laurelwoods (Santa-Maria) (*Calophyllum* spp.) bear spherical, green fruits that weather into a brown or tan, eyeball-like sphere. Most beached fruits are smooth with a small nub surrounded by short fibers. Less worn specimens may have a fibrous covering. Of two similar fruit forms, *C. inophyllum* is slightly larger than *C. calaba*.

Mammee apples (*Mammea americana*) are thumb-size, ovoid, fruit "stones" that have a firm, fibrous-brown covering.

ORIGIN: Box fruit trees are native to Southeast Asia but have been introduced throughout the Tropics. Anchovy pear trees are native to Jamaica, Central America, and Colombia, and grow in groves near rivers and marshes. Mammee apple trees, with 1–4 seeds per tasty fruit, grow throughout the American Tropics.

DID YOU KNOW? The box fruit tree has many parts that are pulverized to release a fish-stunning poison. Its buoyant fruits can be viable after two years at sea and were recorded as being among the first living things to reach the new beaches of Krakatau after the former island went kaboom. A fresh anchovy pear has the appearance and taste of a mango.

Porcupineseeds and Coco Plum

RELATIVES: These plants share the order Malpighiales. Porcupineseeds come from sawari nut trees in the exclusively neotropical family, Caryocaraceae, and coco plums are in the family Chrysobalanaceae.

IDENTIFYING FEATURES:

Smooth porcupineseeds (*Caryocar glabrum*) have brownish-black, oval or kidney-shape endocarps that are typically split on one side. Their surface is covered by dull, woody spines.

Smooth porcupineseed endocarps, max 2 in (5 cm)

Prickly porcupineseeds (*Caryocar microcarpum*) are black or brown, kidney-shape, split, hollow endocarps with clusters of sharp spines.

Other porcupineseed forms (*Caryocar* spp.) have a dense covering of thin spines or a covering worn down to the texture of dull, coarse sandpaper.

Coco plums (*Chrysobalanus icaco*) have fleshy fruits with hard, light brown, teardrop-shape seeds with netlike ridges. Only seeds survive the drift.

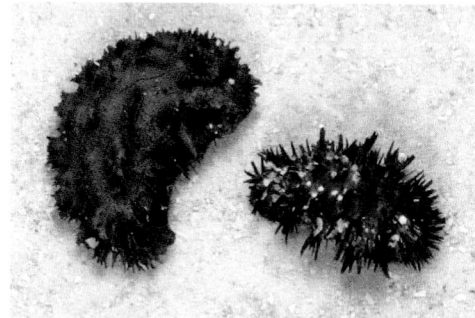

Prickly porcupineseed endocarps, max 1.5 in (4 cm)

ORIGIN: Porcupineseeds come from sawari nut trees that grow in frequently flooded areas of the Amazon and Orinoco River basins. Coco plums are native to coastal areas of the American Tropics, including southern Florida, Mexico, and the Caribbean.

DID YOU KNOW? Sawari nut trees are one of the largest trees in the Amazon rainforest canopy, growing to 160 ft (50 m) tall with a trunk 8 ft (2.5 m) in diameter. Their porcupineseeds protect a large, soft, white kernel that is pleasantly sweet and widely eaten where the tree grows. Edibility rarely survives a sea drift, so don't sample those beach seeds. Coco plums are important tropical dune plants and are widely planted as ornamentals.

Beach-worn porcupineseeds, possibly C. villosum

Coco plum seeds, max 0.75 in (2 cm)

325

Mary's bean seeds, max 1 in (2.5 cm)

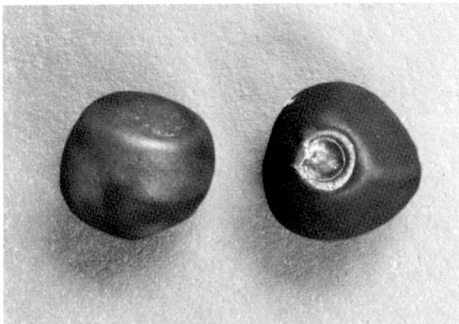

Wood rose seeds, max 0.8 in (2 cm)

Moonflower seeds (top), railroad vine seeds (bottom)

Railroad vine seeds with capsule

Mary's Bean, Wood Rose, Moonflower, and Railroad Vine

RELATIVES: These morning-glories share the order Solanales and family Convolvulaceae with the sweet potato (which is a tuber, not a seed).

IDENTIFYING FEATURES:

Mary's beans *(Merremia discoidesperma)* look like a scorched, hot-cross bun. They are black or brown, stony seeds with an oval hilum (attachment scar) and a distinct, indented cross.

Wood rose seeds *(Merremia tuberosa)* have a circular hilum and flat sides so as to accommodate four tightly packed seeds per capsule.

Moonflower seeds *(Ipomoea alba)*, 5/16 in (8 mm), are similar to wood rose seeds but are only about half the size.

Railroad vine (bayhop) seeds *(Ipomoea pes-caprae)*, 3/16 in (5 mm), mature within a papery capsule that splits to liberate fuzzy, brown, angular seeds with a circular hilum. Those surf-washed enough to lose their fuzz are glossy and hard. The seeds are smaller and darker than moonflower seeds.

ORIGIN: A Mary's bean reaching a Gulf beach likely dropped from a single-seed capsule on a winding vine growing near a Central American river. Wood rose grows throughout tropical America and the West Indies. Many railroad vine and moonflower seeds come from local sources, although these plants grow all over the Wider Caribbean. Each of these seeds may drift for many years and thousands of miles.

DID YOU KNOW? Mary's bean, named for the Virgin Mary, gets its cross from the impression of a strap that secures the seed in its capsule. Holding this seed provides believers with good luck, recovery from snakebites and hemorrhoids, and for expectant mothers, ease of childbirth.

Black Sea Biscuit, Antidote Vine, Candlenut, and Rubber Tree

RELATIVES: Black sea biscuits (Anacardiaceae) are related to cashews, mangos, and poison ivy. Antidote vine is a member of the melon family, Cucurbitacae. Candlenuts and rubber trees share the family Euphorbiaceae with spurges.

IDENTIFYING FEATURES:

Black sea biscuits *(Antrocaryon amazonicum)* are fruit stones from the jacaiacá tree. These endocarps are dark, slightly flattened, and radially symmetrical with four to five divisions (locules, with seeds).

Black sea biscuit endocarps, max 1.2 in (3 cm)

Antidote vines *(Fevillea cordifolia)* produce fruits that reach the beach as brownish, warped discs containing about 10 seeds that do not rattle within.

Candlenuts (kukui nuts) *(Aleurites moluccanus)* have a walnut shape and are typically black with a waxy sheen. Some weathered candlenuts are tan or powdery brown, but all of these nuts are stone hard.

Antidote vine fruits, max 2.25 in (5.7 cm)

Rubber tree seeds *(Hevea brasiliensis)* have a glossy-black to flat-brown endocarp and an elliptical shape with angular sides from sharing a three-seed capsule.

ORIGIN: Black sea biscuits come from the center of a yellow fruit atop a tall tree in the Brazilian rainforest. Antidote vines grow throughout tropical America and the Greater Antilles. Candlenut trees are native to the Indo-Pacific but are widely cultivated in the Tropics. Rubber trees are native to the Amazon region where they are cultivated and tapped for their latex.

Candlenuts, max 1.25 in (3.2 cm)

DID YOU KNOW? Antidote vine seeds are rich in oil used as a purgative and antidote for many kinds of poisoning. The fruits and seeds float to disperse the plant after floods. Candlenuts (kukui in Hawaii, where this Polynesian transplant is the state tree) are polished into black gleaming nuggets suitable for jewelry.

Rubber tree seeds, max 1.2 in (3 cm)

327

Manchineel fruits, max 1.2 in (3 cm)

Monkey pistol fruit carpel, max 1.5 in (3.8 cm)

Cabbagebark seeds, max 2 in (5 cm)

Naval-spurge seeds, max 2 in (5 cm)

Manchineel, Monkey Pistol, Cabbagebark, and Naval-Spurge

RELATIVES: These fruits, fruit parts, and seeds come from plants in the order Malpighiales and family Euphorbiaceae.

IDENTIFYING FEATURES:

Manchineel (poison apple) fruits *(Hippomane mancinella)* drop from trees as yellow, fleshy, flattened globes. They arrive on Gulf beaches as corky, tan fruits in various stages of disintegration, ranging from slightly wrinkled to a skeletal endocarp of radiating ribs.

Monkey pistol (sandbox) trees *(Hura crepitans)* have pumpkin-shape seed capsules with torqued woody carpels, each shaped like a jumping dolphin. The carpels float, but seeds do not.

Cabbagebark seeds *(Andira inermis)* are within a hard, fibrous, round mesocarp encircled by a ridge connected to faint veins. The surface is generally brownish and pitted.

Naval-spurge (cobnut) seeds *(Omphalea diandra)* are blackish-brown with a fine, knobby surface angled on one side. The brittle endocarp is occasionally cracked, allowing an unforgettable experience with the putrid decay of the oily kernel.

ORIGIN: Manchineel trees are native to Central America and the Wider Caribbean, including Florida. Monkey pistol and cabbagebark trees grow throughout the American Tropics. Naval-spurge comes from tidal swamps and rivers of the American Tropics.

DID YOU KNOW? Manchineel fruits are poisonous when fresh but harmless after drifting. The tree is Endangered in Florida. The monkey pistol tree is one of the tallest trees in Central America. Its seeds disperse when the woody fruit containing them ripens, and explodes like a shot, allowing seeds to flutter down to the forest floor.

328

Nickernuts and Little Marble

RELATIVES: These seeds come from legumes, order fabales, family Fabaceae (peas and beans).

IDENTIFYING FEATURES:

Gray nickernuts (nicker or nickar beans) *(Caesalpinia bonduc)* are hard, grayish beans that look like swollen ticks. The endocarp typically has faint fracture lines encircling the seed.

Gray nickernuts, max 0.75 in (2 cm)

Yellow (orange) nickernut *(Caesalpinia ciliata)* are similar to gray nickernuts, but have yellowish lines or are **bleached pale**. Other species of yellowish *Caesalpinia* may drift ashore. These are rare.

Brown nickernuts *(Guilandina major)* are similar to gray nickernuts, including having the faint fracture lines, but are larger, less spherical, and brown. These are common in Texas, rare elsewhere.

Little marbles *(Oxyrhynchus trinervius)* are black, round, glossy, stone-hard beans almost completely encircled by a hilum with a light groove and lipped margins.

Yellow nickernuts, max 0.75 in (2 cm). Bleached (R)

ORIGIN: Nickernuts grow along warm coastlines. Species reaching Gulf beaches likely came from Caribbean Mexico. Little marbles drop from pods grown on a vine from the canopy of Central American rainforests.

DID YOU KNOW? All these seeds can be polished to a lustrous shine. "Nicker" derives from knikker, the Dutch word for a (toy) marble. They are amazingly hard and have floated for more than 30 years in experimental trials. Nickernut beans develop within pods that are bristling with sharp spines that match the recurved prickles covering the rest of the plant. Nickernut plants are a critical host species (food source) for caterpillars of the Endangered Miami blue butterfly *(Cyclargus thomasi bethunebakeri)*.

Brown nickernuts, max 1 in (2.5 cm)

Little marbles, max 0.5 in (1.3 cm)

329

Bloodwood pods, max 2 in (5 cm)

Coin vine, max 1.5 in (3.8 cm) (A), moneybush (B)

Cat's eye pods, max 1.75 in (4.5 cm)

Flame of the forest seeds, max 1.5 in (4 cm)

Bloodwood, Coin Vines, Cat's Eye, and Flame of the Forest

RELATIVES: These seeds are from legumes in the family Fabaceae.

IDENTIFYING FEATURES:

Bloodwood pods (*Pterocarpus officinalis*) are a little larger than a quarter and a few times as thick. Most are brownish or gray, and have a veined texture over a roundish, irregular, winglike shape.

Coin vine pods (*Dalbergia ecastophyllum*) are thin, papery, leaflike fruits about the size of a nickel. Pods larger than 1.5 in (3.8 cm) may be from the **moneybush**, *D. monetaria*, a rarer find on Gulf beaches.

Cat's eye pods (*Millettia pinnata*) drop from the poonga oil tree. The pods are glossy tan, uniformly thick, and have a persistent stem.

Flame of the forest (*Butea monosperma*) seeds are kidney-shape and beach in various stages of decay.

ORIGIN: Bloodwood trees grow in coastal swamps of tropical America and the West Indies. Coin vine is a leggy bush common in coastal habitats around the Wider Caribbean, including southern Florida and Mexico. Moneybushes grow in freshwater marshes of northern South America. Poonga oil trees are native to Malaysia but have been introduced to the Wider Caribbean. Flame of the forest grows as an ornamental all over the Tropics and is native to India.

DID YOU KNOW? Bloodwood, also known as the dragon's blood tree, oozes red juice when the bark is cut. This *"sangre de drago"* is used in the treatment of diarrhea, mouth sores, and diaper rash. The roots and bark of coin vine are crushed for use as a fish poison. Seed oil from the poonga oil tree has been used as feedstock for biodiesel to generate power in remote Indian villages.

Bay Bean, Cathie's Bean, Yellow Flamboyant, and Ear Pod

RELATIVES: These seeds are from legumes in the family Fabaceae.

IDENTIFYING FEATURES:

Bay beans (beach peas) *(Canavalia rosea)* are hard seeds in the classic bean shape. They have an elongate, oval hilum, and a slightly dark mottling to a background of tan or red-brown. The beans tumble from hotdog-shape pods that grow on dune vines.

Bay beans, max 0.75 in (2 cm). Right shows hilum scar

Cathie's bean *(Canavalia nitida)* is a circular to ellipsoid, crimson bean with a black hilum that wraps around about a third of the seed.

Yellow flamboyant (poinciana) fruits *(Peltophorum pterocarpum)* have a flattened flame shape with thin edges and a texture of tiny parallel grooves.

Ear pods *(Enterolobium cyclocarpum)* are seed pods from the guanacaste tree. They have a lumpy, ear shape, are dark brown when fresh, and dull black when weathered.

Cathie's beans, max 1 in (2.5 cm)

ORIGIN: Bay beans grow in dunes around the Caribbean and southern Gulf. Cathie's bean likely drifts from the Greater Antillies. Yellow flamboyant is native to Southeast Asia, but is planted over the Tropics and is invasive in Mexico. Guanacaste trees are common in Mexico and the rest of Central America.

DID YOU KNOW? The rarely found Cathie's bean was named in memory of author, artist, and naturalist, Cathie Katz—the "Seabean Lady." Guanacaste seed pods require animal-assisted dispersal, but are ignored by native critters sharing its Central American range. Evolutionary biologists hypothesize that the pods were formerly exploited by giant ground sloths and other Pleistocene mammals that are now extinct. Today, the seeds are spread by nonnative horses and cattle.

Yellow flamboyant fruits, max 3 in (7.6 cm)

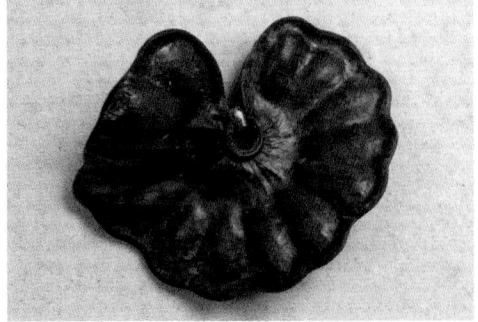

Ear pod from the guanacaste tree, 3 in (8 cm)

331

Coralbean seeds, max 0.5 in (1.3 cm)

Mora seeds, max 7 in (18 cm). Right seed is split

Monkey brain fruit, max 4 in (10 cm)

West Indian locust pod, 7 in (15 cm)

Coralbeans, Moras, Monkey Brain, and West Indian Locust

RELATIVES: These seeds are from legumes in the family Fabaceae.

IDENTIFYING FEATURES:

Coralbeans (*Erythrina* spp.) are scarlet, orange, or (rarely) yellow. Their oval hilum is slightly offset from center. Most are glossy, and all are stone hard. Red seeds less than 1 cm are likely *Erythrina herbacea* (p. 309).

Mora seeds (*Mora* spp., but most commonly, *Mora excelsa*) look like shriveled kidneys. They are brownish, wrinkled, rubbery cotyledons either together as a pair or divided.

Monkey brain fruits (*Andira galeottiana*) look like a fist-size, dark-brown brain, complete with a wrinkled cerebral cortex and divided hemispheres fit for any vegetarian zombie. The most sea-worn fruits have lost their outer coating, revealing a light-brown fibrous layer.

West Indian locust pods (*Hymenaea courbaril*) are rounded, jumbo bean-pods with a tough stem and a rough, thick, brown exocarp (covering). The 2–6, dime-size seeds within are reddish-brown.

ORIGIN: Coralbeans on Gulf beaches likely came from herbaceous coralbean trees growing along the Mexican coast. Mora trees are coastal mangroves from northern South America. Monkey brain fruits are from a small, water-loving tree (the macayo) native to southern Mexico. West Indian locust trees are native to the West Indies and Brazil. Pods may have also come from Central America.

DID YOU KNOW? Coralbean seeds are alluring but toxic, to the extent that they are used as rat poison in Mexico. Mora seeds have the largest plant embryo known. The part that drifts and strands is all embryo.

Sea Purses and Sea Heart

RELATIVES: These seeds are from legumes in the family Fabaceae.

IDENTIFYING FEATURES:

Sea purse beans (*Dioclea* spp., but most commonly, *Dioclea reflexa*) are stone-hard, glossy seeds that are purse-shape due to a flattened side. They are burnt orange, orange-brown, or dark, with some seeds showing a mottled pattern. Their thin, dark hilum is often outlined in orange, wrapping all but the flattened side. Similar beans that are solid dark may be *Mucuna* species.

Sea heart beans (*Entada gigas*) are hard, compressed, circular or heart-shape seeds with an inconspicuous hilum in their indented margin. Although they may have some crusty sea-growth, beneath they are a glossy purple-brown or dark mahogany. The smallest sea hearts tend to be rounder and may have come from the smallest end of the pod.

ORIGIN: Sea purse beans develop with 2–4 sibling seeds in a short, plump sausage-like pod on a tropical rainforest vine. The vine is thought to be native to all of the world's tropics, in part due to the seed's impressive survival at sea. Sea hearts tumble from the long, twisting pods of the monkey ladder vine, a woody canopy-climber from rainforests of the Central and South American Tropics.

DID YOU KNOW? Among the stony, "collectible" seabeans on Gulf beaches, the sea heart is most common, and sea purses are moderately common. The sea heart's monkey ladder vine has the longest bean pod in the world. These seabeans are famous as beach finds as far from their tropical origin as northern Europe. The discovery of a sea heart floating in the eastern Atlantic helped inspire Columbus's search for land to the west.

Sea purse beans, max 1.5 in (4 cm)

Variation in sea-purse-like beans

Sea heart beans, max 2.25 in (5.7 cm)

333

Brown hamburger beans, max 1 in (2.5 cm)

Red hamburger beans, max 1.2 in (3 cm)

Colors of brown (A) and red (B,C) hamburger beans

Hamburger Beans

RELATIVES: These seeds are true beans (legumes) in the family Fabaceae.

IDENTIFYING FEATURES:

Brown hamburger beans *(Mucuna sloanei)* are circular, stone-hard seeds with a thick, black hilum encircling about three-fourths of their circumference. Their texture is slightly roughened away from the hilum. Shapes range from a plump disc to a sphere. Colors are generally on the gray side of brown, and the light border of the hilum is more gray than yellow.

Red hamburger beans *(Mucuna urens)* are difficult to separate from brown hamburgers except for subtle differences in shape and color. Red hamburger beans tend to be flatter (seldom spherical) and redder (on the red side of brown). Their hilum border is more yellow than gray and they often show dark mottling.

ORIGIN: These seeds develop within spiny pods hanging on long stems from woody vines (lianas) native to rainforests of the American Tropics and West Indies. Brown hamburger beans are also known from western Africa.

DID YOU KNOW? Hamburger beans, also known as deer-eye, donkey-eye, horse-eye, ox-eye, or sheep-eye beans, are moderately common on Gulf beaches, and are a regular find in southern Texas. These beans are rich in an alkaloid called L-Dopa, which is used in treating the tremors from Parkinson's disease. L-Dopa is toxic in high doses and is known to cause hallucinations, delirium, and other ills. This chemical defense, along with the seed pod's irritating spines, apparently ensures that few critters eat these beans. Their high survival in the rainforest helps explain why so many hamburger beans end up on distant beaches.

Thick-banded Mucuna and **Black Mucuna**

RELATIVES: These seeds are closely related to hamburger beans, and are legumes in the family Fabaceae.

IDENTIFYING FEATURES:

Thick-banded mucunas (*Mucuna elliptica*) are similar to red and brown hamburger beans except for size and hilum thickness. They are plump discs larger than a quarter and have a wide, black hilum band for most of their thickness.

Black mucunas *(Mucuna holtonii)* are flattened, hard, black seeds with a roughened, dimpled, deflated appearance. They are generally the size of a quarter, and the seed's thin, protruding hilum is easily felt along the edge. Some seeds are more inflated than others.

ORIGIN: Thick-banded mucunas are a mystery. They may come from Caribbean islands or perhaps the Amazon Basin. Black mucunas drop from the velvety pods of "chandelier vines" native to Central America.

DID YOU KNOW? These *Mucuna* species are much less common than brown or red hamburger beans. Black mucunas come to us thanks to Commissaris's long-tongued bat (*Glossophaga commissarisi*), a tiny flying mammal the size of a hummingbird. The seabean's flower has a unique echoreflector that makes it conspicuous to the bat's sonar. Bats landing on the flower are rewarded with an explosion of nectar, and the plant is rewarded with pollen from another flower. Don't know beans? You are not alone. *Mucuna* species are enigmatic finds for beachcombers because they are variable, and only the plant's seeds are seen. Beans with a roughened texture and a wide hilum band are likely ***Mucuna* species**. But some forms of seabeans that look a bit like sea purses are species of **Dioclea**, or could be *Mucuna*.

Thick-banded mucuna beans, max 1.25 in (3.2 cm)

Black mucuna beans, max 1.25 (3.2 cm)

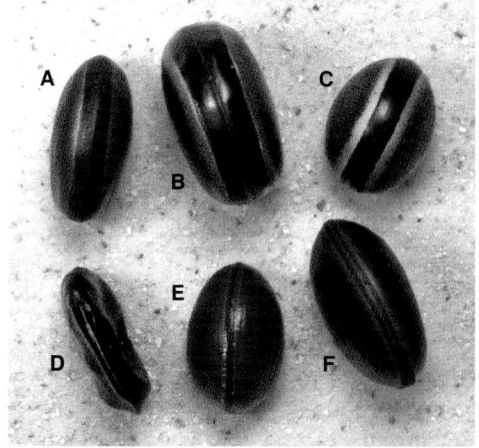

Likely Mucuna *(A–D), and* Dioclea *(E, F) species*

335

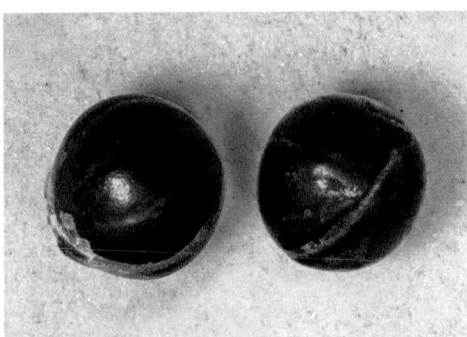

Lantern tree seeds, max 0.5 in (1.3 cm)

Blisterpods, max 2 in (5 cm)

Blisterpods cut in sections showing air pockets

Calatola seed, max 1.5 in (3.8 cm)

Lantern Tree, Blisterpod, and **Calatola**

RELATIVES: Lantern tree (family Hernandiaceae), blisterpod (Humiriaceae), and calatola tree (Metteniusaceae) are distantly related, tropical, flowering plants.

IDENTIFYING FEATURES:

Lantern tree (mago) seeds (*Hernandia sonora*) come from a lantern-like, waxy cupule (fused bracts) that surrounds a black drupe with one large seed. Most beached seeds are hard, dark-brown, marble-size globes with faint longitudinal indentations. Some have tan ribs adhering from remnant, dried fruit-flesh.

Blisterpods (*Sacoglottis amazonica* or *S. guianensis*) look like WWII-era hand grenades because of their woody, air-filled, bulging blisters. Worn pods have open blisters, but internal **air pockets** surrounding the single, inch-long seed remain to keep the pod floating.

Calatola seeds (*Calatola costaricensis*) are a wrinkled, rugby-ball shape with an encircling, longitudinal ridge and many lower, connecting ridges. The rough-coated endocarp is typically light, hollow, and without a viable seed.

ORIGIN: Lantern trees grow in wet forests and along river banks in tropical America and the West Indies. Blisterpods fall from small trees native to the Amazon and Orinoco Basins. Calatola trees are common throughout Central and northern South America.

DID YOU KNOW? The oily seeds of the lantern tree can be burned like tiny candles. The woody, bubble-wrapped, flotation fruit of the blisterpod is exquisitely adapted for aquatic seed-dispersal within the periodically flooded forests of the Amazon. However, the seed does not remain viable after a salty sea voyage.

Hickorys and Pecan

RELATIVES: These nut trees are allied with walnut trees within the order Fagales and family Juglandaceae.

IDENTIFYING FEATURES:

Water hickory nuts *(Carya aquatica)* beach as brownish, lumpy, slightly compressed shells with encircling ridges and variably pointed apexes.

Water hickory nuts, max 1 in (2.5 cm)

Pignut hickory nuts *(Carya glabra)* often come ashore with their charcoal, teardrop-shape, four-part husk intact. The nut within is similar to the water hickory.

Bitternut hickory nuts *(Carya cordiformis)* look like the hickory nuts above but are smoother, without encircling ridges, and have pronounced apex-points if not worn.

Pecans *(Carya illinoinensis)* are similar in shape to the store-bought variety but tend to have a wider range of size and color. Runts may be as small as 1 in (2.5 cm), and color can range from mahogany to tan.

Pignut hickory nuts, max 1.5 in (3.8 cm)

ORIGIN: These nuts fall from trees distributed across the southeastern US. Their upstream source is likely one of the major rivers emptying into the Gulf, especially the Mississippi.

DID YOU KNOW? These are true nuts, defined as a dry fruit with a hardened ovary wall not attached to the seed within. All nuts are seeds, but not all seeds are nuts. Edible, oil-rich hickory nuts have been used by Americans for about 9,000 years. The Powhatan people of Virginia believed that spirits traveling to the rising Sun could only complete their afterlife journey after a drink of *pokahichary* served by a goddess. The drink, made from hickory nuts pounded with water, gives us the name for the tree and its fruit.

Bitternut hickory nuts, max 1 in (2.5 cm)

Pecans, max 1.5 in (3.8 cm)

337

Tropical walnuts, max 1.75 in (4.5 cm)

Black walnuts, max 1.5 in (3.8 cm). Split nut (R)

White walnuts, max 2 in (5 cm)

Mockernuts, max 1.25 in (3.2 cm)

Walnuts and Mockernut

RELATIVES: These nut trees are allied with hickory trees—order Fagales and family Juglandaceae.

IDENTIFYING FEATURES:

Tropical (West Indian) walnuts *(Juglans jamaicensis)* have a hard, wrinkled surface and are indented at one end and bluntly pointed at the other. Their color varies between black and light gray.

Black walnuts *(Juglans nigra)* are brownish-gray with a roughly wrinkled surface. They are bluntly pointed at one end and rounded at the other, and are occasionally heart shape.

White walnuts (butternuts) *(Juglans cinerea)* are deeply wrinkled, elongate, and pointed at one end.

Mockernuts *(Carya tomentosa)* have rounded ridges and a thick shell, often split at the pointed apex.

ORIGIN: Tropical walnut trees have become rare, with most found only in a few forests on the islands of Cuba, Hispanola, and Puerto Rico. Despite its scientific name, the tree is not native to Jamaica. Black walnut, white walnut, and mockernut trees range over much of the eatern US, which is drained by the Mississippi and other rivers that empty into the northern Gulf.

DID YOU KNOW? Few of these nuts are viable, or even contain seeds, following a lengthy drift at sea. Tropical walnut trees are Endangered. Most drift seeds from these trees are probably from Cuba. Elsewhere, few trees have escaped being cut down for housing and agriculture. Tropical walnuts can drift for a long time and travel great distances. Drift-seed experts Charles Gunn and John Dennis observed the nuts to float for about two years.

Tallownut, Madagascar Olive, Seagrape, and **Black Pearl**

RELATIVES: Tallow wood trees (family Olacaceae), Madagascar olive (Oleaceae, olives), seagrape (Polygonaceae, knotweeds and buckwheats), and soapberry trees (Sapindaceae) are not directly related.

IDENTIFYING FEATURES:

Tallownuts (hog plums) fall from the tallow wood tree *(Ximenia americana)* and are light, hard fruits the size of a quail egg or smaller. Weathered fruits are tan, and have four bumps separated by a cross (+) at their roundest end. Some fruits have a brown crust from the formerly yellow fruit flesh.

Tallownut fruits, max 1 in (2.5 cm)

Madagascar olive *(Noronhia emarginata)* fruits are woody and tan with an olive size and shape. Fresh off the bush, they are yellow or purple, but none of this color remains on most beached fruits. Some fruits have a rattling pit.

Madagascar olive fruits, max 1 in (2.5 cm)

Seagrape seeds *(Coccoloba uvifera)* are hard, brownish, fiber-shrouded pits with a rounded end and a pointed end.

Black pearl (western soapberry) seeds *(Sapindus saponaria)* are dark, glossy, stone-hard, little spheres. Fruits grow in clusters, and fall from trees as orange-brown, fleshy berries. Black pearls strand as fruits with a glossy, brown skin or a fibrous net surrounding a papery sack, or as naked seeds.

Seagrape seeds, max 0.75 in (2 cm)

ORIGIN: Tallow wood grows throughout the Tropics. Madagascar olive is native to Madagascar but has been introduced to Florida and Panama. Seagrapes line beaches of the southern Gulf and Caribbean. Soapberry trees range widely from the US to South America.

DID YOU KNOW? Fruits of the western soapberry tree contain abundant saponin, which is a surfactant, soapy enough to lather extensively in water. The fruits were widely used in Mexico to launder clothes.

Black pearl seeds, max 0.5 in (1.3 cm)

339

Crabwood seeds, max 1.75 in (4.5 cm)

West Indian mahogany fruit, max 3.5 in (9 cm)

Nutmeg seeds, max 1 in (2.5 cm)

False mastic seeds, max 0.75 in (2 cm)

Crabwood, Mahogany, Nutmeg, and **False Mastic**

RELATIVES: Crabwood and mahogany are allied within the order Sapindales and family Meliaceae. Nutmeg (order Magnoliales, family Myristicaceae) and false mastic (Ericales, Sapotaceae) are only distantly related.

IDENTIFYING FEATURES:

Crabwood (*Carapa guianensis*) seeds are dull gray-brown, flattened on two sides, and rounded on the other. An elongate, distorted hilum scar is opposite the rounded side.

West Indian mahogany (*Swietenia mahogoni*) fruits are brownish, woody, and round or faintly five-sided in cross-section. A stem or stem scar is opposite the narrow end.

Nutmeg seeds (*Myristica* spp.) come from the center of an apricotlike fruit and are black, textured, and egg shape when they strand on beaches. Larger seeds are probably the commercial species, *M. fragrans*.

False mastic seeds (*Sideroxylon foetidissimum*) are red-brown to dull gray with a light, indented, oval hilum just shy of its least rounded end. Elongate seeds larger than 5/8 in (1.6 cm) may be from the more tropical tempisque tree *(S. capiri).*

ORIGIN: Crabwood is a mangrove tree living coastally from Central America through the Caribbean to Brazil. West Indian mahogany grows in the West Indies and southernmost Florida. Nutmeg is native to the Spice Islands of Indonesia and is commercially grown in Grenada and elsewhere in the Tropics. False mastic has a native range from Mexico and the West Indies to southern Florida.

DID YOU KNOW? West Indian mahogany has been harvested for wood since the 17th century. It was the principal lumber for ship building in England and Spain during colonization of the New World.

340

Eggfruits, Sweetgum, Oak, and **Lotus**

RELATIVES: Eggfruits share the order Ericales and family Sapotaceae with false mastic. Sweetgum (order Saxifragales, family Altingiaceae), oaks (Fagales, Fagaceae), and lotus (Proteales, Nelumbonaceaeae) are only distantly related.

IDENTIFYING FEATURES:

Eggfruit (sapote) *(Pouteria sapota)* seeds are shaped like small, compressed footballs. They are brownish and smooth except for a wide, roughened hilum over their entire length. **Rounder seeds** smaller than 2 in (5 cm) are likely to be other eggfruit species sharing the genus *Pouteria*. All sizes of eggfruit seeds tend to rattle.

Sapote eggfruit seeds, max 3.5 in (9 cm)

Sweetgum *(Liquidambar styraciflua)* fruits are an aggregate of small seed capsules that form a spiked ball. Worn beach specimens have lost their spikes, leaving only the open capsules.

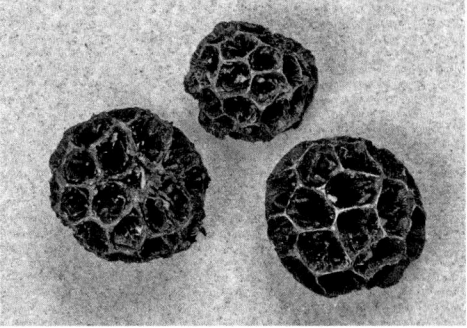

Sweetgum fruit capsules, max 1.2 in (3 cm)

Acorns are the fruits of oak trees *(Quercus* spp.). They are glossy nuts that may or may not have the familiar scaly cap (cupule). Some may have a cap completely enclosing the nut (its pericarp). Shapes vary between species, from spherical to elongate. Naked nuts show the scar where the cap was attached.

American (yellow) lotus *(Nelumbo lutea)* bears fist-size seed pods with open capsules that may have remaining hard, brown seeds.

Acorns from varied oak species, max 1 in (2.5 cm)

ORIGIN: Eggfruits come from sapote trees, which are native to Central America and the West Indies. Sweetgum trees and oaks are common across river drainages of eastern North America. American lotus grows in lakes and sluggish rivers that empty into the Gulf.

DID YOU KNOW? Trees in the eggfruit family make up about a quarter of the Amazon Basin's tall trees.

Yellow lotus pods and seeds, 0.5 in (1.3 cm)

341

Bamboo cane revealing hollow compartments

Bamboo rhizome (A), and culm (B)

Common reed rhizome bundle

Fresh (L) and old (R) smooth cordgrass wrack

Bamboo, Reed, and Cordgrass

RELATIVES: Bamboos are really big grasses in the family Poaceae, shared with their smaller cousins, reeds and cordgrass.

IDENTIFYING FEATURES:

Bamboo (tribe Bambuseae, many woody species) strand as sections of hollow, compartmentalized stem (culm), often with its associated rhizomes, and commonly with cut marks or other signs that it has been used by people. Bamboo canes put to structural use are the commonly introduced Asiatic species of *Bambusa*, although many of species have been transplanted to tropical America. Canes can be up to 6 in (15 cm) in diameter and more than 30 ft (9 m) long.

Common reeds (*Phragmites australis*) beach as large bundles of finger-thick rhizomes bound by dense roots. The bundles stay together after lengthy sea travel even when the mass has sun-bleached to a light gray. Fresher bundles are brownish and may have remnants of the thinner reeds bending at right angles to the rhizomes.

Smooth cordgrass (*Spartina alterniflora*, p. 274) has 3.5-ft (1.1-m) strawlike stems that arrive on the beach like piles of broken hay bales.

ORIGIN: Bamboo has a wide range between North and South America, and introduced species are ubiquitous. Common reeds occasionally grow near beaches (p. 275), but are most common in marshes of the Mississippi Delta, which is eroding away as it sinks into the Gulf of Mexico. Smooth cordgrass dominates saltmarsh behind many barrier islands rimming the Gulf. Cordgrass comes from eroding tidal marshes behind barrier islands.

DID YOU KNOW? Bamboo is used to make houses, furniture, boats, bicycles, bridges, music, and dinner, among thousands of other useful and artistic things.

Tangleballs and Roots

RELATIVES: Tangleballs involve many parts of many kinds of plants. Saw palmetto (Arecaceae), greenbriar (Smilacaceae), and bulrush (Cyperaceae) are distantly related monocots.

IDENTIFYING FEATURES:

Tangleballs are spherical clumps of intertwined plant parts. These parts vary, but commonly include stiff roots, rhizomes, pine needles, and thin twigs. The seawoven balls keep their shape long after being beached.

Beach tangleballs, max 10 in (25 cm)

Saw palmetto (*Serenoa repens*, p. 280) trunks are fuzzy, curved, and composed of densely packed fibers. Some may retain roots and frond boots that give them the appearance of a giant shrimp.

Greenbriar (*Smilax* spp., p. 282) rhizomes are dark-red or black, multi-lobed masses with occasional remaining roots.

Saltmarsh bulrush (*Bolboschoenus robustus*) root nodes are dark, connected chains of three-sided lumps.

Saw palmetto stem, 1–7 ft (0.3–2 m)

ORIGIN: Beach tangleballs are woven from coarse fibers dense enough to collect on the seabottom, which is where these balls are formed. Storms erode plant roots from the dune and wash other plant parts down rivers to the sea. Material that reaches the surf is snowballed together by the rolling action of waves. Saw palmetto and greenbriar wash into the sea when the dune erodes. Bulrush comes from eroding saltmarsh.

DID YOU KNOW? Tangleballs form wherever erosion and waves occur. They are unique conversation pieces and are occasionally sold to tourists as "whale burps." This term also describes the events that produce tangled masses of beached sea stuff. The trunks and roots here are tough, and may have a lengthy drift before returning to a beach. But greenbriar rhizomes don't get far because they tend to sink.

Greenbriar rhizomes, max 24 in (61 cm)

Saltmarsh bulrush root nodes, max 12 in (30 cm)

Turtleweed fruits, max 0.4 in (1.0 cm)

Seeds of the southern swamp lily, max 2 in (5 cm)

River-carried surfwood pile in wrack

Typical surfwood wear patterns

Turtleweed fruits, Swamp Lily Seeds, and Surfwood

RELATIVES: Turtleweed is a dicot in the saltwort family, Bataceae. Swamp lily is a monocot in the family Amaryllidaceae with onions and chives. Surfwood may be from one of hundreds of tree species, including palms, conifers, and hardwoods.

IDENTIFYING FEATURES:

Turtleweed (seaside saltwort) *(Batis maritima)* fruits are greenish-yellow, angular, and the size of garden peas.

Southern swamp (string-) lily *(Crinum americanum)* produces large, fleshy, greenish or gray-brown seeds that often have a sprout.

Surfwood is found as smooth, rounded wood pieces, generally hand-size or smaller. Like driftwood, surfwood is difficult to identify to its tree species. Many pieces are too soft to reveal their grain.

ORIGIN: Turtleweed (p. 289) is common in saltmarsh behind Gulf barrier islands. Swamp lily is a native *Crinum* of southern US, freshwater wetlands. Plants growing along rivers often drop seeds that are carried to the Gulf. Surfwood may be the last remnants of river- and surf-sanded driftwood, or the smoothed pieces of logs pounded into chunks by the sea.

DID YOU KNOW? Both the turtleweed fruits and their small (1 mm) dark seeds float to disperse. This allows the plant to colonize wet, salty areas laid bare by hurricanes. Its fruiting coincides with that season. Fresh, swamp lily seeds planted in wet, slightly acidic soil will develop into a perennial with large, white, frilly flowers; but it will take a few years. Occasionally, plant parts with bulbs are found, and these may flower the following summer. Elongate surfwood likely came from branches that were surf-sanded. The largest piles of surfwood can be found near major river inlets.

Driftwood and Peat

RELATIVES: The Gulf receives driftwood from hundreds, perhaps thousands, of tree species, from both conifers (gymnosperms) like pines and cedars, and from broad-leaved flowering trees (angiosperms) including oaks, hickories, and tropical hardwoods.

IDENTIFYING FEATURES:

Identifying tree species from wood is tricky, even when its origin is known. Determining wood species drifting from unknown sources requires laboratory detective work. On the beach, moderately educated guesses can be made after looking at the wood in a smoothly cut cross-section (end-on slice). Pines often have soft, wide rings with conspicuous resin canals and no vessels. Oaks have numerous vessels in their rings and have conspicuous rays. Most tropical hardwoods have a tight grain, and many are dark or reddish under their sun-bleached exterior. Folks who are haunted by persistent driftwood mysteries can send a 1x3x6-inch sample to the USDA Forest Service, Center for Wood Anatomy Research, 1 Gifford Pinchot Drive, Madison, WI 53726.

Ceder driftwood can be identified by its cedar scent

Peat lumps are hand- to boulder-size masses of dark, compressed vegetable matter. Most lumps that have eroded from the dune are smooth and rounded due to surf-sanding. Occasionally, flat, tablelike formations, many yards across, become exposed (p. 353).

This trunk shows an oak's undulating grain

ORIGIN: Most driftwood comes from rivers, either by bank erosion or from loss of felled timber. Many tropical species are resistant to rot and may drift at sea for years. Peat is a composite of very old, partially decayed vegetation, and forms in swamps and dense marsh. Exposed peat is a clear indication that a beach is advancing on the location where wetlands once backed the former dunes.

Peat lumps from an eroded dune after a storm

345

Kapok bark bits, max 3 in (8 cm) diameter

Bullhorn acacia thorns with ant entrance hole (arrow)

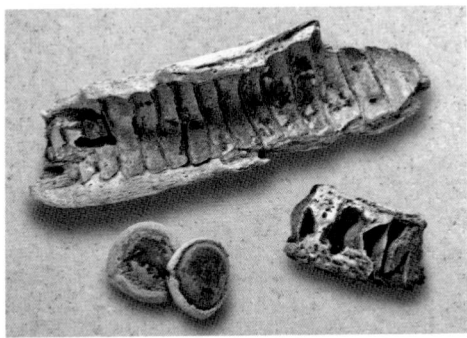

Trumpet-tree trunks, 2 in (5 cm) diameter

Beach brittlestem, 2 in (5 cm) tall

Kapok Bark, Bullhorn Acacia, Trumpet-tree, and Mushroom

RELATIVES: Kapok trees (Malvaceae, the mallow family), bullhorn acacia trees (Fabaceae, legumes), and trumpet-trees (Urticaceae, nettles) are distantly related. Brittlestems are fungi in the order Agaricales, containing gilled mushrooms. Fungi are closer to animals than to plants, sharing a common ancestor with what likely was a single-celled protist living about a billion years ago.

IDENTIFYING FEATURES:

Kapok bark from the silk-cotton (kapok) tree *(Ceiba pentandra)* is studded with cone-shape, corky, bark prickles and knobby or smooth, layered bark chunks. Some prickles may have a spiny tip.

Bullhorn acacia *(Vachellia cornigera)*, 3 in (8 cm), grows dual thorns that resemble the tree's namesake. The hollow thorns begin growing at the base of compound leaves. Older thorns occupy woody branches. **Holes** accommodate ants.

Trumpet-trees *(Cecropia* spp.) have tall, thin trunks with centers formed of hollow compartments. Corky beached portions include entire trunk sections or just the papery compartments.

Beach brittlestem *(Psathyrella ammophila)* mushrooms have pale, domelike caps that gradually turn brown and open upward. The mushroom's crowded, brown gills attach to the stem.

ORIGIN: Kapok trees and trumpet-trees are native to Mexico, Central America, and the Greater Antilles. Bullhorn acacia trees are native to Central and South America. Beach brittlestem grows on beaches all over the Gulf and Atlantic.

DID YOU KNOW? Bullhorn acacia and trumpet trees are each protected by ants that live in the hollow thorns or trunk compartments. The ants receive "bribes" of food from special leaf structures.

BEACH GEOLOGY

What is Beach Geology?

A beach's geology describes its nonliving solid parts and how they came to be. These beach components include sands, other sediments, and rocks of all sizes. Essentially, pebbles, gravel, sand, silt, and clay are all rocks of increasingly smaller size, down to microscopic particles (clay). Rocks and sediments contain minerals, which are naturally occurring, inorganic solids with a uniform chemical composition and an ordered (crystalline) structure. Quartz is the most common mineral on the beach, followed by feldspar and calcite. Calcite makes up most of the shell bits in beach gravel beds (shell hash, p. 26), and the tiniest shell bits become carbonate sand grains. Additional sand minerals include metal oxides like magnetite, semiprecious gemstones like garnet, and others. These pulverized minerals were once within mountains, and their path to the beach may have involved thousands of dramatic events. Some were buried as sediments and turned to stone, only to be eroded back into sediment. All were brought to the beach by water moving in rivers, sea-currents, and waves. If each grain could tell the story of its origin, it would recount volcanoes, landslides, storms, torrents, and countless other fits of Earth's fury over eons.

A beach's geological record also includes tales of former lives told by fossils. Fossils are ancient, buried animal parts that have become mineralized. Most are parts from animals that did not live near a beach, but did live near the place the beach now occupies. Their stories illustrate striking change. Sands, stones, and fossils describe what a beach and its components were like hundreds, thousands, even millions of years ago. The descriptions tell of how beaches were assembled, what forces keep them in a constant state of change, and how their future is likely to unfold.

Sand from Padre Island, Texas, showing grains of quartz, feldspar, shell bits, and gems. Scene is 1/8 in (3 mm)

Gulf Beach Sands

WHAT ARE THEY? Sand is pulverized rock, and includes hard bits of animal shells. The **beach-sand sampler** to the right shows some of the variety in color, grain size, and composition of Gulf-coast sands collected at the high-tide line.

SIZE: Technically speaking, sand is finer than gravel and coarser than silt, with grains between 0.06 and 2.0 mm. Sand grains on most Gulf beaches average 0.2 to 0.5 mm, but there is much variation among and within beaches. Finer sands are found on beaches just west of the Mississippi, during summer, in the dune, and in the offshore bars. Coarser sands tend to occur on Mississippi barrier islands through the Florida Panhandle, during winter, and on the lower beach.

ORIGIN: Most Gulf beach sands came from eroding mountains upstream from coastal rivers. There was a dramatic pulse of erosion as mountains were being formed during the tectonic upheaval of the Late Miocene, about 12 to 5 million years ago. Rivers carried quartz, feldspar, and heavier mineral sands to coastal deltas, which were reworked by Gulf waves and currents during the rise and fall of hundreds of sea level changes. Some sands came from the sea. But these carbonate grains from mollusks, forams, and other skeletal bits make up a small fraction of most Gulf beaches. Although carbonates are constantly being deposited, they are softer than terrestrial sediments and erode into silt and clay more quickly. Shell sediments range from thousands of years old, to freshly ground, whereas quartz sands from inland mountains may have occupied their present granular form for tens of millions of years.

DID YOU KNOW? The whitest sands are nearly all quartz. Light brown and grayish sands have darker quartz, feldspar, and iron-stained shell bits. Dark peppering is from grains of iron or titanium oxides.

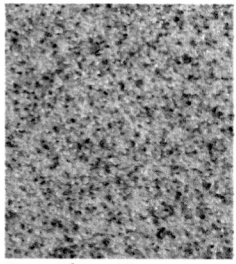

Boca Chica, Texas

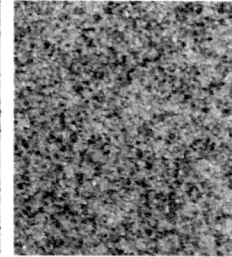

South Padre Island, Texas

North Padre Island, Texas

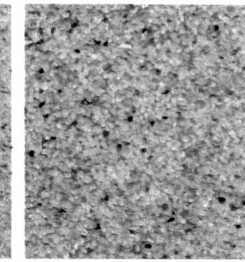

Matagorda Island, Texas

Galveston Island, Texas

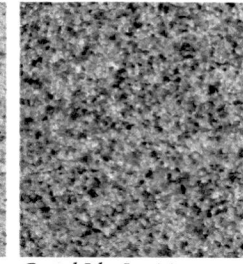

Grand Isle, Louisiana

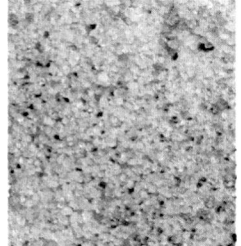

Ship Island, Mississippi

Dauphin Island, Alabama

St. Joe Peninsula, Florida

Alligator point, Florida

Pure quartz sand from Topsail Beach, Florida

Quartz and feldspar grains, South Padre Island, Texas

Beach Sands *(Quartz, Feldspar)*

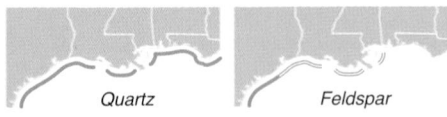

Quartz   Feldspar

WHAT ARE THEY? Quartz and **feldspar** compose 95 percent of Gulf beach sand. Quartz is a mineral of silicon dioxide. Its hardness is 7 on the Mohs scale, between titanium (6) and topaz (8), and its density is light (2.7 g/cm³). Although quartz is the second most common mineral on Earth's surface, it's the most common in beach sand because of its hardness and resistance to wear. The most abundant surface mineral is feldspar. The feldspar of Gulf beach sand is a potassium aluminum silicate, which has a lower hardness than quartz (Mohs 6), but is about the same density (2.6 g/cm³). Feldspar is almost absent from beaches east of the Mississippi, but can make up about 10 percent of sands on some Texas beaches. Under magnification, quartz grains look like rounded glassy crystals—clear, but sometimes milky, gray, or pinkish. Translucent quartz grains make for a sugary white beach. Feldspar grains are squarish-to-rounded crystals with a dull luster and colors that range between gray, pinkish, and yellowish.

SIZE: Each of these minerals in beach sand is variable in size between about 0.1 and 1.0 mm.

ORIGIN: Quartz and feldspar sands originated from silicate igneous (volcanic) rocks, such as granite. Some quartz comes from eroded sandstone (a sedimentary rock), but the original lithified sand came from an igneous source.

DID YOU KNOW? The hills and mountains holding source rocks for Gulf beach sands spanned half of North America.

Beach Sands
(Heavy Minerals, Tiny Gemstones)

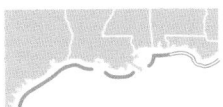

WHAT ARE THEY? "Heavy" sands comprise the densest grains on the beach and are most visible in thin, dark layers (p. 25). Often, these heavy mineral grains include abundant semi-precious **gemstones**. The top image magnifies a small spoonful of olive gray sand that was separated into a layer by rough surf on a south Texas beach. Among the occasional quartz grains (preceding page) and blackish **magnetite** (following page) are **olivine** (glassy olive green), **garnet** (glassy, watermelon pink), **staurolite** (lustrous, partly opaque, dark orange), **zircon** (elongate, rounded, distinctly outlined, and colorless translucent), and **kyanite** (elongate, rectangular, and translucent pale green). These gem grains have densities between 3.5 (kyanite) and 4.8 (zircon) grams per square cm, which is 30 percent greater to twice the weight of an equal volume of quartz and feldspar sand.

SIZE: Diameter averages less than 0.2 mm. A jeweler's loupe is key to exploring this world, but a microscope is needed to truly appreciate these treasures.

ORIGIN: Most of these minerals were components of high-temperature and high-pressure metamorphic and igneous rocks eroded long ago, and were carried to the coast by rivers. These gems are scattered throughout the beach but become concentrated with equally dense grains by heavy wave action.

DID YOU KNOW? Successful micro-gem hunters focus on "placers," which are layers of dense, dark sand (p. 25) that are revealed following rough surf.

Dark olive layer from Boca Chica Beach, Texas

Magnetite   *Olivine*   *Garnet*

Staurolite   *Zircon*   *Kyanite*

Dark gray layer from Ship Island, Mississipppi

Magnetite sand from Boca Chica Beach, Texas

Magnetite grains cling to a bar magnet

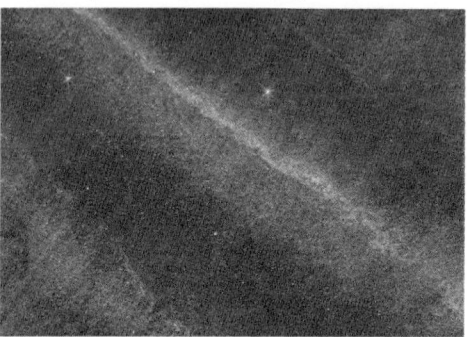
Mica glitters in the trough of an antidune

A closeup reveals tiny mirrorlike mica flakes

Beach Sands
(*Magnetite, Ilmenite, Mica*)

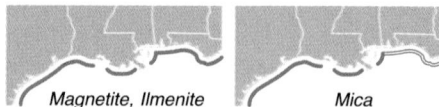

Magnetite, Ilmenite   Mica

WHAT ARE THEY? Beach layers of "heavy" dark sand generally contain a high proportion of metal oxide grains. Chief among these are oxides of iron (**magnetite**) and iron and titanium (**ilmenite**). Both are very dark, opaque, and rounded, with flattened faces and a metallic sheen. Grains of ilmenite weakly move with the swipe of a magnet, but magnetite sand will leap up to meet the bar. These sand grains are some of the heaviest on the beach—between 4.7 and 5.2 g/cm^3. **Mica** is a potassium aluminum silicate mineral similar to feldspar, but mica's crystal arrangement gives it a two-dimensional sheet structure. On the beach, these broken sheets are seen as tiny, glittery, clear or shiny-black flakes. Although at 2.9 g/cm^3, mica is slightly more dense than quartz and feldspar sands, mica's fluttery shape means that it can only collect within wave-protected areas like the troughs of lower-beach anti-dunes (p. 24). Wind-blown dune sands contain little mica.

SIZE: These metal oxide grains are mostly less than 0.1 mm. Mica flakes are often greater than 1 mm.

ORIGIN: Magnetite and ilmenite occur together within igneous and metamorphic rocks. Mica occurs in similar rocks, but is also in sedimentary rock.

DID YOU KNOW? Lightly worn magnetite grains often show their crystalline, octahedral shape. This crystal structure does not allow magnetite grains to be elongate. Because of its reflective properties, mica is one of the most conspicuous minerals in granite.

Firm Sediments *(Clay, Peat)*

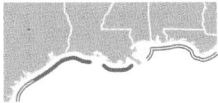

WHAT ARE THEY? Clay is composed of microscopically fine rock particles, each less than 0.05 mm. Compressed particles combine with water to form a soft, sticky, grayish type of soil. Exposed clay formations on the beach break into **clay balls**, and are typically covered by beds of **peat**. Peat is a brownish, solidified accumulation of partially decomposed plant material. It often contains burrows, root casts, and shells of animals from a former time. Clay and peat exposed on retreating beaches show that these formations are slightly more resistant to erosion than the sand that covered them.

SIZE: Clay balls from exposed formations are fist size, whereas clay bluffs stretch for hundreds of yards (meters). Exposed peat formations can be more extensive, and broken pieces are as large as table tops.

ORIGIN: Formations exposed on Gulf beaches include the Beaumont Clay (northeastern Texas) and the Prairie Formation (Louisiana), which slowly accumulated within old river valleys during the Late Pleistocene (about 35,000 years ago) when sea level was hundreds of feet lower. Gulf-beach peat traces accumulated masses of marsh plants that were overtaken by beach sands about 8,000 years ago.

DID YOU KNOW? The Matagorda Peninsula of Texas sits on an old, clay-rich river delta, and is not a barrier island. Most clays comprise feldspar rock particles that weathered into minuscule bits over millions of years. Peat is younger, but burial, compaction, and time can transform it into the flammable rock we call coal (p. 382).

Clay bluffs, Matagorda Peninsula, Texas

Clay balls eroded from an exposed beach formation

An exposed peat formation at Sargent Beach, Texas

Peat with estuarine shells

353

Silty (A) and sandy (B) limestone tumbled by rivers

Chert from central Texas beaches

Clay stones (one broken), Dauphin Island, Alabama

Beach Stones *(Silty and Sandy Limestone, Chert, Clay Stone)*

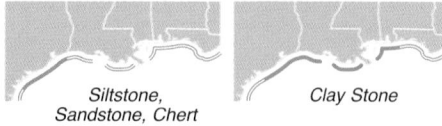

Siltstone, Sandstone, Chert

Clay Stone

WHAT ARE THEY? These sedimentary rocks formed from marine sediments, and were cemented with dissolved-mineral precipitates. Upon striking them with a hammer (use eye protection), **silty/sandy limestone** will cleave through the carbonate cement between grains, leaving a dull, fracture face. Gritty "crumbs" from the break fizz in muriatic acid (aqueous HCL). Silty limestone is darker and has finer grit than sandy limestone. Shapes are disclike or elongate, and colors range light brown to dark gray. **Chert** is a micro-crystalline silica (SiO_2) that is extremely hard (7 on the Mohs scale). These stones are opaque, have a waxy luster, fracture in smooth, curved chunks, and come in grays (flints), pale greens, and sanguine shades (jasper). **Clay stones** are sun-hardened and surf-rounded clay balls (previous page) that can be broken by hand.

SIZE: Most are technically pebbles, which are 0.16–2.52 in (4–64 mm) in diameter.

ORIGIN: Silty and sandy limestone came from eroded formations hundreds of miles up coastal rivers. Chert formed from ooze deep under ancient seas. The hardened ooze later rose into mountains, but chert is essentially the fossilized, silicate remains of diatoms and other tiny sea organisms. Chert was brought to the coast by rivers or by pre-Columbian native people, who worked the stones into spear points. Clay stones likely came from offshore formations, or were pumped onto the beach by dredge-and-fill (p. 393).

DID YOU KNOW? River-carried limestone pebbles tend to be flat like the layers of rock from which they eroded.

Beach Stones
(Lithified Sediments, Coquina Limestone, Ferrous Coquina, Calcite)

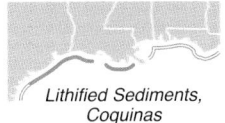

Lithified Sediments, Coquinas

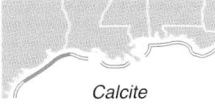

Calcite

Lithified sediment beach stones from western Louisiana

WHAT ARE THEY? These are sedimentary rocks. **Lithified sediment stones** are gritty and grayish with flattened, irregular shapes. **Coquina limestone** contains numerous seashell bits. **Ferrous coquina** occurs in rusty, shelly lumps and attracts a magnet. **Calcite** is calcium carbonate. It forms whitish stones with a waxy feel that break with flat, 60 and 120 degree sides. This mineral is only as hard as a copper penny (Mohs scale 3). All these beach stones have carbonates that will fizz vigorously under a drop of muriatic acid.

SIZE: Mostly 0.3 in to 4 in (0.8–10 cm).

Coquina limestone with oyster shell bits

ORIGIN: Most lithified sediment stones formed where cementing, sandy, carbonate mud interacted with methane under water during the last glacial retreat between 19,000 and 15,000 years ago. Near Louisiana beaches, these stones are still forming at methane seeps in sandy marsh, where the stones envelop modern debris like glass bottles. Coquina can form like this, but also forms where shell beds above sea level have carbonates dissolved by rainwater and re-precipitated as cement. Ferrous coquina forms around iron-rich (e.g., manmade steel) objects surrounded by dissolved calcium carbonate, as from sea shells, making an iron oxide-calcium carbonate concretion. Calcite formed in shallow seas that flooded North America more than 200 million years ago. The calcite formed when calcium carbonate precipitated out of shallow marine waters. Calcite stones reached the beach with other river stones thousands of years ago.

Ferrous coquina from a Texas beach

Calcite pebbles from a central Texas beach

355

Quartzite pebbles (fractured, R) from central Texas

Milky quartz pebbles from Dauphin Island, Alabama

Pink granite fragments from a Texas jetty boulder

Beach Stones
(Milky Quartz, Quartzite, Granite)

Quartzite, Granite

Milky Quartz

WHAT ARE THEY? Quartzite is metamorphic rock, meaning it was once another rock type (sandstone, in this case) but was changed by heat and pressure. The change fused quartzite's original quartz sand grains and made the rock harder. Pure-quartz quartzite is opaque white, and yellowish or reddish versions contain iron hematite. The rock fractures through grains, not around them, leaving lighter colored flakes on a rough, broken surface. **Milky quartz** and **granite** are igneous rocks, which means they crystallized from molten magma. Milky quartz stones are composed of translucent, crystalline quartz, but with numerous imperfections that break up light. Unlike quartzite, quartz (the mineral stone) will shatter like glass. Granite is a mix of many minerals, including feldspar (pink or whitish), quartz (glassy and gray), mica (shiny flecks), and dark minerals like hornblende. All these hard rocks will scratch glass and do not fizz under acids.

SIZE: Most are pebbles, 0.16–2.52 in (4–64 mm). Granite cobbles and boulders occur at jetties.

ORIGIN: These rocks were carried from inland mountains that are as old as hundreds of millions of years. But their journey to the beach was more recent. Rounded pebbles were carried by Pleistocene rivers, less than a half-million years ago. Granite was carried by railcar to build jetties and revetments (p. 392).

DID YOU KNOW? These beach stones were rounded by tumbling down ancient rivers, not by waves and beach wear.

Salts, Pumice, and Fulgurite

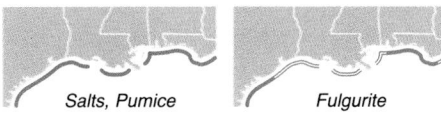

Salts, Pumice   Fulgurite

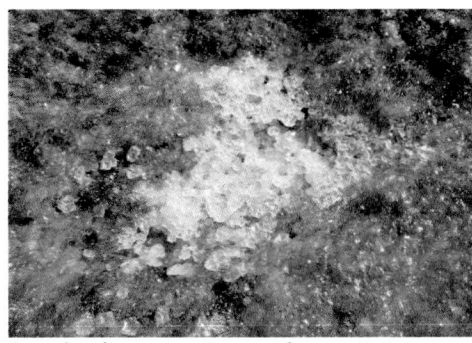

Sea salts after evaporation on a hot granite jetty

WHAT ARE THEY? About 3.5 percent of Gulf water is dissolved salts. Sodium and chloride ions are 85 percent of these, and the remainder is largely magnesium, calcium, sulfate, and bicarbonate ions. **Sea salts** remain dissolved unless they precipitate as minerals where water has evaporated (top image). **Pumice** is an igneous rock that forms when gas- and silicate-rich, frothy lava erupts from a volcano. This bubble-filled volcanic glass is light enough to float at sea. Colors include whitish, gray, brownish, and black. Surf-worn pumice is rounded. **Fulgurite** is also glass, but the silicate-melting energy producing this rock came from lightning striking dune sand. The tubular shape of fulgurite represents a vaporized center surrounded by fused sand (mostly quartz). Beach pieces represent brittle, treelike formations under the sand.

SIZE: Crystallized sea salts are a little coarser than table salt. Pumice stones range from pea-size to over 4 in (10 cm). Fulgurite is mostly finger size, but as large as chunks four inches across.

ORIGIN: Salts in the sea came from minerals dissolved in land runoff. Pumice comes from volcanoes, erosion of their ejected material, and sea currents that carry the floating stones. Possible volcanic sources of Gulf beach pumice over the last 250 years are eruptions in Iceland, Martinique, Guatemala, and southern Mexico. Large pieces may drift for many decades. Fulgurite from dune-crest lightning strikes accumulated over centuries and is revealed by shifting sands.

DID YOU KNOW? Dozens of marine species travel on drifting pumice.

Pumice stones found on Texas beaches

Finger-size fulgurite pieces from a larger formation

Large fulgurite chunk discovered on Padre Island, TX

357

Fossils

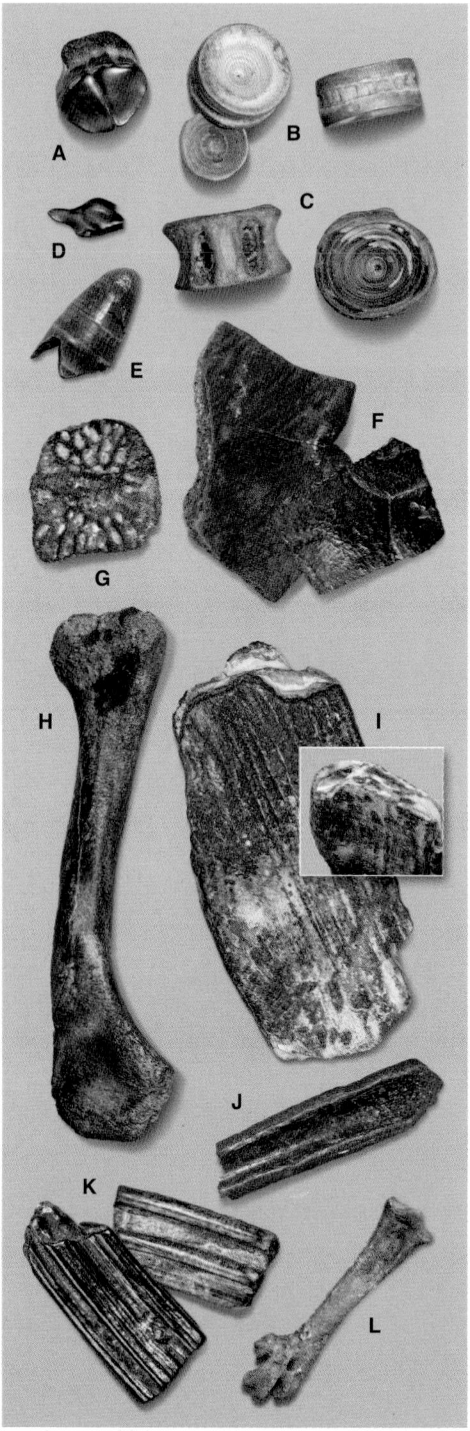

WHAT ARE THEY? Fossils are petrified parts and other traces of living things, changed in composition by permineralization. This process occurs when a bone, tooth, or shell is impregnated by dissolved minerals, hardening the original structure with calcite, iron, and silica. Permineralization makes fossils stonelike, with colors that range from gray to brown. Fossils found on beaches have been uncovered by erosion, and indicate that the site was once a marsh, lake, or estuary where animal parts could be preserved in sediment. Some dark-stained bones may look like fossils, but have not been permineralized. Mollusk shells start out as mostly calcite, and may not change much over time except to change color and become chalky.

Fossils from Gulf beaches include:

A. Pufferfish tooth plate (*Diodon* sp.)

B, C. Shark vertebral centra (spine segments)

D. Garfish scale (*Lepisosteus* sp.)

E. Alligator tooth (*Alligator mississippiensis*)

F. Pond turtle carapace bones (Emydidae)

G. Alligator osteoderm (skin-bone)

H. Alligator humerus (upper foreleg bone)

I. Mammoth molar plate (inset shows crown) (*Mammamuthus* sp.)

J. Deer metatarsal fragment (leg bone) (*Odocoileus* sp.)

K. Horse molars (*Equus* sp.)

L. Bird tarsometatarsus (ankle bone)

Fossil parts of fishes, reptiles, mammals, and birds

Fossils

Shark teeth fossils:

A. Great white shark teeth
   (*Carcharodon carcharias*)

B. Mako shark tooth (*Isurus* sp.)

C. Lemon shark tooth
   (*Negaprion brevirostris*)

D. Sand shark tooth (*Odontaspis* sp.)

E. Tiger shark tooth
   (*Galeocerdo cuvier*)

F. Requiem shark teeth
   (*Carcharhinus* sp.)

G. Shark teeth fragments

SIZE: Whole mammoth teeth, with multiple plates (previous page) are the size of a bread loaf. Broken shark teeth may be only slightly larger than sand grains.

ORIGIN: Because beach dynamics mix many ages of material together, fossils from Gulf beaches may span millions of years. But most beach fossils are from the Pleistocene Epoch (between 1.8 million and 10,000 years ago) when our coastline advanced and retreated with four major changes in sea level, each corresponding to an ice age.

DID YOU KNOW? Because the **tympanic bullae** (ear bones) of whales are so dense, they are one of the more persistent fossils left by whales. **Quahog clams** as large as 6 in (15 cm) lived in estuaries where the beach is now. These may not be fossils, but they are pretty old (hundreds of years, maybe?). Studies show they lived longer lives than modern clams, commonly to 40 years and older. In the Gulf, only Alabama and Mississippi have designated state fossils. Both states chose *Basilosaurus cetoides,* an archaeocete whale from the late Eocene. Their skeletons are approximately 15 meters (50 feet) long.

Fossil shark teeth. Largest here is 2 in (5 cm)

Fossil baleen whale ear bones, 5 in (13 cm)

Large quahog shells. Fossils, or just old?

What's Under a Gulf Beach?

Layers upon layers … Geologists call these layers, facies (FAY-sheez), which are bodies of rock or sediment with unique characteristics. Each facies defines a time in the geologic past. If you dig a hole deep enough, you will see many facies staring back at you—sand, mud, peat, and clay until you get to a depth of hundreds of feet and a couple million years; limestone at thousands of feet representing tens of millions of years past; and 250-million-year-old basement rock only after boring several miles. Can you dig to China from a Gulf beach? Well, no. After penetrating the crust and boring through the searing hot mantle and core, you'll pop up in the southern Indian Ocean off Western Australia.

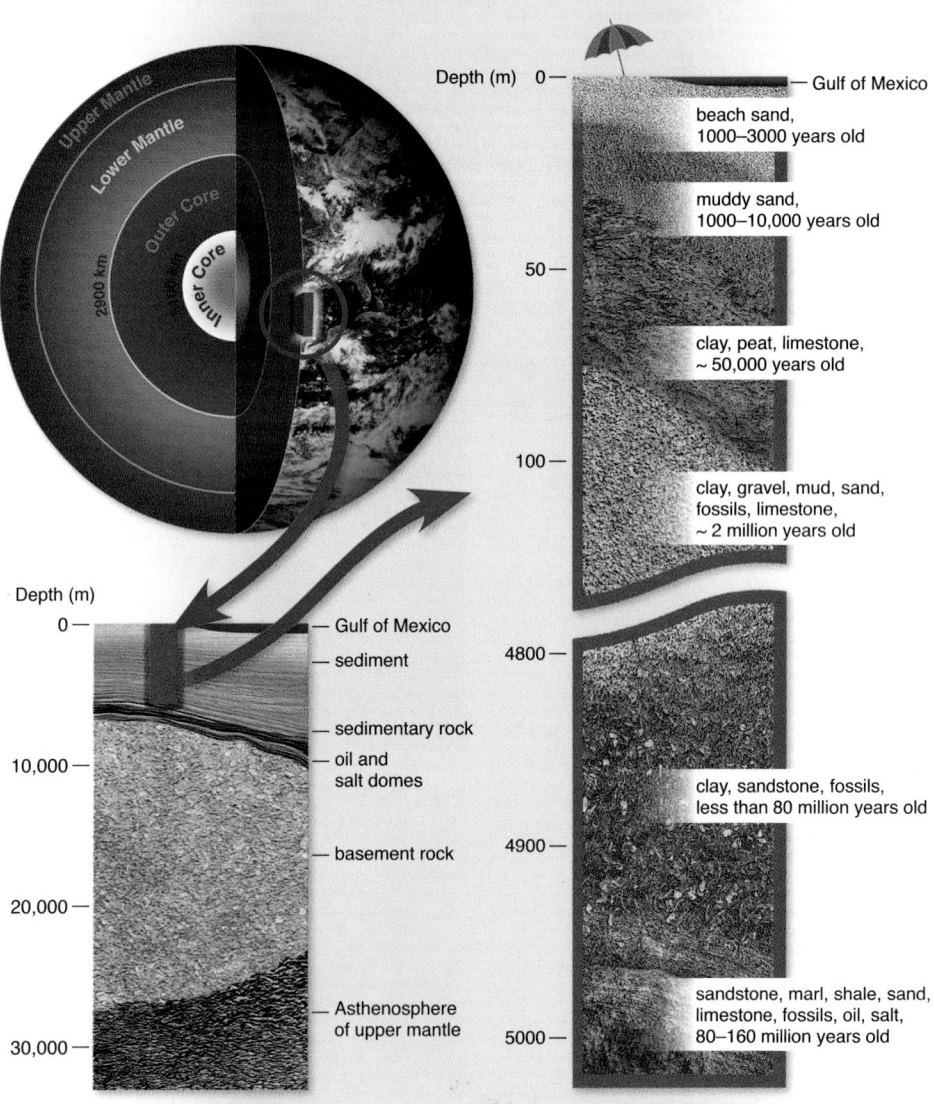

A magnification of the many layers in Earth's crust under a Gulf beach

Gulf Beaches Over Time

Much has happened. A timeline for Gulf beaches shows separating continents, cycles of Gulf filling and drying, species evolving and disappearing, and dramatic events like the Earth-altering Chicxulub (CHIK-su-lube) asteroid. Humans are recent arrivals. But even in our short, 30,000-year history with the Gulf, we've experienced vast changes in climate and sea level. Our current coastline has been moderately stable for about 8,000 years.

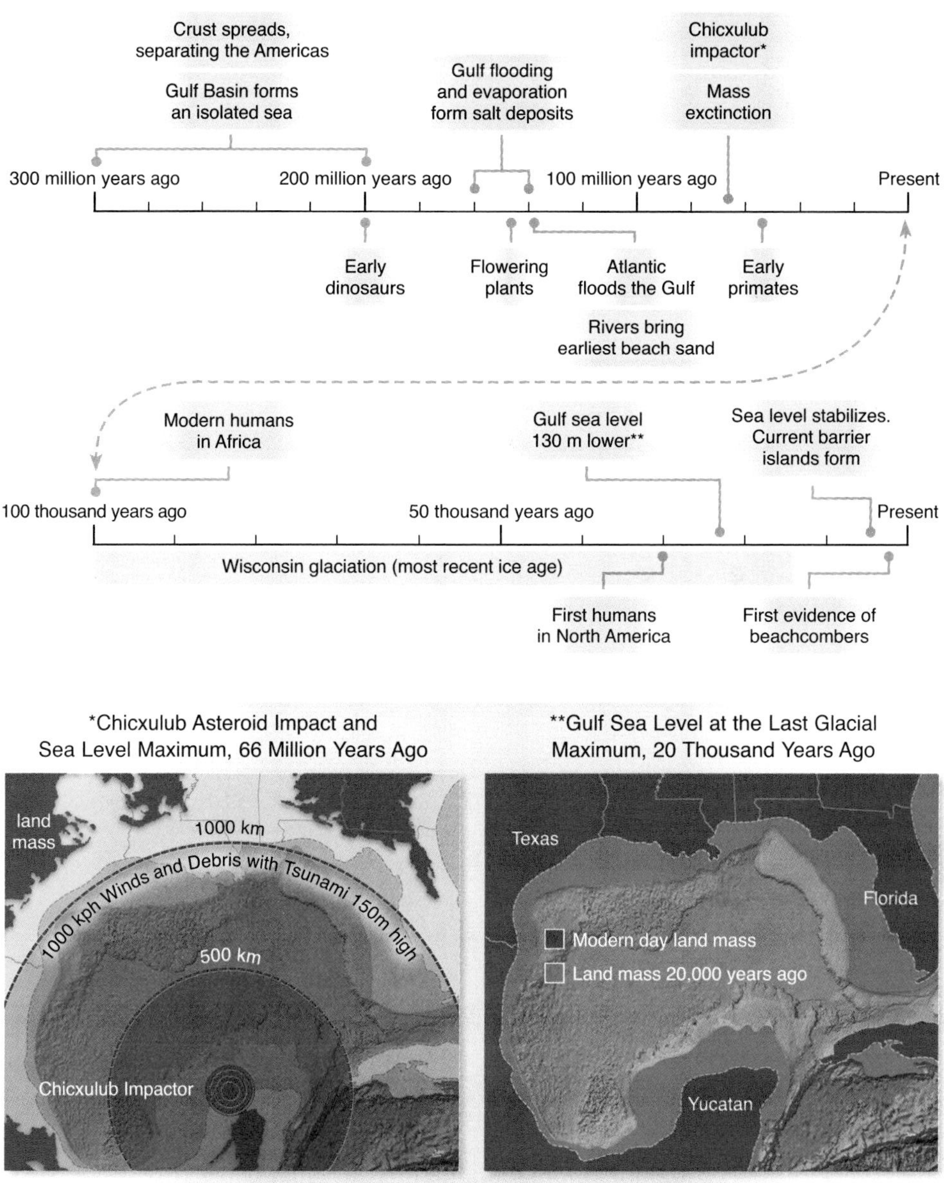

Crust spreads,
separating the Americas

Gulf Basin forms
an isolated sea

Gulf flooding
and evaporation
form salt deposits

Chicxulub
impactor*

Mass
exctinction

| 300 million years ago | 200 million years ago | 100 million years ago | Present |

Early
dinosaurs

Flowering
plants

Atlantic
floods the Gulf

Early
primates

Rivers bring
earliest beach sand

Modern humans
in Africa

Gulf sea level
130 m lower**

Sea level stabilizes.
Current barrier
islands form

| 100 thousand years ago | 50 thousand years ago | Present |

Wisconsin glaciation (most recent ice age)

First humans
in North America

First evidence of
beachcombers

*Chicxulub Asteroid Impact and
Sea Level Maximum, 66 Million Years Ago

**Gulf Sea Level at the Last Glacial
Maximum, 20 Thousand Years Ago

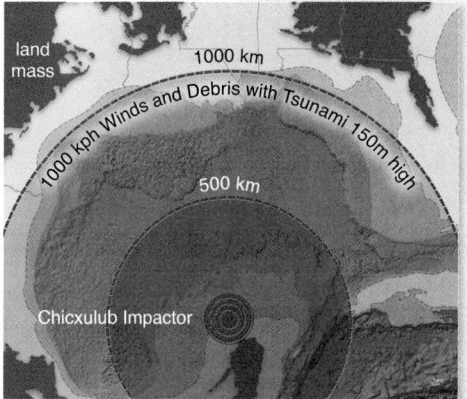

land
mass

1000 km

1000 kph Winds and Debris with Tsunami 150m high

500 km

Chicxulub Impactor

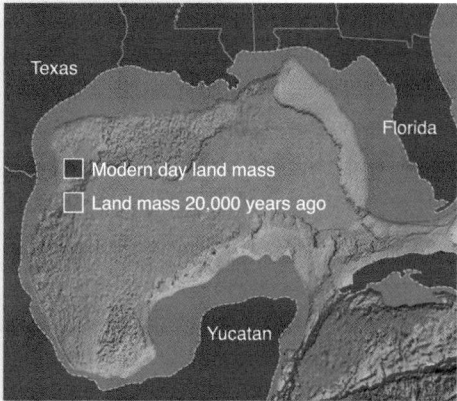

Texas

Florida

Modern day land mass

Land mass 20,000 years ago

Yucatan

A timeline of geological events shaping Gulf beaches

361

School of Rock

ROCK, OR NOT A ROCK? Many rock-like objects don't have a strictly geological story. Here are some common "rocks" one might encounter on Gulf beaches.

A. Quahog fragment. Surfworn, thick pieces may look like rocks.

B. Oyster fragment. Old oysters have thick, rocklike shells.

C. Shell aggregate asphalt. Old, eroding, coastal roads were paved with this rock-like material.

D. Sponge-bored quahog shell. (p. 150).

E. Broken sea turtle carapace bone. Lighter than a fossil bone. The beach is a final resting place for many sea turtles.

F. Fossil rib bone. Surfworn, but has the texture and feel of a fossil bone (many fossils are porous and will stick to a damp finger). This was likely from a marine mammal.

G. Iron plate. Rocklike, flat, and heavy. A magnet will confirm, it's iron.

H. Fossil bone fragment. This piece is darkly mineralized. Broken, worn, fossil bits are distributed on the beach similar to seashells.

I. Fired clay from a brick. Red, hardened clay remains after the surf tumbles a broken brick.

J. Concrete. Surfworn, but a recognizable composite of limestone pebbles and cement.

K. Dark seaglass. A tap on metal will verify it's glass (seaglass, p. 383).

L. Airplane skin. Aluminum with a composite honeycomb core, not geological.

"Rocks" of varied provenance found on Gulf beaches

HAND OF MAN

What Have We Had a Hand In?

Almost everything. Our touch is ubiquitous … so much that we are among the most influential of the Gulf's coastal forces. We consume, manage, and shape beach landscapes, and we leave lots of traces. Even on remote beaches, the synthetic crumbs of our existence are everywhere. Just as beaches record connections between the "natural" components of continents and seas, coastal sands also receive our own well-traveled discards.

Our familiarity with many of the items featured on the following pages makes this section different from the others in this book. But even recognizable things can have unfamiliar stories. Who made it? Where did it come from? Why are there so many of such an odd item? How did this strange thing get here?

In many ways, the human influence on beaches is heavy-handed, and even, counter to our own interests. One of the most profound effects we have on beaches stems from our demand for permanence within a system that is ever-changing. Living persistently near beaches has required that we do battle against the perpetual forces that shape the shore. These skirmishes have brought about drastic measures, conspicuous losses, and a host of unintended consequences.

Yet, it's the peaceful, grand beauty of beaches that seduces, sooths, and inspires us. Visitors drawn to the beach and overtaken by joy may feel compelled to leave testament to their spiritual experience—art. Although an emotional representation of this art may last, the creation itself does not. In this way, beach art is much like the elaborate Tibetan Sand Mandalas, which by Buddhist tradition are painstakingly created, then, ceremonially destroyed and released back into nature, thus symbolizing the ephemeral essence of life.

Quarried, granite boulders of the Freeport Jetty (1896). This structure and the associated Brazos River diversion (1929) dramatically altered the Texas shoreline.

Beach Shrines and Beach Art

WHAT ARE THEY? Beach shrines and beach art comprise human traces that commemorate beach visits and other important events. Constructions typically use beachcombed materials. Included are sand castles and sand sculptures of all shapes, using either the bucket-mold, wet-drip, or hand-smoothed methods. Shrines portray a sampling of local flotsam and sea shells, incorporated into artistic collages. Evidence from memorial ceremonies at sea, such as flowered wreaths, occasionally reach the beach. Artistic expressions range from small to large, and are limited by day-length, work ethic, availability of materials, and location relative to the tide. Some are added to by subsequent beach visits and visitors, and can become iconic monuments.

HOW COME? Why did the Druids construct Stonehenge? Because they could? Or perhaps, because they had time to kill? Like Stonehenge, the cultural significance of many beach shrines may remain a mystery. One hypothesis is that a trip to the beach can bring about a sense of whimsy in just about anyone.

DID YOU KNOW? Shrines we've seen have included the creative use of pig bones, underwear, sponges, doll heads, fish heads, mismatched flip-flops, mummified sting-rays, and a plethora of colorful plastic drift toys. All are common beach-finds. Nearly every surface of the locally famous, **"UFO Beach" pod** on South Padre Island, Texas, is bejeweled with shells, beer cans, and drift plastic.

A shrine to lost hard hats, South Padre Island, Texas

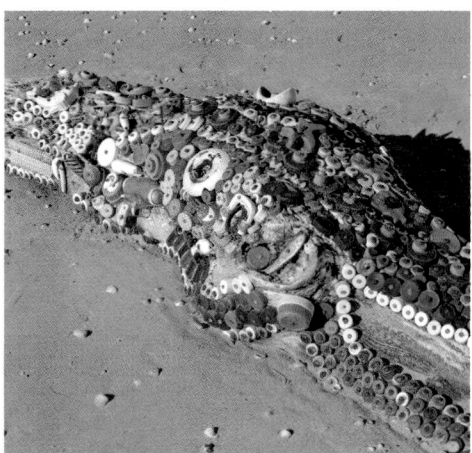

Driftwood+bottlecaps+whimsy=beach art

Oil rig escape pod at "UFO Beach," Texas

Finding one precious object can start a lifetime search

Finders keepers? Be aware of laws protecting antiquities

A barnacle bedecked bale of waterlogged weed

Discarded dunnage snapped by heavy cargo

Metal, Drugs, and Dunnage

Metal, Dunnage

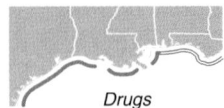
Drugs

WHAT ARE THEY? Some beach treasures are of silver or gold. But, **metal detectorists** take care—Gulf states have antiquities laws to protect archaeological artifacts. Chances are, much of the beach you search is state or federal property. Ask about detecting permits at parks. Most federal lands prohibit this practice. Know what else is prohibited? … that **bale** of marijuana or cocaine you found. How do you know it's drugs? Anything in meticulously wrapped bricks or bales is suspect. Take a photo, tell the cops, and share the story with your friends. Unlike treasure and drugs, **dunnage lumber** is jetsam (jettisoned) cargo. These packaging boards serve as pallets, skids, braces, or chocks for heavy freight and are generally rough, North American or tropical hardwoods of high compression strength.

HOW COME? In 1554, three Spanish ships, transporting silver and gold plundered from the Mēxihcah people, wrecked in a storm off Padre Island, Texas. A business from Indiana began removing artifacts from the site in the 1960s, which alarmed Texas officials, who identified the artifacts as property of Texas citizens. In response, Texas adopted the 1969 Texas Antiquities Code. Other Gulf sates have similar laws. Demand for narcotics in the US incentivizes a steady supply from the south. Dunnage is tossed from the many international cargo ships plying the Atlantic and Gulf of Mexico.

DID YOU KNOW? Archaeologists are keenly interested in learning from beach-found artifacts. If you find something really old, take a photo and location notes, and contact a state museum.

Boat Parts

WHAT ARE THEY? Beaches punctuate the final sea voyages of many vessels and their parts. Simple **wooden boats** show that in many upstream locations, craftsmanship exceeds access to materials. Other boat parts include **fenders** and **foam flotation**. Fenders are hung or floated to protect boats and moorings from bumps. They range from ship- to small-boat-size. Polyurethane foam flotation is used in many boats, breaks apart when boats do, and has a long life at sea.

HOW COME? Even after a boat founders, many of its parts keep afloat and can drift for thousands of miles. Primitive wooden boats may have come from Mexico and Central America. The bow section in the top image shows a plank-on-frame design that has been in use for hundreds of years. Fenders are meant to float between boats and their dock, but lines securing them occasionally break. Synthetic foam is one of the most common floating remnants of foundered vessels.

DID YOU KNOW? Marine lumber, which is often from tropical hardwoods, can provide beautifully figured wood for woodworking projects. Even "wormy" drift lumber shows interesting character and can highlight beachcombing-art designs. When new, a large yacht fender can cost many hundreds of dollars. Simple woven-rope fenders have been in use for more than a thousand years.

Old wooden boats show basic boat-building skills

A floating fender once cushioned a docked ship

Rope fender—old-school protection for docked boats

Foam polyurethane is used as flotation in boats

Man-overboard radio beacons

Lighted buoy used in overboard emergencies

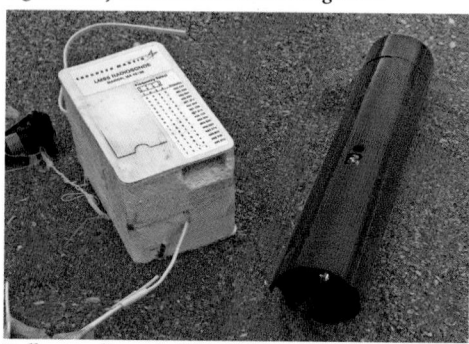

Balloon radiosonde (L) and dropsonde (R)

This PSAT popped off a billfish and is worth $500

Beacons, Sondes, and Fish Tags

WHAT ARE THEY? Man-overboard radio beacons are used to mark the drift paths of people or objects in the water during rescue operations. **Buoy lights** have a similar function and are deployed with life rings as soon as the emergency occurs. **Balloon radiosondes** are expendable instrument packages that measure and broadcast position, altitude, and atmospheric data on balloon flights to more than 100,000 ft (31,000 m). They are foam-encased atmospheric computers that often have shreds of the balloon that carried them aloft before bursting in the rarefied upper atmosphere. **Dropsondes** broadcast similar data during their fall from hurricane-hunter aircraft. **Pop-off Satellite Archival Tags** (PSATs) were attached to a large fish by fisheries researchers.

HOW COME? Man-overboard beacons on vessels occasionally wash overboard even when there is no distress. Radiosondes and dropsondes are disposable instruments deployed by meteorologists to forecast hurricanes and other weather. PSATs contain electronics that biologists use to track fish movements using daylight-period position estimates.

DID YOU KNOW? Although beacons and sondes contain relatively expensive electronics, they are not worth much on eBay. But the National Weather Service will gladly take them back for refurbishment, which saves taxpayer money. Some have postage-paid mailbags enclosed within. Rewards given for PSATs ensure a more efficient return rate. It's the detailed fish-migration data they contain that are so valuable, enough to warrant rewards of $100–$500.

Military Flares and Sonobuoy

WHAT ARE THEY? Military phosphorus flares (Marine Location Markers) are flare and smoke generators used in military operations to track submarines and mark floating survivors during search and rescue operations. They are bare aluminum cylinders labeled with hazard warnings and instructions to contact police or military. **Sonobuoy shipping containers** are large black (or faded) plastic octagonal tubes with screw threads on the open end. The tube once contained an expendable temperature sonobuoy probe that was ejected from an aircraft or ship conducting war games or marine research. **Decoy flare end caps** are jettisoned when the flares are deployed.

HOW COME? Military flares float ashore or roll along the bottom after sinking. Formerly sunken flares are often fouled, with important hazard messages difficult to read. These messages are serious. Call authorities and DO NOT TOUCH! Sonobuoy tubes are often discarded at sea. Decoy flares are defensive mechanisms used by military aircraft to avoid targeting by heat-seeking missiles. Each of these military materials may originate from practice maneuvers throughout the Gulf and Atlantic.

DID YOU KNOW? Aerial signal flares only burn for a few minutes, but are as bright as a 3,000-watt incandescent lightbulb. Because decoy flares are intended to deceive infrared missiles, the flares reach temperatures of thousands of degrees.

Military phosphorus flare, 18 in (47 cm)

A fouled flare, covered with encrusting animals

Sonobuoy shipping container, 3.8 ft (1.1 m)

End caps of decoy flares show maneuvers at sea

369

Steel, lighted, starboard-hand lateral buoy

Radar reflectors top this steel buoy

Green port-hand channel buoy

A beached, offshore mooring buoy, 3 ft (0.9 m)

Buoys *(Navigation, Mooring)*

WHAT ARE THEY? Steel buoys provide rustic and rusty reminders of impermanence at sea, even for our most robust maritime constructions. Steel, **lateral buoys** as tall as 12 ft (3.7 m) mark the edges of Gulf shipping channels. Green buoys typically mark the boat's left (port) side when returning from sea, and red marks the vessel's right side (starboard). These offshore buoys often have **light platforms** and right-angled plates that serve as **radar reflectors**. **Mooring buoys** are hollow steel spheres with rings for chains or lines.

HOW COME? Buoys that are aids to navigation are deployed and maintained by the US Coast Guard. There are hundreds of these structures around the Gulf region, and storms often tear them free of their mooring chains. The air in the hull of a navigation buoy is confined by half-inch (1.3 cm) thick steel plate. This makes them persistent at sea, and on the beaches where they land.

DID YOU KNOW? The US Coast Guard is interested in cataloging aids to navigation that have gone astray. They can be reached by marine radio, phone, and email (search: <your state> Coast Guard Aids to Navigation Office). These navigation buoys cost many thousands of dollars new. But, a typical channel-marking buoy will weigh about 13,000 pounds (5,900 kilos), so think twice about plans to salvage one.

Buoys *(Buoys in Depth)*

WHAT ARE THEY? Floats that sink endure great pressures in deep Gulf waters. These buoys include inflatable **mooring buoys** that are meant for use at the surface, but were dragged down and crushed at great depth. **PVC foam floats** retain their original buoyancy down to about 1,300 ft (400 m), but are permanently squished beyond that. Special, deepwater buoys filled with **syntactic foam** and ping-pong size spheres, and **glass spheres** encased in "hard hats," are buoyant to a depth of over 3.7 miles (6,000 m).

An inflatable mooring/fender ball crushed by depth

HOW COME? Crushed floats were likely used on fishing or exploration equipment that was used too deep, and were discarded as damaged. Even crushed, they still barely float and will drift and strand with other debris. Syntactic foam could be from broken, deepwater buoys used in petroleum exploration. **Glass spheres** and their hard hats likely tore free from the equipment they were supposed to keep buoyant. Common equipment deployed at great depth in the Gulf include oceanographic sensors and ocean-bottom seismology instruments used in oil and gas exploration.

Depth-compressed, PVC foam net floats

DID YOU KNOW? A buoy sunk to a depth of 6,000 meters experiences 594 times the pressure it is under at the sea surface. A basketball brought to this depth would shrink to about an inch (2.5 cm) in diameter.

Syntactic foam with plastic spheres. Colors (inset)

Glass ball buoys with protective "hardhat" cases

Alabama oil drilling platform with lighted flare stack

A V-gard, full brim hard hat lost by an oil worker

Fresh crude oil in a wave during the 2010 BP Spill

A melted tar ball on a Gulf beach

Petroleum Industry

WHAT ARE THEY? Offshore drilling platforms visible from a Gulf beach are typically anchored to the bottom by large supporting legs. The largest rigs have living accommodations for workers, ship moorings, and helicopter landing pads. Many rigs have flaming **flare stacks**, which are used for burning off unwanted, flammable petroleum gas during production. **Hard hats** are worn by workers on offshore platforms to reduce injuries. **Tarballs** are the black, sticky, semi-solid leftovers from weathered **crude oil.**

HOW COME? The Gulf supplies about 18 percent of US oil production and 4 percent of the country's natural gas. This energy extraction makes $30 billion annually. About 10,000 drilling and extraction rigs are erected within Gulf areas leased by the Bureau of Ocean Energy Management (BOEM). The rigs occupy coastal marshes, out to more than 1.5 miles (2,450 m) deep. Rig workers wearing hard hats are outside, often under windy conditions. The hats are fitted with a head-harness made of webbing and plastic, but for some reason, the hats seldom have chin straps. A wind-tossed hard hat falling from a rig is unlikely to be retrieved until it reaches a Gulf beach. BOEM reports that most beached oil in the Gulf comes from natural seeps and messy transfer of petroleum from the thousands of Gulf rigs. Known spills have a significant contribution to beach tar only following large events.

DID YOU KNOW? The Gulf has seen the two largest oil spills in history—BP Deepwater Horizon (2010, 4.9 million barrels) and Ixtoc I (1979, 3.3 million barrels).

Petroleum Industry

WHAT ARE THEY? Petroleum from undersea deposits is a yellow-black liquid containing gaseous, liquid, and dissolved solid hydrocarbons. After this crude oil is leaked or spilled, the lightest hydrocarbons evaporate, leaving heavier liquids and solids. The heaviest solids, mostly asphaltenes, make up about 6 percent of Gulf crude. These solid compounds are refined for use as paving materials and roof shingles. The black tarballs found on beaches are similar to the sticky, black, highly viscous, semi-solid "asphalt" used in road construction. The freshest tarballs stick readily to a beachcomber's foot. More weathered forms stick only when sun-heated. The most weathered tarballs have the consistency of a roadway "blacktop." Some asphalt slabs are as large as a car.

HOW COME? Beach tar origins include industry leaks and spills, and natural seeps. Oil seeps are fractures in the sea-floor that leak crude oil. In pre-Columbian times, long before the practice of oil extraction, the Karankawa Indians of Padre Island decorated pottery and water-proofed boats with beach-found tarballs. Today, seeps are thought to produce over half of Gulf beach tarballs. The remainder may have a responsible party. Tarball samples sent to the US Coast Guard lab in Connecticut for "oil fingerprinting" can have their source identified … sometimes. Newsworthy spills are rare, but Gulf oil production leaks thousands of barrels a day in obscurity.

DID YOU KNOW? Sunscreen assists a good first effort in removing tar from feet.

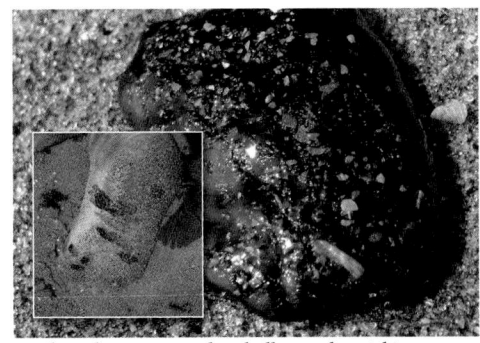

Fresh and sun-warmed tarballs are the stickiest

A weathered, asphalt "cookie" in the wrack

A large asphalt slab on a Texas beach

Spilled oil becomes news when it reaches a beach

373

A section of VIV strake sheath in the surf

VIV strake sheath for a small-diameter pipe

Wing from seismic streamer control "bird"

A spilled emergency oil spill kit, 3.5 ft (1.1 m) tall

Test ball plugs are used to test pipes under pressure

Petroleum Industry

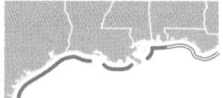

WHAT ARE THEY? Petroleum exploration and extraction in the Gulf is a complex operation involving strange equipment that often goes astray. **VIV strake sheathes** are half-inch thick, table-size sheets with stegosaurus-like diagonal fins. They are yellow or dark and are made of tough polyethylene. **Smaller strake sheathes** have longer, spiral fins. **Seismic streamer control "birds"** are used to control multiple, ship-towed "streamer" cables bearing hydrophones, which pick up the bottom reflections of onboard, seismic air cannons. The ships towing this equipment are searching for petroleum deposits. **Spill kits** lost from oil rigs and ships resemble giant plastic, screw-cap jars. They contain absorbent boom, disposal bags, and other supplies. **Test ball plugs** are inflatable, rubber fittings designed to test pipes for leaks under pressure.

HOW COME? VIV strake sheathes cover the extra deep pipes between oil platforms and the sea floor. The fins keep the pipes from wagging in the current, reducing metal fatigue and pipe leaks associated with this wear. Streamer birds are controlled in the instrument rooms of seismic survey ships looking for places to drill. They occasionally tear free in rough seas. Spill kits are common on many vessels and rigs. The petroleum industry uses a lot of pipe, and because leaks are bad, they must be pressure-tested with ball plugs.

DID YOU KNOW? The US Gulf has over 50,000 miles (80,500 km) of underwater petroleum pipeline.

Fishing Discards *(Floats)*

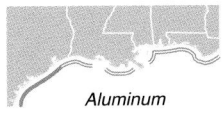

Aluminum

Other Floats

WHAT ARE THEY? Almost every fishing activity involves some sort of float. **Aluminum floats** may be vintage, collectible buoys from European fisheries. Commercial **stone crab trap floats (A)** bear an X-number, **spiny lobster trap floats (B)** have a C-number, and **blue crab floats** have a V- or F-number identification. Recreational traps have an "R." Most are foam polystyrene and painted colors unique to fishers who used them. Trap floats formerly had, or may still have, a rope that attached them to the trap. PVC-foam **bullet buoys** are used for surface longlines (thousands of hooks on a long horizontal line). These typically have strong monofilament line leading through a center hole to a large hook in the bottom. Primitive, **hand-made floats** are often fashioned from flotsam like polystyrene blocks, plastic bottles and jugs, or even used flipflops.

HOW COME? Floats that reach beaches may be casualties of broken lines and stormy seas. But many floats and other fishing gear at the end of a useful life simply get tossed. Because floats float, they are among the most common types of fishing gear on beaches. Hand-made floats are probably from developing countries where beach flotsam is essential raw material.

DID YOU KNOW? Aluminum floats may be old and valuable. Floats embossed with "Coquille France," "Phillips Trawl Grimsley England," or "La Coruna Spain" and a lighthouse icon, may be from the middle 20th century. Glass fishing floats formerly used by the Portuguese are rare now. Most sold in curio shops are reproductions.

Aluminum floats from France and Spain, 8.5 in (22 cm)

Stone crab (A) and lobster (B) floats, 8 in (20 cm)

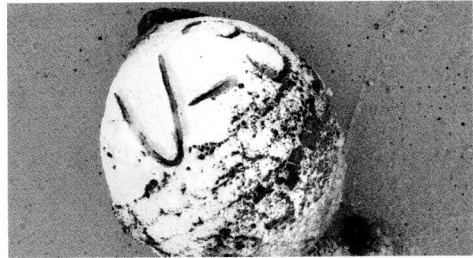

Blue crab V-float, 8 in (20 cm)

A longline "bullet" buoy, 18 in (46 cm)

Homemade foam-and-rope float

375

Deepwater "football" floats, 6 in (15 cm)

Spherical deepwater floats, 8.5 in (22 cm)

Pontoon oyster-bag float, 32 in (81 cm)

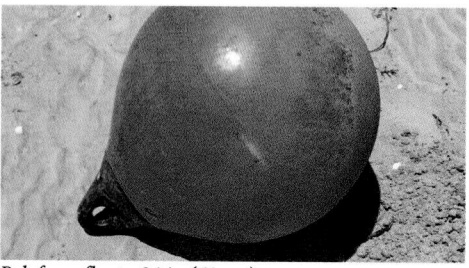

Polyform floats, 24 in (61 cm)

A beached high flier, 12 ft (3.6 m)

Fishing Discards *(Floats)*

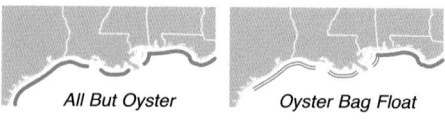

All But Oyster   *Oyster Bag Float*

WHAT ARE THEY? Deepwater floats are hollow, hard-plastic football, pill-capsule, or sphere shapes with rope-attachment rings. Most of these are used for deep longlines and trawl (vessel-towed) nets. **Pontoon oyster floats** are used in coastal oyster farms to keep mesh bags of living oysters at the surface. Inflatable **Polyform floats**, are used as markers for longlines, pots, and traps. Larger buoys of this type are used as fenders (boat bumpers). A **high flier** is a float with a long pole through it, typically with flags or an angular aluminum radar reflector. Those from developing countries may be no more than a bamboo pole with colored rags, attached to a polystyrene block. The makeshift high flier in the lower image had a cast-iron window-sash weight as ballast.

HOW COME? Floats used in deepwater fisheries are made of rigid plastic that is resistant to compression. Floating oyster bags are used in aquaculture operations to keep oysters in better water quality near the surface. Mesh envelops valuable floats to reduce the chance they would be lost by a single line breaking. Frayed lines indicate when this tactic did not work. High fliers fly high to make them easier to see in the open ocean. Storms greatly increase the stranding likelihood of this floating fishing equipment.

DID YOU KNOW? Harvest of wild oysters in the northern Gulf is being replaced by aquaculture. Oyster reefs have greatly diminished due to pollution from runoff and alteration of freshwater inputs to coastal estuaries.

Fishing Discards *(Traps)*

WHAT ARE THEY? These **traps** catch crustacean seafood. Pots are traps that are set and left unattended. Beached traps are constructed of **wood**, plastic, or **wire mesh**, and are designed to catch spiny lobsters, stone crabs, or blue crabs. Each trap has (or had) a funnel entrance, latchable lid, marker float, and rope (trap line). Blue crab traps are required to have plastic **escape (cull) rings,** or rectangular escape vents, which allow juvenile crabs to exit. **Latch hooks** on elastic cords secure the access door. As old traps corrode, these plastic parts can float free.

HOW COME? Most traps probably end up on beaches by accident and are evidence of the power of storms to move even deep objects over long distances. Some old traps have become miniature reefs with a wide variety of bottom creatures growing on them. Although stone crab and spiny lobster traps come from offshore waters, most blue crab traps come from coastal estuaries.

DID YOU KNOW? Traps set for blue crabs often entrap and drown terrapins (p. 204), which have suffered severe declines across the Gulf. A Bycatch Reduction Device (BRD) or "terrapin excluder" is known to prevent this mortality. The device is a plastic rectangle affixed to the trap's entrance funnel, which allows entry of crabs, but not terrapins. Studies show that traps with BRDs catch crabs at rates equivalent to traps without BRDs. The regulatory body overseeing coastal fisheries, the Gulf States Marine Fisheries Commission, has yet to require these devices.

A partially buried, wooden, spiny lobster trap

A blue crab trap in the swash zone

Blue crab traps are common near coastal estuaries

Crab-trap cull ring and latch hook (inset)

377

A stone crab trap showing the lid latch (lath) in use

Spiny lobster and stone crab lid latches

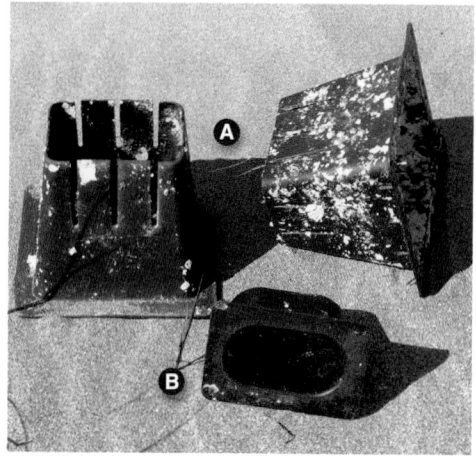

Spiny lobster (A) and stone crab (B) trap funnels

Fishing Discards
(Lid Latches, Trap Funnels)

WHAT ARE THEY? These black-plastic trap parts are some of the most persistent floating fishing litter. **Lid latches** (laths) hold the lids closed on both plastic and wooden traps for lobster and stone crabs in southern Florida. **Trap funnels** are entrances to traps that allow crustaceans to enter but not escape. Spiny lobster funnels end in a rectangle, and stone crab funnels end in an oval.

HOW COME? Most of these trap parts drift from fisheries in southern Florida. Lid latches are required to be fastened with a corrodible nail that eventually releases the lid after the trap is long abandoned. This design is an attempt to reduce the time that lost traps continue ghost fishing—the needless entrapment of marine life. Lost and abandoned traps are common. Wooden traps slowly disintegrate, freeing their floating parts, including latches, bait cups (following page), and funnels.

DID YOU KNOW? Stone crabs in Texas are caught recreationally with a generic, "chicken-wire" trap (previous page). The commercial season for trapping spiny lobster in Florida waters is early August through March. Stone crab season is open October 15 through May 15. Most commercial stone crab traps are set between the Keys and Florida's Big Bend. Lobster traps are set mostly in the Florida Keys. Currents carry parts from these traps far and wide. Plastic fishing gear is black because of the color's resistance to degradation in sunlight. The persistence of this material at sea may be greater than a hundred years.

Fishing Discards
(Trap Tags, Bait Cups, Octopus Pot)

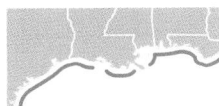

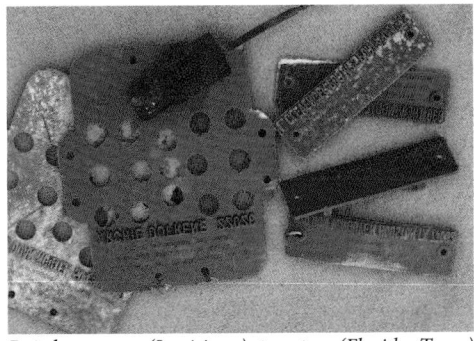

Bait-box covers (Louisiana), trap tags (Florida, Texas)

WHAT ARE THEY? Gulf states require commercial traps to be identified with the owner's name and license number. In Texas, **trap tags** are generally orange plastic. In Louisiana, the trap's bait-box cover is identified. Tags in Mississippi, Alabama, and Florida range widely over the years. Tags were clasped, nailed, or stapled. **Stone crab bait cups** hold the bait (fish heads, pig feet) within traps. **Octopus pots** are black or brown, open plastic jugs with a flat side, set to catch octopuses. Many have no markings. Others are embossed "NIBER" or "H·O·T–P·P·A"

Spiny lobster pot annual tags, Florida

HOW COME? The commercial "endorsement numbers" on tags and floats identify trap owners to fish and wildlife officers. Bait cups are one of many trap parts freed when traps break apart. The plastic octopus pots that beach are without the concrete or stone that originally lined their flattened side. Un-baited and lidless, they attract octopuses only as a hiding place. The pots likely have come from the eastern Atlantic, including fisheries off Spain, Portugal, Morocco, and Mauritania. Spain has recently banned plastic pots. In the other countries, lost octopus pots number up to 9,000 pots per month.

Stone crab trap bait cups

DID YOU KNOW? Derelict crab traps are no longer useful to fishers, but often continue catching and killing crabs as "ghost traps." Given the number of commercial blue crab traps in use, and the rate of trap loss, estimates are that more than 250,000 derelict blue-crab traps are added each year in the Gulf of Mexico.

Octopus pots are usually found battered and worn

Shrimp trawl netting

Derelict gill net with ensnared animals

Beached FAD. A common FAD tracking device (inset)

Trap line (rope)

Fishing line with reel coils

Fishing Discards *(Nets, FAD, Line)*

WHAT ARE THEY? Fishing nets and lines are made of synthetic twine or nylon monofilament. They are designed either to strain animals from the water, tangle them, or secure implements that catch them. **Shrimp trawl nets** are of the straining variety. Their meshes of stiff twine are generally too small to get a fist through. Fish trawl nets are similar but may have larger meshes. **Gill nets** are either thin twine or monofilament. These nets typically have a distinct float line (with floats) and a lead-cored line called a leadline. **FADs** (Fish Attracting Devices) include bamboo rafts with branches and palm leaves, often bound by ropes and netting. Fancier FADs have a tracking **radio beacon**. **Trap ropes** attach buoys to traps and are typically black. **Fishing line** is monofilament nylon.

HOW COME? These nets are illegal within Florida's state waters, extending nine nautical miles into the Gulf. Shrimp and fish trawls are heavy nets dragged along the bottom by steel cables. Although the nets are strong enough to scrape up corals and sponges, they occasionally encounter items too heavy to tear away from the sea floor. Nets torn free of their tow vessels are subsequently moved by storms. Trap line (rope) is made of braided polypropylene, chosen because it floats in seawater.

DID YOU KNOW? Concerns over indiscriminate entanglement mortality of marine life prompted bans of gill net use in the coastal waters of Texas and Florida. In 2010, an estimated 10,000 FADs were circulating in the Atlantic and Gulf. The devices attract tropical tunas for purse-seine vessels.

Fishing Discards *(Miscellaneous)*

Seafood basket, 24 in (61 cm)

WHAT ARE THEY? Seafood baskets are used on fishing and shrimping vessels to temporarily hold the catch. Latex **shrimp-heading (picking) gloves** are used by shrimpers as they pick shrimp from their trawl catch. Colors are orange, yellow, and blue. **Fishing glowsticks** (lightsticks) are transparent plastic tubes containing cylume chemicals and fluorescent dyes. The sticks glow for several hours after an internal vial of hydrogen peroxide is broken. They have various attachment clips and rings. Lost fishing lures, typically "**teasers**" and **plugs**, come in a variety of fisher-attracting color combinations.

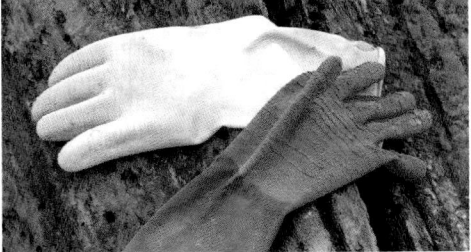

Shrimp-heading gloves

HOW COME? Most seafood baskets are damaged, indicating that they were discarded rather than lost. Shrimp-heading gloves and glowsticks are disposable items with a limited life. Glowsticks mark fishing buoys at night and are used as fish attractors on commercial longline hooks set to catch swordfish at night. Because they last only one night of fishing, and because the Atlantic longline fleet uses millions of baited hooks, glowsticks are common fishery items on the beach. Teasers are hookless lures towed behind trolling vessels to attract gamefish to hooked baits. Saltwater fishing plugs used by recreational fishers for casting or trolling cost 10–20 bucks or more. This makes us think that most were lost, and are now part of an incredible fish story.

Fishing glowsticks, 3–6 in (8–15 cm)

Teaser lures mimic flying fish, 4–9 in (10–23 cm)

DID YOU KNOW? Glowsticks may attract swordfish that mistake the light for bioluminescent squid. Reusable, battery-LED lights are now available, which may hopefully reduce this marine litter.

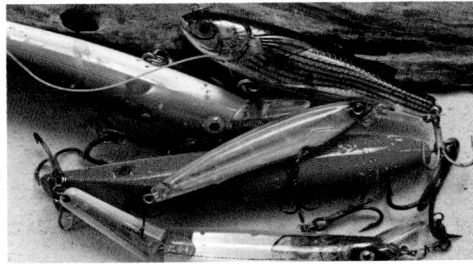

Lost fishing plugs, 4–9 in (10–23 cm)

381

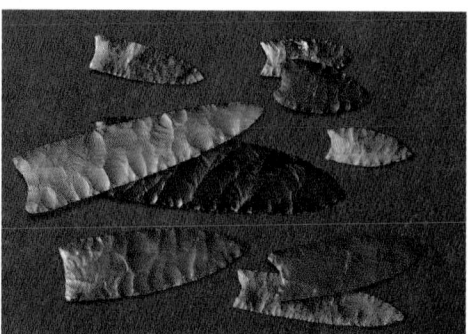

Chert Clovis points from northeastern Texas beaches

Coal from a steam ship's bunker

Bale of rubber sheets, Padre Island, Texas

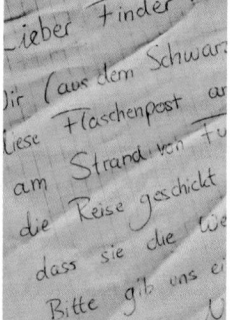

An Atlantic-spanning message, Matagorda, Texas

Messages from the Past

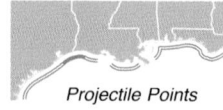
Projectile Points

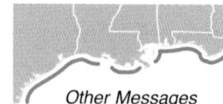
Other Messages

WHAT ARE THEY? With careful detective work, artifacts help tell the story of human experiences near Gulf beaches. **Stone projectile points** reveal the presence of Clovis and Folsom people, who were named for the distinct stone projectile points they used. These people lived between 13,000 and 9,000 years ago when the current beach was marsh, which was many miles from the Gulf shoreline. **Coal** on the beach comes from the disintegrating coal bunkers of sunken steam ships that last sailed in the early 1900s. **Bales of crepe (raw sheet) rubber** are a mystery. Some may be from a transport wrecked off Brazil in World War II. These bales have the words "Product of French Indochina." **Messages in bottles** range old to recent. The message in the image drifted onto a Texas beach from German tourists in the Canary Islands off Morocco.

HOW COME? Beach artifacts represent the human record of change—from stone tools, to steam ships, to international trade and warfare, to leisure travel. These traces were brought to beaches by the same forces that erase them.

DID YOU KNOW? Clovis people are the ancestors of most indigenous people throughout the Americas. A concentration of Clovis artifacts, along with fossils from Pleistocene animals, occur near McFaddin Beach, Texas. There, a salt dome just off the beach has pushed up Pleistocene sediments, making them available for erosion and onshore transport. The most tragic time for steam-ship wrecks off Gulf beaches was the American Civil War, when more than two dozen were lost. To deliver your sentiments without littering the sea, just give a note to a random person.

Sea Heroes (Drift Toys) and **Seaglass**

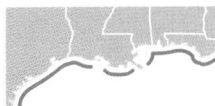

WHAT ARE THEY? Seaglass pieces are broken bottle shards that have been rounded, smoothed, and etched by sand and surf. **Sea heroes** are plastic, once-cherished, childhood friends set adrift on the open sea. These little synthetic figures often bear evidence of their lengthy voyage, such as sun-fading, lost limbs, and accumulations of encrusting creatures.

HOW COME? Seaglass shows how this select aspect of discarded debris can be formed by the sea into interesting and beautiful objects. Although seaglass has human origins, it ends up on beaches the same way seashells do—by tumbling within waves and currents, often for many decades. Sea heroes float and drift after being lost or discarded.

DID YOU KNOW? Seaglass colors from most common to least common are (generally) brown, medium green, clear, light green, light blue, dark blue, yellow, red, lavender, and black. Lavender is from pre-WWI glass made with manganese. Black glass may be from the 1700s. Held up to strong light, this glass looks dark green or amber. Variations on the classic plastic army man, comic-book characters, and baby-doll parts, seem to be the most common sea heroes. We've collected hundreds of plastic figures and no two are the same. Many figures may be from developing countries upstream, having long ago fallen out of favor with jaded American kids. Drifters from Mexico include **Luchadores** (wrestlers) who have been produced as tiny plastic figurines since the 1980s.

Seaglass gathers with shell hash

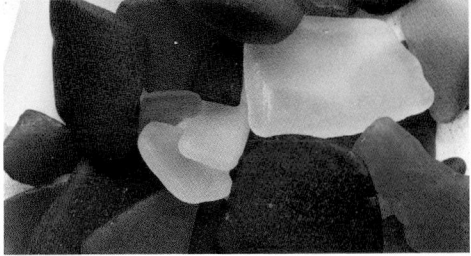

Brown seaglass is common, blue is rare

A sampling of sea heroes—no two are alike

Facsimiles of Mickey (L), rubber (plastic) duckies (R)

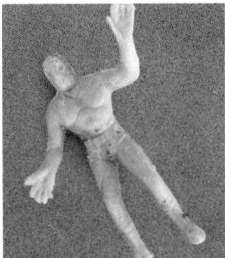

Plastic piggy banks (L), Luchador (R)

Foil balls tell of human foibles

Practice golf balls are cheap enough to leave as litter

Roll-on balls, typically 0.4–1.4 in (1.0–3.6 cm)

Nurdles, max 3/16 in (5 mm)

Balls and Nurdles

WHAT ARE THEY? Aluminum **foil balls** are buoyant elements of galley waste. **Practice golf balls** are like the originals, but made of brightly colored, cheap, synthetic foam. **Roll-on balls** are hollow plastic spheres that top the increasingly less common underarm deodorant applicator. **"Nurdle"** is the entirely-too-cute, colloquial name for an industrial resin pellet of HDPE (high-density polyethylene). The pellets are precursors for a plethora of plastic products.

HOW COME? Foil balls are symptoms of two common habits—wadding up used cooking foil and tossing trash overboard. Trapped air allows them to float. Practice golf balls are inexpensive versions of standard course balls. The common practice of hitting them into the sea leaves the floating balls to wash up somewhere. Roll-on deodorant containers are part of the worldwide circulation of discarded plastic. The balls are the most persistent part of the deodorant package and may float for decades. Nurdles are spilled at coastal production sites and during mass transport at sea.

DID YOU KNOW? Little turtles eat nurdles, although the plastics are bad for them. Nearly all yearling sea turtles that wash ashore in Florida have ingested plastics, including nurdles. Petrochemical companies in coastal Texas are major producers and transporters of nurdles, and they spill a lot. Estimates are that there are hundreds of billions of nurdles circulating in our seas. Want to help solve the mystery of where this plastic waste comes from? Visit: www.nurdlepatrol.org

Container Seals and Packaging

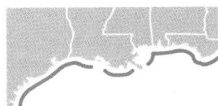

WHAT ARE THEY? Container seals recorded that shipboard containers were unopened. Containers (isotainers) are the international, semi-trailer-size boxes that haul almost every product on the planet, products that also have their own plastic **packaging**. Varieties span many languages. **Butter tubs** from Haiti are common, as are **thin-walled plastic containers** without reusable screw caps. These inexpensive containers sold in the Dominican Republic once held bleach (white plastic labeled CLORO), vinegar (yellow plastic, labeled VINEGRE), detergent (white or yellow), or other consumables. Those water bottles with yellow liquid? A micturition convenience for litterbug boaters. Yuk!

Plastic container seals, about 6 in (15 cm)

HOW COME? Plastic floats for decades. Beached container seals reveal both the maritime tradition of tossing trash overboard and the astounding volume of goods shipped in containers. Beached plastic packaging is multinational, which may indicate its most common source—international shipping. Island landfills are another important source of plastic litter. People of limited means who live on islands have few resources to manage their waste, and much of this plastic washes into the sea.

An international array of beached trash

DID YOU KNOW? Household (and shipboard) plastic packaging enters the ocean at an estimated rate of 6.4 million tons per year, reports the National Academy of Sciences. This is 10 times what gets recycled. We use plastic because it seems cheap and convenient. Yet, over a third of all plastic packaging becomes litter, where it is about 20 times more expensive to clean up than its original cost.

Ti Malice butter tubs from Haiti

Thin-walled plastic containers from the DR

385

Plastic, yarn packing cones, 8 in (20 cm)

Plastic confetti (microplastic) including nurdles

A plastic bag's knot is the last to break into pieces

Yarn Cones, Plastic Confetti, and **Plastic Knots**

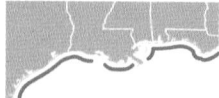

WHAT ARE THEY? Plastic, yarn packing cones are wound with the yarns (textile threads) used in clothing manufacture. "**Confetti**" describes the abundant, tiny pieces of plastic (microplastics) that were once larger litter. **Plastic bag knots** remain after a knotted bag has shredded due to sunlight exposure and time at sea.

HOW COME? Yarn cones? Really? How obscure is that? There's a lot of obscure plastic litter floating out there. This example may have resulted from a broken shipping container spilling the cones at sea. Many weird things end up on beaches this way. Sunlight makes plastics brittle after years of exposure. Waves break the brittle litter into tiny floating shards that collect at the sea surface in the same places where fishes, young sea turtles, and seabirds search for food. Because these open-sea animals live in a virtual desert of rare opportunity where any novel item might be food, they mistakenly eat a lot of plastic. Plastic knots remain after the rest of the bag breaks into pieces because the knotted plastic is thicker, with outer plastic layers shielding inner layers from the sun. And, plastic bags are everywhere.

DID YOU KNOW? On the tiny, remote archipelago of the Cocos Keeling Islands in the Indian Ocean, plastic covers the beaches. A study estimated the litter to include over 14 million plastic items—bottle caps, straws, almost a million shoes, and 400,000 toothbrushes. The litter drifted there by sea. The mass was equivalent to what the local community would produce as waste in roughly 4,000 years.

Mystery Plastics and Wax

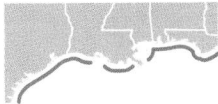

WHAT ARE THEY? Burned or melted plastics may be from trash-burning piles of Caribbean islands. Plastic spirals that look like drift fusilli are the one-piece **drill shavings** from marine, high-density polyethylene (HDPE) sheets. **Carnauba wax** is beached as hard, grayish or brownish chunks. The wax is harvested from fronds of the wax palm *(Copernicia cerifera)*, a native fan palm of Brazil. **Paraffin** (petroleum) wax comes ashore rough or flat, colorless chunks. This petroleum product begins to melt at 99 °F (37 °C).

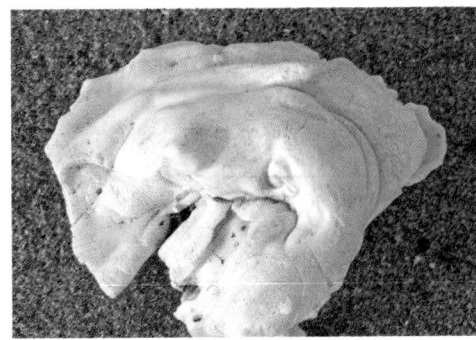

A melted lump of white mystery plastic

HOW COME? To reduce landfill waste volume, trash is often burned. Incineration is often incomplete, resulting in plastics that are only scorched or melted. Inhabitants of islands often have landfills next to the sea, which may be why this waste escapes to drift. HDPE is a marine structural plastic that has replaced teakwood on modern vessels. When drilled, the shavings go overboard as puzzling spirals. Carnauba wax is used in making varnishes, polishes, car wax, and candy. Chunks of this wax fall off ships all over the Atlantic and Gulf. Paraffin wax is used in the manufacture of candles, crayons, and food additives. The wax is transported by tanker ships, which carry it as a heated liquid. The ships commonly clean their tanks of the buoyant leftovers at sea to cut down on operational costs.

Drill shaving from marine HDPE

DID YOU KNOW? The World Economic Forum reports that, at predicted rates, the mass of plastics in our oceans will outweigh the fish by 2050. Marine industry plastics are resistant to sunlight degradation and will probably outlive us all.

Carnauba wax, 1–10 in (2.5–10 cm)

Paraffin wax chunks, 1–10 in (2.5–10 cm)

387

Toy hand grenade smoke bomb (A), firework tips (B)

Mylar® balloons, shredded latex balloon (inset)

Shotgun shells (A) and plastic wadding (B)

A mallard drake decoy that escaped to sea

Fireworks, Balloons, and Hunting Discards

WHAT ARE THEY? Plastic tips of many colors come from launched fireworks. **Black-plastic grenades** are spent, smoke bombs. Beached balloons were formerly filled with enough helium to keep them aloft. **Latex balloons** are often shredded. **Mylar® plastic balloons** are generally intact, but sun-faded. Both typically have a ribbon attached. **Plastic shotgun shells** missing their brass head are merely a plastic casing. The shells fire **plastic wadding** along with the lead shot. Hunters set **duck decoys** in coastal waters to attract waterfowl. A bottom keel keeps them upright.

HOW COME? What goes up must come down. Balloons burst or leak their helium in the thin upper atmosphere and then fall to Earth (or to sea). This way, the joy of an inland birthday party is transmitted hundreds of miles. Unfortunately, the sentiment is lost on the sea life harmed by this litter. Tentacle-like shreds of a popped latex balloon develop when the balloon freezes at high altitude before bursting. Fireworks end up in the ocean due to the common belief that water provides a safe location for ballistic revelry. Shotgun shells are discarded by waterfowl hunters in bays and sounds. At sea, shooting is at clay pigeons (skeet targets) launched from cruise ships. Both the shot wadding and the ejected casings float for years. Mallard drakes are common decoys because they attract many other species.

DID YOU KNOW? Under Florida law, it is illegal to release 10 or more balloons in a day. Environmental concerns and passenger casualties have begun to curtail the use of shotguns on cruise ships.

Bite Marks on Litter

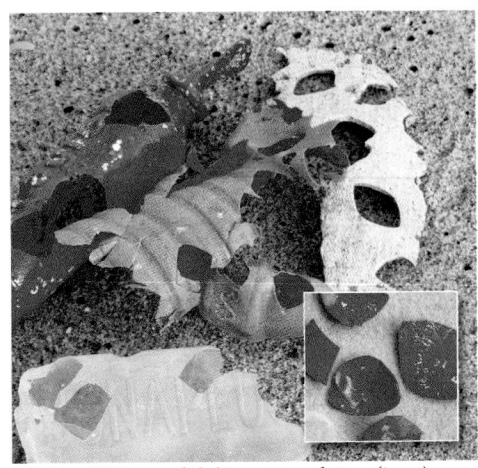

Plastic with triggerfish bites, pieces bitten (inset)

WHAT ARE THEY? Bite marks show that plastic at sea is an ingestion hazard to marine life. Triggerfish (family Balistidae) leave **diamond-shape holes** with close-set teeth marks at upper/lower vertices of the mark. These bites also result in V-shape edge bites. Turtles bite plastic too, but have no teeth. Green turtles can leave **finely crenulated bite marks** from their serrated jaws. Typically, only larger juvenile turtles, with bites more than 1.5 in (3.8 cm) wide, are able to bite through plastic. Bites from larger sea turtles, like loggerheads, are occasionally seen in **foam floats.** Sharks bite floats too, and leave ragged crescent-shape marks.

HOW COME? Plastic drifts within the world of surface-feeding marine animals. Their indiscriminate feeding style evolved in a world without plastics. Because plastics are tough, only bites from animals with strong, sharp mouths are recorded. The most common identifiable bite marks we've seen on Gulf beach plastics are likely from the ocean triggerfish *(Canthidermis sufflamen).* This fish's bite has a combination of sucking and sharp teeth that draws in and snips neat diamonds out of thin sheets and containers.

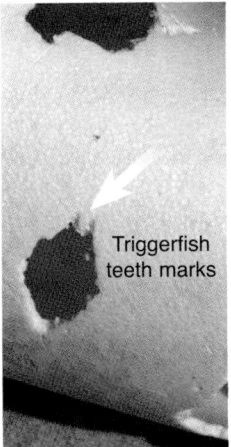

Triggerfish teeth marks

Green turtle, no teeth marks

Ttiggerfish bites in foam   *Turtle bites in plastic*

DID YOU KNOW? Young sea turtles are well known to ingest hazardous amounts of plastic. But rather than biting pieces out of whole litter, the turtles most often swallow sun-brittled plastics that were broken into bite-size shards. Most of the plastic bites in the sea go unobserved. This synthetic waste is eaten by animals as small as larval fishes and plankton, which puts our potentially toxic trash into the marine food web that produces our seafood.

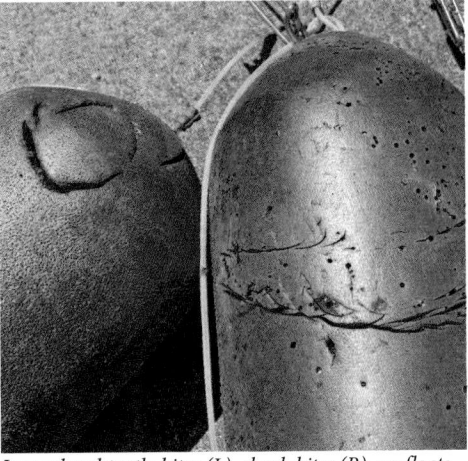

Loggerhead turtle bites (L), shark bites (R), on floats

A beach cleaning tractor, Mississippi

Beach driving occurs on most Texas beaches

Ruts on a busy stretch of Padre Island, Texas

Rows of wooden-slat sand fencing, Surfside Beach, TX

Beach Grooming, Driving, and Sand Fences

Grooming, Fences

Driving

WHAT ARE THEY? Grooming (beach cleaning, beach raking) flattens the beach and mechanically removes the wrack. **Beach driving** describes the use of beaches by private vehicles for recreation. **Sand fences** are posted lengths of plastic mesh or wired wooden slats.

HOW COME? The beach-grooming debate outlines divergent expectations. One is that beaches should be clean surfaces free of litter and biological material, even if the removal of this material means the loss of shorebirds, beachcombing finds, and other aesthetics. An alternative expectation is that true litter can be easily plucked by hand, leaving beaches as functioning habitats where beachhoppers, seabeans, bird life, and natural sand sculpturing are preserved. As a compromise, some communities only remove wrack when seaweed piles become ponderously large. Public beach driving is legal in most of Texas, Holly Beach (western Louisiana), and parts of Gulf and Walton Counties, Florida (by permit only). Where traffic is dense, the character of the beach is drastically changed. Rules regulating speed and vehicle-free zones to protect nesting sea turtles, shorebirds, and human pedestrians are enforced on a scale ranging from strict to negligent. Sand fences function to prevent human access, to defend beach property lines, and to keep wind-driven sand on the beach and out of the dune. The fences are only slightly poorer at accumulating sand than are clumps of seaoats and other beach plants.

Coastal Armoring

WHAT ARE THEY? Coastal armoring includes **seawalls**, **rockpiles** (revetments), and **geotextile tubes** (sandbags) placed so that they reduce movement of sand on the upper beach and dune. Seawalls are steel or aluminum sheetpile, concrete, or wood, and may be vertical or inclined. Geotextile tubes, also called geotubes, are giant bags of polypropylene mesh. The bags are filled by pumping a sand-water slurry into them. Water drains out, leaving the swollen bag with a rocklike firmness.

The 8-mile granite revetment at Sargent Beach, Texas

HOW COME? Armoring is constructed to retain the existing sand beneath beach-front yards, roads, and buildings. Unintended consequences of armoring include acceleration of erosion in front of and beside the structure. The **Galveston Seawall** was built following the Galveston Hurricane of 1900, to defend the city from future storms. It has worked so far, but as an unintended consequence, it caused erosion that almost completely scoured away the sandy beach. Now, Galveston's beach-sand loss is abated only through frequent dredge-and-fill projects. Texas and Mississippi now prohibit shoreline armoring. Alabama and Florida permit armoring following frequent storm emergencies. Geotextile tubes were tried as "soft" armoring and "dune restoration," but the tubes proved to be as detrimental as other armoring, and less resilient to wave action. Shreds of old geotextile tubes are evident following erosion on many Gulf beaches.

A rock revetment, Grand Isle, Louisiana

DID YOU KNOW? Rising sea level is likely to accelerate creative engineering to stop beaches from moving beneath our coastal buildings in dangerous locations.

Galveston Seawall, Texas, constructed in 1904

A geotextile tube groyne traps updrift sand

Beach breakwater with a tombolo, Dauphin Is., AL

Fishers on a granite jetty

A derelict, steel, sheetpile jetty, Rollover Inlet, Texas

Groynes, Breakwaters, and Jetties

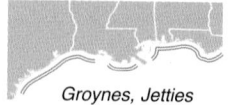

Groynes, Jetties

Breakwaters

WHAT ARE THEY? Groynes (groins) are constructed perpendicular to shore so that they interrupt longshore drift of surf sand. They may be of sheetpile, geotextile tubes, wood, or more commonly, rock. **Breakwaters** are rock structures in the surf zone aligned parallel to shore. **Jetties** are typically piles of heavy granite boulders lining the channel into an inlet, and may extend offshore for more than five miles (8 km).

HOW COME? Groynes promote accretion (p. 30) on their updrift side, and cause net erosion (p. 29) on their downdrift side. This is due to capture of sand transported by the longshore current (p. 33). They are meant to stave off local erosion, but because of the "robbing Peter to pay Paul" aspect, groynes have fallen out of favor for beach-sand management. Breakwaters steal sand as groynes do, and build **tombolos** of sand connecting the breakwater to the shoreline. Jetties have sand-theft drawbacks similar to groynes and breakwaters, but remain a popular option for keeping sand out of deep inlet channels. Granite is the rock of choice for jetties because the rock's density (50 percent greater than limestone) helps keep the boulders in place during big wave events. Jetties of **steel sheetpile** have a shorter life than rock.

DID YOU KNOW? Inlets have pairs of jetties—one on the updrift side, and one downdrift. The updrift side is the direction of the strongest average winds, which drive the sand-carrying longshore current. Updrift jetties extend the longest because that is the side with the greatest potential for longshore-drifting sand filling the inlet channel.

Artificial Nourishment

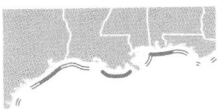

WHAT IS IT? Artificial beach nourishment is the accretion of beaches by artificial means. Most commonly, this involves trucking sand from inland quarries or pumping sand from offshore sources. This process is also known as "beach renourishment," "beach restoration," and "dredge and fill."

HOW COME? Most projects take place where there is an unsatisfactory sand buffer between the sea and valuable structures built on the dune crest. Where seawalls, revetments, or buildings occupy the beach, artificial nourishment provides a beach that would otherwise only be possible if the armoring and structures were removed. Benefits to tourism and natural resources are also cited as justifications for nourishment projects, with the debatable assumption that there is a correlation between beach-width and use by tourists and wildlife.

DID YOU KNOW? Creating a living beach is difficult. Artificial beach projects smother existing swash-zone animals and beach plants/seeds with sterile sand, and tend to create berms that are wider, flatter, and harder than natural beaches. After about two years of surf action and colonization by plants and animals, many artificial beaches can be as alive as natural beaches. Federal, state, and local government funding for artificial nourishment of Gulf beaches costs tens of millions of dollars per year. Half of all Mississippi mainland beaches are artificial. The 26-mile (42-km) stretch between Pass Christian and Biloxi was once marsh, but is now the longest artificial beach in the world.

Projects involve heavy equipment on land and sea

Pipes carry pumped offshore sand in a seawater slurry

The sand slurry settles and is bulldozed into place

A newly engineered beach

Río Bravo, Mexico

Aransas Pass, Texas

Matagorda Island, Texas

Point Bolivar, Texas

Biloxi, Mississippi

Sand Island, Alabama

Cape St. George, Florida

St. Marks, Florida

Historic Structures *(Lighthouses)*

WHAT ARE THEY? Lighthouses are towers of brick or steel topped with a rotating light-source. **Faro Río Bravo**, Mexico (unknown date), is a 59 ft (18 m) light just south of the Rio Grande and is the southernmost visible light in Texas. **Aransas Pass Light Station**, Texas (1857), 60 ft (18 m), marks the pass to Aransas and Corpus Christi Bays. **Matagorda Island Light**, Texas (first lit in 1852, current structure lit in 1873), is a 91-ft (28-m), cast-iron lighthouse. **Point Bolivar Light**, Texas (first lit in 1852, current structure lit in 1872), shines its light at 116 ft (35 m). **Biloxi Light**, Mississippi (1848), provides a 60-ft (18-m) beacon on Mississippi Sound. **Sand Island Light**, Alabama (first lit in 1838, current structure lit in 1873), shines at 131 ft (40 m) off Dauphin Island, Alabama. The **Cape St. George Light**, Florida (1852), collapsed and was moved (2005–2008) from Cape St. George to St. George Island. **St. Marks Light**, Florida (1842), stands 82 ft (25 m) at the eastern range of this book.

HOW COME? Before modern instruments, lighthouses aided navigation around reefs, sandbars, capes, and inlets.

DID YOU KNOW? Many Gulf lighthouses have had multiple incarnations. After its second post-storm rebuild in 1852, the Cape St. George Light stood for 153 years before crumbling into the sea. It was originally built hundreds of yards from the beach. The current structure was moved and reconstructed, brick-by-brick. The Matagorda Island lighthouse was damaged in 1861 when confederate soldiers tried to blow it up, and was moved away from the retreating beach in 1866.

Historic Structures *(Forts)*

WHAT ARE THEY? These beachfront military installations are monuments to an obsolete coastal defense system. **Forts** (fortifications) are permanent defensive military facilities, generally with accommodations for personnel. **Artillery batteries** are components of multiple, large gun (canon) emplacements.

HOW COME? Following the War of 1812, the US Congress funded seacoast-defense structures known as the Third System. Third System forts on Gulf beaches include Fort Pickens (Pensacola, Florida), Fort McRee (Pensacola, Florida, in ruins buried by sand), Fort Morgan (Mobile Point, Alabama), **Fort Gaines** (Dauphin Island, Alabama), **Fort Massachusetts** (Ship Island, Mississippi), and **Fort Livingston** (Barataria Island, Louisiana, sinking ruins). The development of penetrating, explosive shells in the late 1800s favored reliance on defensive batteries of long-range artillery. One example of this is the **Battery Laval** (Galveston, Texas) and other gun emplacements that were part of the harbor defense of Galveston.

DID YOU KNOW? Confederate forces seized Fort Gaines in 1861 at the start of the Civil War, and kept it until August of 1864, when the fort was surrendered to the Union fleet following the Battle of Mobile Bay. The fleet was commanded by Admiral David Farragut, who during the battle gave the famous order to "Damn the torpedoes, full speed ahead!" He was referring to the mines (torpedoes) strung across the Mobile Bay channel. Most of these sites saw their only fighting during the Civil War, although all are currently in a desperate battle with the rising and frequently savage sea.

Battery Laval, Texas (1897–1920)

Fort Gaines, Alabama (1821–1946)

Fort Massachusetts, Mississippi (1859–1875)

Fort Livingston, Louisiana (1835–1889)

The skeletal pier at Camp Helen, Florida

The wreck of "Boca Chica No. 2," southern Texas

Memorial for the 1900 Galveston Storm

Historic Structures
(Wooden Piers, Shipwrecks, Memorials)

WHAT ARE THEY? Wooden piers on Gulf beaches have historically accommodated strolling, fishing, restaurants, and even amusement parks. Original wooden piers were supported by pine logs treated with coal-tar creosote. The oldest standing Gulf piers include the Bob Hall fishing pier (1950, Corpus Christi, Texas), Galveston Island Pleasure Pier (1943, Galveston, Texas), and St. Andrews State Park Pier (1959, Florida). You may notice that there are not very many, and they are not that old. Storms are unrelenting. Even these survivors have been through several cycles of rebuilding after hurricanes. Other piers are mere remnants, like the **pier at Camp Helen** State Park, Florida (built in the 1930s). Wooden shipwrecks are occasionally uncovered by storm erosion. Most have unknown identities, such as the **Boca Chica No. 2 wreck** on Boca Chica Beach, Texas. Its construction suggests it was built between 1790 and 1840. Memorials mark tragic events not to be forgotten, like the **1900 Galveston Hurricane**, which killed approximately 8,000 people and destroyed the city. On the 100-year anniversary of the disaster, the memorial was set on the seawall that now defends the city of Galveston.

HOW COME? The prospects for piers are enticing—sea breezes and beautiful scenery. The reality for structures built on the shoreline is that all will eventually be destroyed. The stormy sea also takes ships. Most of those grounding near the beach have reached their final resting place.

DID YOU KNOW? Remembering lessons taught by nature can save lives.

Conservation *(Sea Turtles)*

Texas sea turtle nest patrol, ridley season flag (inset)

WHAT IS IT? Because sea turtles are threatened with extinction and depend on sandy shorelines for reproduction, many **sea turtle conservation** activities take place on beaches. Each nesting season (spring–summer), **morning nest-counts** occur on almost every beach from south Texas through the Florida Panhandle. In Texas, **Kemp's ridleys** nest in daylight, when the public can help surveyors find nests. The nests are commonly relocated to **hatcheries** to protect them from beach driving. Some **hatchlings** are released with an audience in efforts to educate the public and promote conservation. To take part, call the Hatchling Hotline (361-949-7163). In Florida, loggerhead and green turtle nests are left in place and identified with **signs**. After, a 2-month incubation, surveyors assess hatching success.

HOW COME? Sea turtle populations remain depleted from their historical abundance. Their recovery hinges on humans making some small changes, such as keeping artificial lighting off the beach at night. Morning nest-counts serve to keep track of sea turtle reproduction and population trends. Nest surveyors often use all-terrain motorcycles with low-pressure balloon tires. The soft tires leave minimal ruts without harming advancing beach plants.

DID YOU KNOW? The Kemp's ridleys that nest on Gulf beaches feed on crustaceans across coastal waters of the northern Gulf. Loggerheads nesting on our beaches grew up around the entire North Atlantic. Throughout their range, the number one threat to sea turtle survival is mortality from incidental capture in fisheries. To identify safe and sustainable seafood, visit: www.seafoodwatch.org

Citizens aid TX surveys   *Florida nest sign*

Hatchery, South Padre Island, Texas

A hatchling release is an education opportunity

397

Reporting banded shorebirds helps track movements

Biologists monitor important shorebird areas

Not all nesting colonies are conspicuously marked

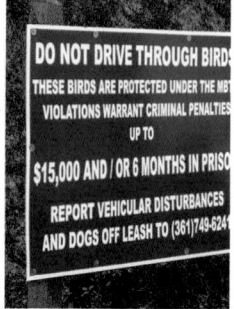
Signs to limit disturbance, which can be lethal to birds

Conservation *(Beach Birds)*

WHAT ARE THEY? Bird researchers apply numbered and colored **leg bands** to track movements and fates. Some bands can be read with binoculars. Helpful information about banded birds includes location, species, band types (metal, colors, flag), and attachment (left, right, upper and lower leg). Beach-nesting birds are either colonial seabirds, or solitary shorebirds. Seabirds include terns and skimmers (pp. 243–249), which nest in large groups. Shorebirds like plovers and oystercatchers (pp. 226–230) nest in isolated pairs. **Posted shorebird nesting areas** are sections of beach where entry by humans or dogs could cause birds to abandon their eggs and chicks. Many nesting areas may not be posted. Circling or agitated birds on the upper beach is a sign that one's presence is unappreciated. Even if a beach has no **signs** recommending good behavior to keep birds undisturbed, please let them have their rest.

HOW COME? Banded birds reveal important information to resource managers trying to reduce threats to bird populations. Posted shorebird nesting areas protect birds from foot traffic and disturbance, as long as the signs are heeded. Beach birds have a stressful life, especially when nesting. Disturbances can mean added stress and death to eggs and chicks.

DID YOU KNOW? Careful behavior and providing space can allow successful shorebird nesting even in developed areas. But disturbance is common. Over half of our least terns and black skimmers are relegated to nest on gravel rooftops of stores and warehouses. Have you seen a banded shorebird? Report the bird's information at: www.bandedbirds.org

Conservation *(Beach Clean-ups, Artificial Reefs, Dune Re-planting)*

Clean-ups, Planting   *Artificial Reefs*

WHAT ARE THEY? Beach clean-ups are organized efforts to remove trash from the beach. Nearly every Gulf community has a "Keep <your location> Beautiful" chapter that organizes local clean-up efforts. The Ocean Conservancy's International Coastal Cleanup involves hundreds of Gulf locations, with principal events held in April and September. Aligning **yellow posts** on the dune will locate four **artificial snorkeling reefs**, arranged in the shape of sea animals, just outside the surf zone between Destin and Panama City, Florida. **Dune re-planting** efforts follow dune erosion, sand placement, and attempts to make artificial dunes less artificial. Beachfront property owners are offered free (or reduced-cost) dune plants, and public areas are re-planted by volunteers. Planted species are predominantly seaoats, but some projects strive for more plant diversity (see *Beach Plants* section).

HOW COME? Big clean-ups are sponsored by nonprofit conservation groups, who tabulate what items are collected and how much. Other important efforts take place on an individual scale when each of us fills a bag with trash during a beach walk. It is rumored that the experience gives one a unique satisfaction. Piles of bare sand don't look like dunes, and can't function like them. To bind and accumulate sand, plants need a head start.

DID YOU KNOW? The Ocean Conservancy reports that the most frequent beach-trash items are cigarette butts, plastic bottles, food wrappers, plastic bottle caps, plastic straws and stirrers, and plastic bags.

Clean-ups are a great excuse for beachcombing

Markers for a snorkeling reef, Grayton Beach, Florida

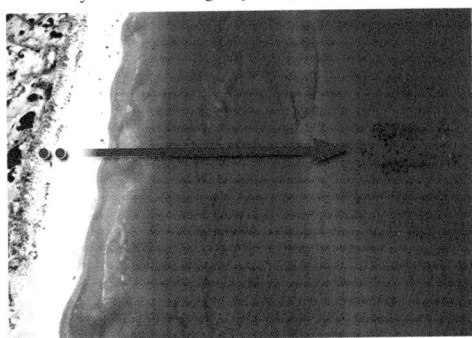

Aligning yellow markers finds the "turtle" reef

Volunteers help re-plant a dune

399

Beach Quests

Many beachcombers delight in the beach itself. Others are driven by quests. These are the searches for uncommon finds whose prospects reinforce the beachcombing obsession. The rarest items can prompt legendary pursuits and become the symbolic excuse for a lifetime of beach adventures.

This list briefly outlines tactics for where, when, and how to target some suggested quest incentives on Gulf beaches. The tactics describe likely times and places among many opportunities that are more fully described in the referenced pages. Even more research may be helpful. Knowing the "habits" of one's quest item, and having a complete "search image," are critical to realizing rare opportunities. But the most important element in fulfilling a quest is persistence. Beaches are places where every footstep and every tide brings something different. An adequately persistent and prepared beachcomber will be able to fulfill all of these beach quests, and even casual combers will find a few. Seek your quest, but don't stray too far from the philosophy of a happy beachcomber—the best beach is the one closest to you, and the best time is every chance you get.

Gemstone Sands

Boca Chica State Park, Texas, and Ship Island, Mississippi, low beach after strong onshore winds, see **page 351**

Sea Turtle Hatchlings

Malaquite Beach PINS, Texas, mid-June through August, call ahead and arrive before dawn, see **page 397**

Blue Button

Central and south Texas, winter following east winds, search the recent tide line, see **page 62**

Black Skimmer

North Padre Island, Texas, May through July in calm surf, bring binoculars and admire from a distance, see **page 249**

Common Sundial

San José Island, Texas, spring tide low or on a calm day following a storm, see **page 102**

Spotted Ground Squirrel

Padre Island National Seashore, Texas, look for dune burrows, see **page 260**

Stone Points

Beaches near the Texas/Louisiana border, following rough seas, record location and report to a local museum, see **page 382**

Sand Island Lighthouse

Three miles off Dauphin Island, Alabama, use binoculars or take a boat tour, see **page 394**

Round-rib Scallop

Santa Rosa Island, Florida, on a calm day following a storm, search high and low, see **page 119**

Ram's Horn Squid

Any beach after strong onshore winds, search the recent tide line and upper beach wrack, see **page 147**

Migrating Shorebirds

Grand Isle, Louisiana, April, bring binoculars and admire from a distance, see **page 254**

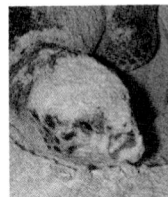

Nesting Sea Turtles

St. Joseph Peninsula State Park, Florida, June and July, you may only see tracks, but those are cool, see **pages 205–207**

Texas Pricklypear

Central and south Texas, blossoms in spring, see **page 287**

Mary's Bean

Matagorda Island, Texas, during and after windy spring days, after strong east winds, turn the wrack, see **page 326**

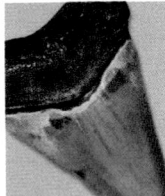

Shark Tooth

Bolivar Peninsula, Texas, spring tide low or on a calm day following a storm, see **page 359**

Blue Seaglass

(Big and Little) Shell Beaches, Padre Island National Seashore, Texas, search the shell hash, see **page 383**

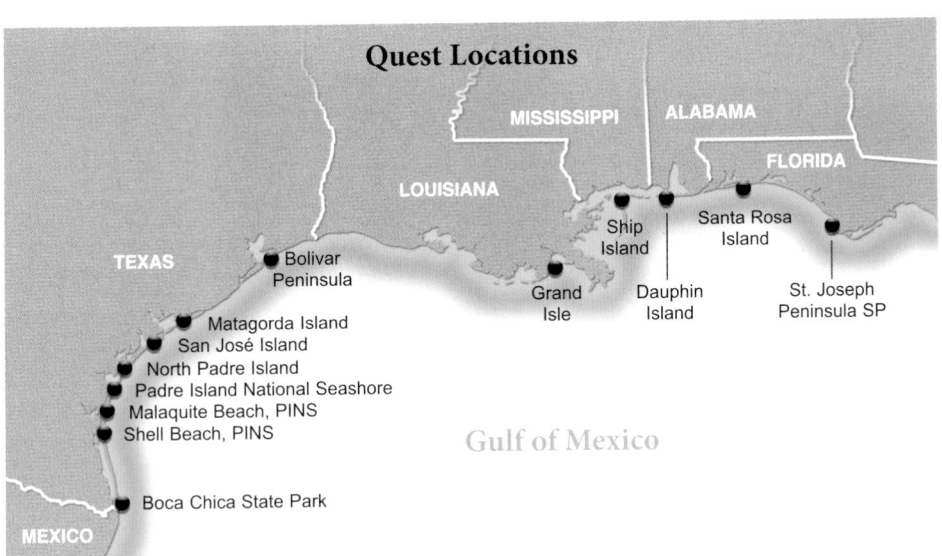

Quest Locations

MISSISSIPPI ALABAMA

FLORIDA

LOUISIANA

Ship Island

Santa Rosa Island

TEXAS

Bolivar Peninsula

Grand Isle

Dauphin Island

St. Joseph Peninsula SP

Matagorda Island

San José Island

North Padre Island

Padre Island National Seashore

Malaquite Beach, PINS

Shell Beach, PINS

Gulf of Mexico

Boca Chica State Park

MEXICO

The Future of the Gulf's Living Beaches

The Living beaches rimming the Gulf of Mexico have changed, are changing, and will continue to change. Change … it's the only constant. But how will beaches change? And, what is our role? In *The Tempest*, William Shakespeare's Antonio said, "Whereof what's past is prologue; what to come, in yours and my discharge." That is, although the past has set the stage for the next act, we still can shape our destinies. To the extent that this keen and dramatic observation by Billy Shakes applies to our relationship with beaches, we have some serious implications to consider.

Beaches are a focal point for the human imprint on Earth. Our contribution to climate change and sealevel rise, our habitation of the coast, and our incomplete control of waste, have obvious consequences at the shore. These effects contribute to the drama of an already dynamic system. The Gulf beaches we now see were shaped by changing sea levels and by the sinking and growth of sandy shorelines. Future changes will be similar, but accelerated. They will be both gradual and sudden, driven by steady forces, but punctuated by acute shifts caused by tumultuous storms. Of changes to come, the most straightforward to predict is a rising sea. Reliable forecasts by the Intergovernmental Panel on Climate Change are that global seas will rise about two feet (60 cm) by the end of the century. In the northern Gulf, the apparent rise is even greater because many coastal lands are sinking. By 2100, northern Texas and Louisiana may see a 4-ft (1.2 m) rise, which would move the shoreline many miles landward.

A fisher contemplates his path ahead as waves wash over his jetty. Access to natural resources is one of the challenges we will face as seas rise and shorelines change.

Beach changes create scenes of stark beauty. As the Gulf of Mexico returns to reclaim former beach from this existing pine forest, there is exquisite drama to behold.

Whether predictions of extreme shoreline change are shadows of things that *will* be, or are shadows of the things that *may* be, depends in no small part on our behavior. A principal effect we have on sea level is how our burning of carbon fuels and removal of forests heats up the planet. But we have additional effects on beaches, including management of sand transport by rivers, trapping of sand by jetties, armoring of coastal structures, and decisions to build where structures are destined to require an armored defense. As shorelines move, we will adapt, but hopefully in anticipation of the most startling movement. Humility will be key. There may be no more humbling experience than witnessing the power of a major hurricane as it reshapes a beach. With immense waves piled upon a storm surge made higher by a rising sea, future tempests will melt dunes into the surf, fill inlets and form new ones, and flood sand through remnants of beachside communities. It's grim, but the good news is that these lessons in humility won't happen everywhere all at once, and that policies on strategic retreat can be shaped by what we learn from the experience of others. We may even sufficiently temper our effects on climate so that only the lower estimates of sealevel rise come true. More good news for the beachcombers who follow us is that our beaches, as they are now, will be alive. Living beaches do not merely change; they are born of change. They are both phenomenon and entity, intertwining processes and cycles of life like few other aspects of our world. Ahead, as in the present, every moment on a beach will be different and amazing.

403

Resources and Suggested Reading

Beach Features

Davis, Richard A. *Beaches of the Gulf Coast.* Texas A&M University Press, 2014.

Beach Animals

Farrand, John Jr., editor. *The Audubon Society Master Guide to Birding.* Alfred A. Knopf, Inc., 1983.

Ruppert, Edward E., and Richard S. Fox. *Seashore Animals of the Southeast.* University of South Carolina Press, 1988.

Hoese, Dickson, and Richard H. Moore. *Fishes of the Gulf of Mexico, Texas, Louisiana, and Adjacent Waters.* Texas A&M University, 1998.

Tunnell, John W., Jean Andrews, Noe C. Barrera, and Fabio Moretzsohn. *Encyclopedia of Texas Seashells.* Texas A&M University Press, 2010.

Witherington, Blair E., and Dawn E. Witherington. *Florida's Living Beaches.* 2nd ed. Pineapple Press, Inc., 2017.

Cornell Lab of Ornithology, bird identification and information. **www.allaboutbirds.org**

iNaturalist, species records and distributions. **www.inaturalist.org**

World Register of Marine Species, primary source for species names. **www.marinespecies.org**

Beach Plants

Lehman, Roy L. *Marine Plants of the Texas Coast.* Texas A&M University Press, 2013.

Nelson, Gil. *The Shrubs and Woody Vines of Florida.* Pineapple Press, Inc., 1996.

Perry, Ed, IV, and John V. Dennis. *Sea-Beans from the Tropics.* Krieger Publishing Company, 2003.

North American Plant Atlas, plant distributions and species names. **www.bonap.net/napa**

Beach Geology

Gale, Bob, and Pam Gale. *A Beachcomber's Guide to Fossils.* University of Georgia Press, 2020.

Weise, Bonnie, and William White. *Padre Island National Seashore: A Guide to the Geology, Natural Environments, and History of a Texas Barrier Island.* Virtual Landscapes of Texas, 1980.

Hand of Man

Ebbesmeyer, Curtis, and Eric Scigliano. *Flotsametrics and the Floating World.* Harper Collins, 2009.

Pilkey, Orrin H., Linda Pilkey-Jarvis, and Keith C. Pilkey. *Retreat from a Rising Sea: Hard Choices in an Age of Climate Change.* Columbia University Press, 2016.

Beach conservation. **www.surfrider.org**

Photo Credits

Photographs, illustrations, and graphics are © Blair and Dawn Witherington unless listed:

p. 6 second, Google Earth
p. 41 top, courtesy of NOAA
p. 41 bottom, courtesy of USGS
p. 45 bottom, Google Earth
p. 47 bottom, Fitzsimmons/Shutterstock
p. 48 second, courtesy SpaceX
p. 60 bottom, © Linda Ianniello
p. 147 second, © Carly DeMay
p. 156 top, © Hans Hillewaert
p. 171 third, © Roger Birkhead
p. 174 top, © Dennis Bonal
p. 204 top, © Ashleigh Holden
p. 207 top, © Adrienne McCracken
p. 207 middle, © Shigetomo Hirama
p. 208 top right, Donna Shaver, NPS
p. 212 top, Joe Farah/Shutterstock
p. 212 bottom, © Ryan Chabot
p. 227 second, © Kevin Edwards
p. 228 center, © Kevin Edwards
p. 240 top, Agami Photo/Shutterstock
p. 246 third, © D.J. McNeil

p. 247 bottom, © Kevin Edwards
p. 251 second, Almost heaven/Shutterstock
p. 258 top, Steve Bower/Shutterstock
p. 259 top, USFWS
p. 259 2nd, 3rd, 4th, 5th, Florida FWC
p. 260 top, © Robert Deans
p. 260 third, inset, © Terry Ross
p. 261 center, © Cullen Hanks
p. 262 top, © Steve Johnson
p. 263 bottom, Randimal/Shutterstock
p. 264 top, Bormozaya/Shutterstock
p. 264 middle, © Steven Pinker
p. 266 top, Tory Kallman/Shutterstock
p. 266 bottom, TMMSN
p. 300 bottom, Captain's Travels/Shutterstock
p. 373 bottom, Cheryl Casey/Shutterstock
p. 380 third and inset, © Tom Pitchford
p. 394 third right, Kraig Anderson
p. 396 middle courtesy of TX Historical Com.
p. 397 bottom, courtesy of PINS NPS
p. 399 bottom, © Donna Lee Crawford

Entries in **bold** indicate photos and illustrations.